Advances in Single Cell Analysis

Vijai Singh
Editor

Advances in Single Cell Analysis

From Fundamentals to Translational Research

Springer

Editor
Vijai Singh
Department of Biosciences, School of Science
Indrashil University
Mehsana, Gujarat, India

ISBN 978-981-95-9788-8 ISBN 978-981-95-9789-5 (eBook)
https://doi.org/10.1007/978-981-95-9789-5

This Springer imprint is published by the registered company Springer Nature Singapore Pte Ltd.
The registered company address is: 152 Beach Road, #21-01/04 Gateway East, Singapore 189721, Singapore

Foreword

I am delighted to write a foreword for the book on *Advances in Single Cell Analysis: From Fundamentals to Translational Research*, which represents a timely and valuable contribution to the rapidly expanding fields of single-cell research with its strong focuses on precision and translational applications. Single-cell approaches enable the study of individual cells within complex populations. It can be measured through time-lapse microscopy or confocal microscopy. Single-cell approaches offer a way to investigate cellular heterogeneity that exists among individual cells within a population. Currently, single-cell research has become a vital and transformative area across biology, medicine, healthcare, and biotechnology.

This book is organized into two sections. *Section A* presents the fundamentals of single-cell analysis, covering emerging technologies, single-cell genomics, single-cell RNA sequencing, transcriptomics, proteomics, metabolomics, and the role of microfluidics in single-cell studies. *Section B* focuses on applications of single-cell approaches in bacterial infections, cancer, neurological and cardiac disorders, precision and regenerative medicine, animal models of metabolic and blood lipid disorders, nanotechnology, and the economic aspects of single-cell research.

I am happy to recognize the valuable efforts of Prof. Vijai Singh, who has tirelessly worked to conceive an excellent book under Springer Nature. This book covers a rich literary text of excellent depth, clarity, and coverage. This book is an excellent source of information from which scientific knowledge can grow and widen the field of single cell. I personally believe that this book may play a key role in better understanding of single-cell analysis and its role in translational research with greater future outcomes. It is useful not only for beginners but also for students, researchers, scientists, clinicians, practitioners, policymakers, and stakeholders who are interested in accessing the potential of single cell in several areas.

Keshav K. Singh

Joy and Bill Harbert Endowed Chair in Cancer Genetics,
Professor of Genetics, Dermatology, Pathology,
and Environmental Health, Founding Editor-in-Chief,
Mitochondrion Journal, Director, Cancer Genetics Program,
Department of Genetics
Heersink School of Medicine, The University
of Alabama at Birmingham,
Birmingham, Alabama, USA

Preface

The cell is the fundamental building block of all living organisms. Millions of cells interact and function together to process biological information while maintaining life. However, it cannot evolve the cellular heterogeneity and molecular dynamics in a certain cell population. Single-cell enables the study of an individual cell within a complex population. It assists in the investigation of genomics, transcriptomics, proteomics, metabolomics, and other molecular layers at single-cell resolution. Single-cell analysis refers to the analysis or manipulation of an individual cell in a population of cells under the particular condition using an optical or electronic microscope, time lapse or confocal microscope. Single cell is being used to screen for high-throughput drugs and their effects. Recent developments in emerging technologies allow analysis of the genome and transcriptome of single cells as well as quantification of their proteome and metabolome. By revealing cellular heterogeneity within populations, single-cell analysis becomes a powerful tool and is used in biology, medicine, and biotechnology.

This book is organized into two sections. Section A presents the fundamentals of single-cell analysis, covering an overview of single-cell analysis, emerging technologies, single-cell genomics, single-cell RNA sequencing, single-cell transcriptomics, single-cell proteomics, single-cell metabolomics, and the role of microfluidics in single-cell analysis and its future potential. Section B focuses on applications of single-cell approaches in bacterial infections, cancer research and discovery, neurological and cardiac disorders, precision medicine and regenerative medicine, animal models of metabolic disorders and blood lipid disorders, nanotechnology, and the economic aspects of single-cell research and innovations.

This book focuses on a fundamental aspect of single-cell analysis from basic applications to applied biological sciences. This book covers a wide range of topics for better understanding and expanding research. This book uses a rich literary text of excellent depth, clarity, and coverage. This book is an excellent basis from which scientific knowledge can grow, widen, and accelerate single-cell research and innovations.

Mehsana, Gujarat, India

Vijai Singh

Acknowledgments

I am delighted to thank all the authors for their excellent contributions to this book. I would like to thank Dr. Soumya Basu, Associate Editor-Clinical Medicine, Biomedicine and Life Sciences, Asia, Springer, and Vijayalakshmi Dayalan, Production Editor from Springer, for their excellent management of this project.

I would like to give thank Prof. Rakesh Rawal, Prof. Bharat Maitreya. Dr. Poonam Bhargava, Dr. Pablo Carbonell, Dr. Satya Prakash, Dr. Vimal C. Pandey, Prof. Suresh Ramakrishna, Dr. Dinh-Toi Chu, and those whose names do not feature here but have directly or indirectly contributed in shaping this project.

I greatly appreciate the support of my students Khushal Khambhati, Dharmisha Solanki, Karan Murjani, and Khushbu Panchal, whose discussion and comments helped to shape this book.

I wish to express my gratitude to my beloved wife, Pritee Singh, for her endless support, patience, and inspiration. Lots of affection for my kids, Aaradhya and Ayush, who missed me during this project.

I am aware that, despite our best efforts, the first version always contains some errors that may have crept into the compilation. I would be delighted to receive feedback from readers to further improve the book. Last but not least, my sincere thanks to GOD for his supreme POWER for endowing me to live with joy and victory in the shape of this book.

Vijai Singh

Contents

About the Editor

Vijai Singh is a Professor and Dean (R&I) at Indrashil University, Rajpur, Mehsana, India. He has worked as Associate Professor in the Department of Biosciences, School of Science at Indrashil University, Rajpur, Mehsana, India. He was an Assistant Professor in the Department of Biological Sciences and Biotechnology at the Institute of Advanced Research, Gandhinagar, India, and also an Assistant Professor in the Department of Biotechnology at the Invertis University, Bareilly, India. Prior to that, he was a Postdoctoral Fellow at the Institute of Systems and Synthetic Biology, Paris, France, and School of Energy & Chemical Engineering at the Ulsan National Institute of Science and Technology, Ulsan, South Korea. He received his Ph.D. in Biotechnology (2009) from the ICAR-National Bureau of Fish Genetic Resources, Uttar Pradesh Technical University, Lucknow, India, with a research focus on the development of molecular and immunoassays for diagnosis of *Aeromonas hydrophila*. His research interests are focused on building novel biosynthetic pathways for production of medically and industrially important biomolecules. Additionally, his laboratory is working on CRISPR-Cas9 tools for genome editing. He has more than 15 years of research and teaching experience in synthetic biology, metabolic engineering, bioinformatics, microbiology, and industrial microbiology. He has published 113 articles, 91 book chapters, 36 books, and 3 patents. He serves as an associate editor, editorial board member, and reviewer of several peer-reviewed journals. He is also a member of the Board of Study and Academic Council of Indrashil University and is the Member Secretary of the Institutional Biosafety Committee (IBSC) at the same University.

Overview of Single-Cell Analysis

1

Dharmisha Solanki, Karan Murjani, and Vijai Singh

Abstract

Single-cell analysis has revolutionized biology by revealing the enriched cellular heterogeneity that is masked by traditional bulk measurements. It provides a detailed view of individual genomes, epigenomes, transcriptomes, proteomes, and more. Over the course of its development from simple microscopic approaches and flow-cytometry in the 1970s–1990s to the pivotal development in 2009, scRNA-seq demonstration was rapidly developed. Platforms such as 10× chromium and Drop seq, using microfluidics and combinatorial barcoding, allow profiling of millions of cells that are also cost effective. Methodological workflows include controlled tissue dissociation, isolation of single cells with the help of fluorescence-activated cell sorting (FACS), magnetic-activated cell sorting (MACS), or nano-wells. These advances are supported by sophisticated library preparation techniques such as Smart-seq2 for full-length transcripts or scATAC-seq for chromatin accessibility and robust computational pipelines that features normalization (e.g., scran), doublet detection (e.g., Scrublet), and trajectory inference (e.g., UMAP). These advances serve as basis for transformative applications in oncology including identification of tumor subpopulations, progression of circulating tumor cells, immunotherapy biomarkers; in neuroscience: brain cell types mapping system, lineage relationship and functional and pathological states with molecular identity. Despite its high resolution, single-cell analysis is limited by technical noise, data sparsity, and loss of spatial context during tissue dissociation, which complicate biological interpretation.

Keywords

Single-cell analysis · Genomics · Proteomics · Transcriptomics · Screening · FACS

D. Solanki · K. Murjani · V. Singh (✉)
Department of Biosciences, School of Science, Indrashil University, Rajpur, Mehsana, Gujarat, India
e-mail: vijai.singh@indrashiluniversity.edu.in

V. Singh (ed.), *Advances in Single Cell Analysis*,
https://doi.org/10.1007/978-981-95-9789-5_1

1.1 Introduction

The cell is the core component of all living organisms. The individual cell profile has become indispensable for understanding biological complexity. Bulk measurements, average signals across heterogeneous mixtures, and unknown rare cell types and transient states; single-cell approaches resolve these issues by measuring molecular features at single-cell resolution, revealing previously hidden subpopulations, developmental trajectories, and cellular responses to perturbations (e.g., drug or infection) (Wu et al. 2024).

Single-cell analysis is the study of genome, epigenome, transcriptome, proteome, metabolites, and functional states at the resolution of individual cells rather than bulk populations, that are developed using technologies that isolate individual cells, measure DNA, RNA, or protein simultaneously. It helps to decode the cellular heterogeneity, and includes interactions between the cells (Lim et al. 2024). Traditional cell-based assays primarily quantify mean response from group of cells; thus, it does not indicate actual response from single cell. However, by measuring the average response of the cells may result in missing significant information, such as in the case of cancer, where tumor microenvironment is composed of tumor cells and its surrounding noncancerous stromal cells. Thus, these two different kinds of cells possess different consequences under the pathological conditions (Schor and Schor 2001).

In the previous study, researchers have deployed limited high throughput techniques to analyze single cell including single-cell PCR, immunofluorescence, and fluorescence in situ hybridization (FISH), to identify various biomarkers (Taniguchi et al. 2009; Citri et al. 2011). These methods enable assessment of only a small number of factors in a single cell. While the other method includes sequencing of DNA and RNA, these methods are also dependent on millions of cells for the collection of data that are independent of their cellular function due to variation in cells (Hu et al. 2016). Thus, single-cell sequencing was developed as there was a need to study the cells separately for elucidating cellular function, and in 2013 the Nature methods was awarded the "single-cell sequencing" as the "methods of the year" ("Method of the Year 2013" 2014).

Through physical isolation of single cell and quantification of these data, single-cell analysis enables the generation of high-dimensional representation of cell types, partially differentiated states, and time-resolved cellular trajectories within heterogeneous tissues (Cheng et al. 2025). Recent technological development in droplet microfluidics, high throughput sequencing, combinatorial barcoding, high-dimensional single-cell cytometry, and spatially resolved capture arrays led to the development of measurement of cells from 10–12 cells per experiment to millions of cells. On the other hand, per cell cost is also reduced, which enables the multiomics profiling of transcriptome, genome, epigenome, and proteome within the same cell (Lim et al. 2024).

1.1.1 Evolution of Single-Cell Analysis

Early work in 1970s–1990s was focused on characterizing the cells using microscopy and low parameter flow cytometry primarily to study the morphology and electrophysiological analysis of the cells. In addition to this, flow cytometry and fluorescence-activated cell sorting (FACS) allowed quantitative analysis of intracellular markers, and physical parameters of single cells, and physical isolation of defined subpopulations for downstream assays (He et al. 2015). In the 1990s, single-cell qPCR and linear amplification methods were pioneered by James Eberwine and Norman Iscove, who showed that, mRNA from single cells could be amplified to adequate levels for multigene expression analysis that provides the conceptual bridge between single-cell biology and transcriptomics (Kolodziejczyk et al. 2015).

A major turning point came with the advancement of next generation sequencing and the first validation of single cell transcriptomics in the 2009 when Tang et al. sequenced cDNA from single blastomeres and oocytes, which proved that single-cell RNA sequencing was technically feasible (Tang et al. 2009). Over the years, full length scRNA-seq protocols such as smart-seq, and smart-seq2, as well as tag-based methods including STRT-seq, Quartz-seq, and CEL-seq, have improved sensitivity and wider coverage (Picelli 2017).

1.2 Methodological Approaches in Single-Cell Analysis

The methodological strategies in single-cell analysis form a complete workflow that are designed to transform mixed cell populations into interpretable high-dimensional data. It characterizes difference between cells and how they change over the time. These processing steps are modular, which allows sample specific customization, targets DNA, RNA, or protein simultaneously, however, it consistently includes sample preparation, single-cell isolation, immobilization of molecule, preparation of library, sequencing, and computational processing (Cristinelli and Ciuffi 2018). Throughout each step specific technical challenges which are related to lower concentration of sample, systematic distortions, and noise, that have been systematically addressed over time with the help of development in microfluidics, statistical modeling, and barcoding (Duncombe et al. 2015).

1.2.1 Sample Preparation and Dissociation

Sample preparation and dissociation are the fundamental step, in which complex tissues, biofluids, cell cultures, or microbial communities are converted into suspension of viable single cells or nuclei suitable for downstream acquisition. For solid tissues such as brain, liver, or tumors, enzyme-mediated digestion using collagenase, dispase, trypsin, or papain which are combined with mechanical dissociation such as pipetting, grinding, or vertexing. It disrupts extracellular matrices and intercellular junctions to release individual cells, while DNase I prevents aggregation from release of genomic DNA (Reichard and Asosingh 2019).

Optimization of enzyme concentrations, incubation temperatures (often 37 °C), time periods (minutes to hours), and agitation were required to balance the complete dissociation with preservation of RNA integrity, surface proteins, and chromatin structure, as excessive proteolysis that can trigger the stress responses such as upregulation of heat-shock genes or degradation of epitopes, which are required for FACS gating (Gupta and Lazzaro 2022). For samples that are cryopreserved, single-cell RNA sequencing (snRNA-seq) protocols uses mild detergents such as NP-40 or IGEPAL in isotonic buffers that helps to lyse plasma membranes, which resulted in release of intact nuclei. These intact nuclei are then purified by density gradients such as fluorescence-activated nuclear sorting (FANS), and this method leads to the degradation of mRNA, although this approach loses mRNA. It is suitable for human tissues where fully dissociating cells is difficult or to introduce bias (Schmokel et al. 2023). In the case of microbial samples, they represent distinct challenges because of rigid cell walls and lower RNA concentration, that often require lysozyme, achromopeptidase. It is a type of lysyl endopeptidase, or probe-based permeabilization used to degrade the cell wall in Gram-positive or Gram-negative bacteria (Heiniger et al. 2016).

1.2.2 Isolation of Single Cell

Single-cell isolation and capture methodologies ensure that each molecular composition of a cell which is processed independently, to prevent any cross-contamination and to enable multiplexing. Fluorescence-activated cell sorting (FACS) remains a core technique for cell sorting, where dissociated cells are stained with fluorescent antibodies against surface markers or intracellular targets. These samples then hydrodynamically focused into a narrow stream and deflected into wells or tubes based on forward/side scatter and fluorescence parameters; advanced sorters support 10–18 colors and index sorting, which records intensities of per-cell marker for later correlation with omics data (Komoda and Matsunaga 2015). Magnetic-activated cell sorting (MACS) provides a gentler, and more accessible alternative specifically for routine applications, that uses antibody-coated superparamagnetic beads to retain target cells in a magnetic field before elution, often as a prestep for rare populations (Maes et al. 2020).

Another method is low-throughput methods such as laser capture dissection (LCM) uses infrared or ultraviolet lasers, that precisely isolate a single cell from tissue, preserving spatial context for histology-guided sampling. On the other hand, micromanipulation uses glass micropipettes under inverted microscope to isolate a single cell (Jin et al. 2020). However, high-throughput microfluidic platforms have revolutionized this step into droplet-based systems (Drop-seq, inDrop, 10× Chromium), which encapsulate cells automatically and generate picoliter oil droplets, each droplet function as an individual reaction chamber that are followed by Poisson-distributed single cells. It includes most droplets that are selected empty, some exactly contains one cell, which are desired, a small fraction contains doublets cell. The concentration of the cells are carefully tuned to maximize the number of single droplets while minimizing the doublets. In addition to this, cells and barcoded beads are flowed together via oil-aqueous phase in microfluidic channels, this oil phase separates the aqueous stream into individual droplets, each droplet contains one cell and one barcoded bead where lysis occurs in situ and transcripts are immediately captured (Zilionis et al. 2017). Microwell arrays and valve-based chips such as, fluidigm C1 and microwell-seq, trap cells in nanoscale wells for controlled lysis and reaction. In fluidigm C1, microfluidic valves actively trap the single cells and lysis reaction are precisely controlled. In microwell-seq, cells settle into tiny wells by gravity thus, each well holds one cell. These capture methods are evaluated on throughput 10^2 to 10^6 cells with >95% purity of single occupancy, for recovery, fraction of input cells are captured and analyzed, and stress minimization, with droplets dominating large-scale studies due to their scalability (Yuan and Sims 2016).

In single-cell genomics, molecular assay and library preparation are essential steps that ensure the compatibility for sequence through the conversion of cellular biochemical contents into stable DNA molecules 19. Whole genome amplification (WGA) is being used to amplify the whole genome of individual cell. PCR-based WGA techniques, such as degenerate oligonucleotide-primed PCR (DOP-PCR) and primer extension pre-amplification (PEP-PCR), use random primer and often result in nonuniform genome coverage and loss of one allele during amplification (Volozonoka et al. 2022). To address these limitations, multiple displacement amplification (MDA) utilizes high fidelity phi29 polymerase to generate the long DNA fragments; however, strand displacement during amplification can lead to the chimeric DNA fragments. An alternative approach, multiple annealing and looping-based amplification cycles (MALBAC), enhances amplification uniformity by using quasi-linear preamplification, which form self-looping amplicons that prevent reamplification. The amplified DNA was subsequently subjected to fragmentation and adapter ligation for illumina sequencing (Chen et al. 2014; Wang et al. 2022).

Description of single-cell isolation also focuses on transcriptomics and epigenomics library preparation and analytical methods were applied after cells that have been isolated. Transcriptomics start with lysis and reverse transcription, where RNA was converted into complementary DNA (cDNA). Some full-length transcriptome method such as smart-seq, has unique property of reverse transcriptase called template-switching ideal for isoform analysis. In contrast, droplet-based 3′ or 5′ scRNA-seq protocols such as 10× genomics, capture only 3′/5′ transcripts end. This was achieved by using barcoded oligo-dT primers and unique molecular identifier (UMI), supporting reliable transcript quantification after amplification (Picelli 2017; Ramsköld et al. 2012). Moreover, single-cell epigenomics adapt bulk protocol by using different biochemical method (bisulfite conversion, immunoprecipitation, or enzymatic digestion) to detect chemical and structure changes on DNA at the level of individual cells. There are other methods: (1) DNA methylation profiling (*scBS-seq* and *scRRBS*), (2) histone modification (*scChIP-seq*), (3) Tn5 transposase tags into open chromatin or accessible DNA (*scATAC-seq, scDNase-seq*), (4) chromation conformation (*scDamID, scHiC*) (Buenrostro et al. 2015; Clark et al. 2016).

In other approaches, proteomics is study of proteins content within cells after isolation where proteomics mainly relies on fluorescently labeled/metal-tagged antibody panels for flow/mass cytometry or advanced MS workflow. However, hybrid method such as cellular indexing of transcriptomes and epitopes by sequencing (CITE-seq), combines antibody-based protein profiling with single-cell droplet RNA sequencing. In this method, cells were encapsulated with barcoded primers that were coated microbeads and lysis buffer into oil-separated nanoliter droplets using a microfluidic device. Ideally each droplet contains a single cell and bead, that enables in situ lysis and immediate mRNA capture. They were used for droplet-based platforms such as Drop-seq and 10× Genomics Chromium (Birhanu 2023; Kirsher et al. 2025; Timp and Timp 2020). In single cell, multiomics methods used to measure multiple molecular layers from the same cell by using cell barcodes system. There are many other multiomics methods such as scTrio-seq technique used for separation of nucleus (DNA/ATAC/methylation) and cytoplasm (RNA) of single cell at physical level, while GT-seq method employed in separation of RNA from gDNA by using magnetic oligo-dT beads. Methods such as sci-CAR/SNARE-seq jointly measure RNA transcription and chromatin tagmentation (ATAC) in the same nuclei, whereas ECCITE-seq method extend RNA-protein profiling to include clonotype and CRISPR perturbation readouts, with all modalities linked by shared cell barcodes (Lee et al. 2020).

In single cell analysis, computational techniques is analytical backbone that converts raw data into interpretable biological signals. Raw FASTQ data reads cell by feature count matrix through demultiplexing based on index sequence of specific sample, alignment (STAR), or pseudoalignment (Kallisto) where each read was linked to gene, transcript, or genomic region (Du et al. 2020). By using quality control methods, cells of low quality can be removed that are based on minimum requirement of genes which were <200 or >3500 genes, mitochondrial genes can also be eliminated at expression >30% for cells and >5% for nuclei. In addition to these doublets can be predicted by using computational tools such as Scrublet or DoubletFinder which is based on python and R language, respectively (Hong et al. 2022; Quatredeniers et al. 2023). A diverse set of normalization techniques have been established to decrease the technical noise in scRNA-seq data which includes single-cell tagged reverse transcription (SAMstrt), gamma regression model (GRM), scran, bayesian analysis of single-cell sequencing data (BASiCS), and robust normalization of single-cell RNA-seq data (SCnorm) (Lytal et al. 2020).

Highly variable genes (HVG) which are also known as highly variable features (HVF) are fundamental step in RNA sequencing of single-cell analysis. HVGs are those genes whose expression vary significantly across cells, which makes them informative for distinguishing different types of cells. In analysis methods such as seurat, HVG selection occurs after quality control normalization. Once the selection of genes has been done for all downstream analysis, dimensionality reduction method such as principal component analysis (PCA), visualization techniques such as uniform manifold approximation and projection (UMAP) and biological analysis which includes cell clustering and trajectory interference all of these are dependent on HVG. This computational workflow ensures that raw data yields robust and reproducible biological conclusions as shown in Table 1.1 (Squair et al. 2021; Wang et al. 2025; Zhu et al. 2019).

1.3 An Overview and Applications of Single-Cell Analysis

The use of single-cell analysis in scientific investigations is increasing exponentially. It has been influenced by many scientific fields such as cancer biology immunology, neurology, and many more. Since single-cell analysis applications is widely used, it is not feasible to describe each one individually. Therefore, major fields such as cancer biology and neuroscience frequently use this approach, and a brief overview of other applications is shown in Fig. 1.1 (Hu et al. 2016).

1.3.1 Application of Single-Cell Analysis in Cancer Biology

Recently, analysis of single cell is being used in a wide range of cancers isolated from tumor cells such as pancreatic adenocarcinomas, muscle invasive bladder cancer, lung adenocarcinoma cancer, acute myeloid leukemia, renal cell carcinoma, colon cancer, and many (Hu et al. 2016). Furthermore, technologies of single cell have become important tool in cancer research by supporting the mapping of tumor heterogeneity, reconstruction of cellular evolution tracking, and guiding of precision therapies. In single-cell sequencing (SCS) technology is applied in oncology, such as scRNA-seq, scDNA-seq, and scATAC-seq, have revealed previously unusual cell population within tumor. These include damaged activated regenerative stem cell in intestinal tumor and distinct aberrant subtypes that predominate in lung adenocarcinomas with distant metastases. Such discoveries facilitate the lineage tracking of tumor progression. Another important use of SCS is the analysis of circulating tumor cells (CTCs) that shows early diagnostic, prognostic stratification, and assessment of metastatic potential (Chen et al. 2023). Studies comparing the positive and negative CTC test of patients with non-small cell lung cancer (NSCLC) have shown that patients with CTC-positive that have greater possibility of cancer disease reappearance. In stage II prostate cancer CTCs frequently exhibit stem cell related mutation that mainly assists in the tumor survival and therapy resistance. Overall, single-cell approaches assist the scientist to identify the specific cellular and molecular processes that enables cancer cells to survive therapy (Wankhede et al. 2022).

Furthermore, single-cell approaches help to identify the molecular and cellular mechanisms that drive drug resistance. In multiple myeloma, for example, coexisting transcriptional state reveal that different cancer cell within same patient that can exhibit distinct gene expression simultaneously. For example, CXCR4, a chemokine receptor associates with these states because it may not respond uniformly to therapy. They contribute to treatment failure and relapse. In renal cancer stem cells, it is observed that the spatiotemporal shift means cancer stem cells that can change their location within the tumor, which is called spatial and alter their molecular

Table 1.1 Overview of methodologies for single-cell analysis

Category	Method/platform	Principle	Reference
Single-cell isolation	FACS (fluorescence-Activated Cell Sorting)	Fluorescently labeled cells are hydrodynamically focused and electrostatically deflected	Komoda and Matsunaga (2015)
	MACS (Magnetic Activated Cell sorting)	Antibody-coated magnetic beads retain target cells in magnetic field	Maes et al. (2020)
	Laser Capture Microdissection (LCM)	IR/UV lasers isolate cells from tissue sections	Jin et al. (2020)
Microfluidic-based Single Cell Isolation	Droplet-based systems (Drop-seq, inDrop, 10× Chromium)	Poisson-based droplet of cells into oil water droplet with barcode beads	Zilionis et al. (2017)
	Microwell-seq	Gravity-based settling of cells into nanowells	Yuan and Sims (2016)
	Fluidigm C1	Valve-controlled microfluidic trapping	Yuan and Sims (2016)
Genomic library prep	DOP-PCR/PEP-PCR	Random primer-based PCR amplification and allele loss at amplification step	Volozonoka et al. (2022)
	MDA	Phi29 polymerase with strand displacement led to the chimeric DNA fragments	Chen et al. (2014) and Wang et al. (2022)
	MALBAC	Quasi-linear pre-amplification with looping amplicons that block re-amplification.	Chen et al. (2014) and Wang et al. (2022)
Transcriptomics Epigenomics	Smart-seq/Smart-seq	Full-length cDNA synthesis via template switching for analysis of isoform	Picelli (2017) and Ramsköld et al. (2012)
	Droplet-based scRNA-seq (10× Genomics)	Barcoded oligo-dT primers capture transcripts 3′/5′ ends	Picelli (2017) and Ramsköld et al. (2012)
	scBS-seq/scRRBS	Bisulfite conversion for DNA methylation	Clark et al. (2016)
	scChIP-seq	Antibody-based histone mark profiling	Clark et al. (2016)
	scATAC-seq	Tn5 transposase tags open chromatin	Buenrostro et al. (2015) and Clark et al. (2016)
	scHi-C/scDamID	Chromatin interaction mapping	Clark et al. (2016)

(continued)

Table 1.1 (continued)

Category	Method/platform	Principle	Reference
Proteomics	Flow/Mass cytometry	Fluorescent or metal-tagged antibodies	Birhanu (2023), Kirsher et al. (2025), and Timp and Timp (2020)
	CITE-seq	Encapsulated cells with primer coated barcode microbead system followed by mRNA capture and + scRNA-seq	Birhanu (2023), Kirsher et al. (2025), and Timp and Timp (2020)
Multi-omics	scTrio-seq	Physical separation of nucleus and cytoplasm	Lee et al. (2020)
	sci-CAR/ SNARE-seq	Combinedly measure tagmentation chromatin and RNA profiling	Lee et al. (2020)
	ECCITE-seq	RNA + protein + clonotype/ CRISPR followed by shared cell barcode	Lee et al. (2020)

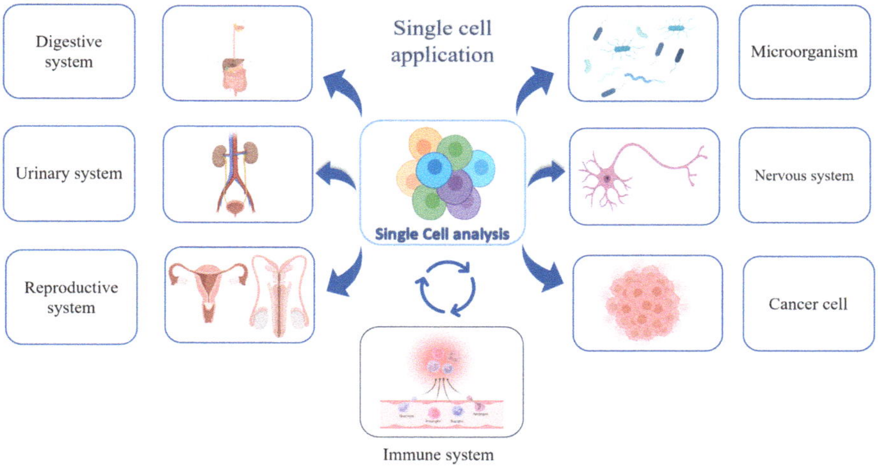

Fig. 1.1 Application of single-cell analysis in different fields

profile over time called temporal shift. These dynamic changes allow cancer stem cells to adapt to therapy (Frede et al. 2021).

In the context of immunotherapy, tumors are diverse immune cells that include tumor infiltrating lymphocytes (TILs), whose complex immune ecosystem can be mapped by SCS. SCS profiles each individual immune cell at high resolution. There are not all cells respond equally against immunotherapy; therefor, SCS allows patients to be classified based on their immune cell composition. For instance, T-cell profiles in hepatocellular carcinoma and esophageal cancer are classified based on their clinical response. Single-cell data can further predict responsiveness to anti-PD-1 therapy by identifying specific immune cell states, making SCS a powerful biomarker tool. Additionally, single-cell analysis of CART-T therapy reveals cytokine heterogenecity among engineered T-cell, providing insights for improving therapeutic efficacy and safety. Whatever, it is taken into account that SCS changes tumor microenvironments into useful information, which includes customized drug screening on primary cells to CTC liquid biopsies (Chen et al. 2023; Krieg et al. 2018).

1.3.2 Role of Single-Cell Analysis in Neurosciences

The brain is the highly interconnected neural network in human body. It contains more than 100 billion neurons, and each neuron is well connected with other and form approximately 10,000 direct connections. Technologies of single-cell sequencing provide individual cells analysis one by one through transcriptomics and epigenomics assays instead of averaging across thousands or millions of cells. In the context of neuroscience, these technologies can exhaustively define the cellular-type diversity of the brain. These analyses reveal the mechanism of regulatory control of cells which shapes the cellular identity and enhance understanding of development and brain cell population evolution. Furthermore, data of single- cell sequencing facilitates the development of functional tools for analyses components of brain circuit, connects molecular identities with anatomical connectivity, morphological, and physiological characteristics (Armand et al. 2021).

In the course of development, *Mus musculus* mouse has been used as foundational model for studying neurodevelopment and understanding of how genetic programs control neurogenesis and cellular differentiation. For instance, the *Tsc2* gene was involved in the control of neuroprogenitor differentiation. Loss of *Tsc2* gene disrupts translational control and impairs neuronal maturation, resulting in developmental defects that parallel neurological features seen in human tuberous sclerosis. Currently, introduction of single-nucleus multiomics technologies such as sn-m6A-CT enables integrated analysis of transcriptomes and m6A methylomes at single-cell resolution which has provided understanding of how epitranscriptomic modifications influence lineage specification and developmental trajectories in murine systems. Through the integration of RNA methylation states to gene expression programs in individual cells undergoing differentiation, these approaches provide insight into previously unresolvable layers of regulatory control at the time of brain development. Together, these advances highlight the continued importance of murine models in biomedical research, especially for dissecting the cellular and molecular mechanisms underlying neurodevelopmental disorders and other human diseases (Woo and Eyun 2025).

Recent work has further revealed how sequencing of single-cell methods that may be employed to trace the cellular lineages in the adult brain. For instance, a novel nuclear sequencing of single-cell technique has been applied to track the adult brain of second-generation single cell of (Lake et al. 2016). Similarly, others studies investigated the application of single-cell sequencing to reconstruct developmental trajectories in the cerebellum. These analyses help to identify the major cerebellar cell populations and revealed additional subtypes that facilitates to cerebellar development in mouse models. Together, these provide evidence that the significance of single-cell sequencing for providing insight into cerebellar development and advancing research into neurobiology and neurological diseases (Desta and Birhanu 2024).

1.3.3 Applications of Single-Cell Analysis in Diverse System

There are several other applications present in single-cell analysis in different fields such as urinary system, digestive system, reproductive system, and variety of micro-organisms that characterize the key insights from non-invasive diagnostic sampling, disease mechanisms, and host–microbial interaction. At the time of injury or disease in urinary tissue, scRNA-seq of urinary cells offers noninvasive approach to interrogate kidney and bladder pathology by profiling cells naturally exfoliated into urine (Klocke et al. 2022). Evidence indicates its value under conditions of acute kidney injury (AKI), such as those related to COVID-19, by identifying immune infiltration patterns and elevated stress pathways in tubular cells that differentiate patients with AKI (Klocke et al. 2022). Moreover, in chronic kidney condition such as membranous nephropathy (MN), urinary macrophages display inflammation-associated transcriptional programs in therapy responders while nonresponders show resident like profile, helping in predicting treatment response and improve prognostic stratification (Liu et al. 2023).

Another application of scRNA-seq in the digestive system has significantly advanced to resolve the cellular complexity of the digestion system at a depth unmatched by earlier approaches. It has exposed functional diversity and earlier previously unrecognized variability in populations of stromal, immunological, and epithelial cells. It also discovered the analyses of disease with disease specific cell and immune pathways in condition such as inflammatory bowel disease (Crohn's disease and ulcerative colitis), celiac disease, and colorectal carcinoma. This helped in understanding of how the gut maintains homeostasis and how intestinal cellular mechanism leads to disease (Bigaeva et al. 2020). Furthermore, SCS application in microbiology field shows genetic makeup and functional activity of individual micro-organisms. As compared to traditional microbial studies, single-cell study of microbial techniques such as microbe-seq and split-pool ligation-based transcriptomics can analyze bacterial cell at individual level. Therefore, it is used in different areas including antibiotic resistance, microbial evolution, and deciphering the microbial dark matter (Wu et al. 2024).

Another application of SCS is in the reproductive system, which provides a rigorous transcriptional framework for detailed analysis of endometriosis disease. Ovarian endometriotic cysts may develop into clear cell and endometrioid ovarian cancers, emphasizing the molecular heterogeneity of endometriosis and the requirement for subtype-specific targeted therapies. Therefore, scRNA-seq has characterized the cellular and molecular level endometriosis. Scientists have created a cell-type reference map for identifying stromal cells and endometrial epithelial cells, dysregulated inflammatory pathways and upregulated complement signaling. However, by combining scRNA-seq results data from 63 women, both having and not having endometriosis, Marečková et al. develop method like human endometrial cell atlas, which made it possible for reliable comparison between healthy and pathological condition (Qi et al. 2025). Overall, these applications highlight the adaptable single-cell technologies for more clarifying disease mechanisms, host–microbe interaction, and tissue specific biology among several other organ systems.

1.4 Challenges and Limitations

Advancement in single-cell technology and computational methods have shown cellular heterogeneity within tissues, providing novel findings into cellular states, dynamics, and regulation processes in both development and disease. Despite significant advancements, substantial challenges persist in the evaluation, synthesis, and analysis of single-cell biological data. Here, we evaluate the present limitations of the discipline and explore future possibilities (Yuan et al. 2017).

1.4.1 Technical and Experimental Challenges

There are various technical challenges faced during single-cell analysis due to trace quantity of DNA, RNA, protein present in single cell. These require amplification steps which can cause systematic error, background noise, and irregular genome/transcriptome representation (Ning et al. 2014). At the time of sample preparation, low RNA input and poor mRNA capture showed high technical noise and variance with missing value in single-cell RNA sequencing often called dropout event. Dropout create false zero, where cell transcripts were not detected and caused systematic underestimation of expression levels (Malhotra et al. 2021). Moreover, isolation method of single-cell analysis such as fluorescence-activates cell shorting (FACS), droplet-based platform, and microfluidics do not always capture all cells equally. These methods mainly capture selectively enriched cells with specific size, morphologies, or physiological states can cause bias in cell type representation (Xu et al. 2016). Additionally, dissociation of cell and preparation of single-cell suspension can cause cellular stress, damage their membranes leading to clumping of cell. These changes can alter gene expression level and reduced the reliability and biological insights from single-cell analysis (Jovic et al. 2022).

1.4.2 Clinical Translation Barriers

In single cell analysis, reliability and consistency are two major limitations because of variability in laboratory protocol, different technologies, and data analysis methods. These make less reliable data for clinical diagnosis or application of predicting patient outcomes. In addition, researchers have limited understanding of technical factor and biological difference leading to variability in data of single cell resulting misinterpretation and false positive finding. Ideally, sample processing time and preservation media after collection met not strict condition in routine clinical practice that causes changes in gene expression pattern and introduce cellular stress in sampled cell (Ascensión et al. 2022). In addition, single cell technology involves complex laboratory procedure, expensive instruments, and large datasets that leads to extra time consuming in silico analysis. As a result, despite the strong potential of accurate diagnosis in single-cell analysis, it is difficult to use with long turn around time in situation where fast clinical decision that are required, such as patient treatment planning (Wang et al. 2025). Despite the higher resolution of single-cell data and its powerful capabilities, clinical application remains slow due to regulatory and ethical challenges. These datasets raise concern about patient privacy, standardization, and data ownership. Moreover, the large public databases such as human cell atlas, European genome-phenome, and gene expression omnibus allow scientists to share and access single-cell profile. However, highly detailed nature of these datasets increases the risk of re-identification and misuse of personal information. By managing these issues appropriately, researcher and institutions get benefits of single-cell data, sharing in a responsible and safe way (Smith et al. 2024).

1.4.3 Limited Spatial and Temporal Context

As highlighted by Yuan et al., a primary difficulty in single-cell analysis is the loss of spatial information, since most traditional methods require tissue dissociation, thereby eliminating knowledge of tissue architecture, cell–cell interactions, and microenvironmental context. Single cell DNA-based and scRNA-based assays encompass the following steps such as cell isolation; cell sorting; and library preparation followed by sequencing (Yuan et al. 2017). Cells have been isolated from local environment and destroyed before profiling. These static "snapshots" obscure critical native information related to cellular spatial context and developmental along dynamic biological trajectories. (Skylaki et al. 2016). In this case, both sources of information are important for interpreting precise state of a cell during its isolation.

1.5 Conclusion and Future Remarks

Single cell analysis has revolutionized scientific studies in biology by unlocking the cellular heterogeneity and rare cell states that are hidden at the macrolevel assessments. Isolation development techniques such as, rapid sequencing, and quantitative analysis, which have facilitated detailed multiomics profiling of individual cells across biological sciences. These approaches have substantially improved understanding of development, infection mechanisms, and clinical improvements, particularly in cancer and neuroscience. Despite these limitations such as technical noise, high price, loss of contextual data, and barriers to clinical translation, persistent technology and analytical breakthroughs continue to address these limitations. Overall, analysis of single-cell is a revolutionary model that is shaping the future of precision biology, medicine, and healthcare.

References

Armand EJ, Li J, Xie F, Luo C, Mukamel EA (2021) Single-cell sequencing of brain cell transcriptomes and epigenomes. Neuron 109(1):11–26

Ascensión AM, Araúzo-Bravo MJ, Izeta A (2022) Challenges and opportunities for the translation of single-cell RNA sequencing technologies to dermatology. Life 12(1):67

Bigaeva E, Uniken Venema WT, Weersma RK, Festen EA (2020) Understanding human gut diseases at single-cell resolution. Hum Mol Genet 29(R1):R51–R58

Birhanu AG (2023) Mass spectrometry-based proteomics as an emerging tool in clinical laboratories. Clin Proteomics 20(1):32

Buenrostro JD, Wu B, Litzenburger UM, Ruff D, Gonzales ML, Snyder MP, Chang HY, Greenleaf WJ (2015) Single-cell chromatin accessibility reveals principles of regulatory variation. Nature 523(7561):486–490

Chen M, Song P, Zou D, Hu X, Zhao S, Gao S, Ling F (2014) Comparison of multiple displacement amplification (MDA) and multiple annealing and looping-based amplification cycles (MALBAC) in single-cell sequencing. PLoS One 9(12):e114520

Chen S, Jiang W, Du Y, Yang M, Pan Y, Li H, Cui M (2023) Single-cell analysis technologies for cancer research: from tumor-specific single cell discovery to cancer therapy. Front Genet 14:1276959

Cheng J, Anne R, Chen YC (2025) Transforming microfluidics for single-cell analysis with robotics and artificial intelligence. Lab Chip 25(23):6100–6125. https://doi.org/10.1039/d5lc00216h

Citri A, Pang ZP, Sudhof TC, Wernig M, Malenka RC (2011) Comprehensive qPCR profiling of gene expression in single neuronal cells. Nat Protoc 7(1):118–127. https://doi.org/10.1038/nprot.2011.430

Clark SJ, Lee HJ, Smallwood SA, Kelsey G, Reik W (2016) Single-cell epigenomics: powerful new methods for understanding gene regulation and cell identity. Genome Biol 17(1):72

Cristinelli S, Ciuffi A (2018) The use of single-cell RNA-Seq to understand virus-host interactions. Curr Opin Virol 29:39–50. https://doi.org/10.1016/j.coviro.2018.03.001

Desta GM, Birhanu AG (2024) Single-cell RNA sequencing, current progresses and future perspectives. Open Biotechnol J 18(1)

Du Y, Huang Q, Arisdakessian C, Garmire LX (2020) Evaluation of STAR and Kallisto on single cell RNA-Seq data alignment. G3 Genes, Genomes, Genetics 10(5):1775–1783

Duncombe TA, Tentori AM, Herr AE (2015) Microfluidics: reframing biological enquiry. Nat Rev Mol Cell Biol 16(9):554–567. https://doi.org/10.1038/nrm4041

Frede J, Anand P, Sotudeh N, Pinto RA, Nair MS, Stuart H, Yee AJ, Vijaykumar T, Waldschmidt JM, Potdar S (2021) Dynamic transcriptional reprogramming leads to immunotherapeutic vulnerabilities in myeloma. Nat Cell Biol 23(11):1199–1211

Gupta V, Lazzaro BP (2022) A robust method to isolate drosophila fat body nuclei for transcriptomic analysis. Fly (Austin) 16(1):62–67. https://doi.org/10.1080/19336934.2021.1978776

He JL, Chen AT, Lee JH, Fan SK (2015) Digital microfluidics for manipulation and analysis of a single cell. Int J Mol Sci 16(9):22319–22332. https://doi.org/10.3390/ijms160922319

Heiniger EK, Buser JR, Mireles L, Zhang X, Ladd PD, Lutz BR, Yager P (2016) Comparison of point-of-care-compatible lysis methods for bacteria and viruses. J Microbiol Methods 128:80–87. https://doi.org/10.1016/j.mimet.2016.07.007

Hong R, Koga Y, Bandyadka S, Leshchyk A, Wang Y, Akavoor V, Cao X, Sarfraz I, Wang Z, Alabdullatif S (2022) Comprehensive generation, visualization, and reporting of quality control metrics for single-cell RNA sequencing data. Nat Commun 13(1):1688

Hu P, Zhang W, Xin H, Deng G (2016) Single cell isolation and analysis. Front Cell Dev Biol 4:116. https://doi.org/10.3389/fcell.2016.00116

Jin Q, Yang Y, Jackson JA, Yoon C, Gracias DH (2020) Untethered single cell grippers for active biopsy. Nano Lett 20(7):5383–5390. https://doi.org/10.1021/acs.nanolett.0c01729

Jovic D, Liang X, Zeng H, Lin L, Xu F, Luo Y (2022) Single-cell RNA sequencing technologies and applications: a brief overview. Clin Transl Med 12(3):e694

Kirsher DY, Chand S, Phong A, Nguyen B, Szoke BG, Ahadi S (2025) Current landscape of plasma proteomics from technical innovations to biological insights and biomarker discovery. Commun Chem 8(1):279

Klocke J, Kim SJ, Skopnik CM, Hinze C, Boltengagen A, Metzke D, Grothgar E, Prskalo L, Wagner L, Freund P (2022) Urinary single-cell sequencing captures intrarenal injury and repair processes in human acute kidney injury. *bioRxiv*, 2022.2002. 2015.479234

Kolodziejczyk AA, Kim JK, Svensson V, Marioni JC, Teichmann SA (2015) The technology and biology of single-cell RNA sequencing. Mol Cell 58(4):610–620. https://doi.org/10.1016/j.molcel.2015.04.005

Komoda T, Matsunaga T (2015) Chapter 6 – Biotechnological study. In: Komoda T, Matsunaga T (eds) Biochemistry for medical professionals. Academic, pp 75–92. https://doi.org/10.1016/B978-0-12-801918-4.00006-2

Krieg C, Nowicka M, Guglietta S, Schindler S, Hartmann FJ, Weber LM, Dummer R, Robinson MD, Levesque MP, Becher B (2018) High-dimensional single-cell analysis predicts response to anti-PD-1 immunotherapy. Nat Med 24(2):144–153

Lake BB, Ai R, Kaeser GE, Salathia NS, Yung YC, Liu R, Wildberg A, Gao D, Fung H-L, Chen S (2016) Neuronal subtypes and diversity revealed by single-nucleus RNA sequencing of the human brain. Science 352(6293):1586–1590

Lee J, Hyeon DY, Hwang D (2020) Single-cell multiomics: technologies and data analysis methods. Exp Mol Med 52(9):1428–1442

Lim J, Park C, Kim M, Kim H, Kim J, Lee DS (2024) Advances in single-cell omics and multiomics for high-resolution molecular profiling. Exp Mol Med 56(3):515–526. https://doi.org/10.1038/s12276-024-01186-2

Liu X, Zhao Y, Niu Y, Xie Q, Nie H, Jin Y, Zhang Y, Lu Y, Zhu S, Zuo W (2023) Urinary single-cell sequence analysis of the urinary macrophage in different outcomes of membranous nephropathy. Clin Kidney J 16(12):2405–2416

Lytal N, Ran D, An L (2020) Normalization methods on single-cell RNA-seq data: an empirical survey. Front Genet 11:41

Maes E, Cools N, Willems H, Baggerman G (2020) FACS-based proteomics enables profiling of proteins in rare cell populations. Int J Mol Sci 21(18). https://doi.org/10.3390/ijms21186557

Malhotra A, Das S, Rai S (2021) Analysis of single-cell RNA-sequencing data: a step-by-step guide. Biomedinformatics 2:43–61. In: s Note: MDPI stays neutral with regard to jurisdictional claims in published …

Method of the Year 2013 (2014) Nat Methods 11(1):1–1. https://doi.org/10.1038/nmeth.2801

Ning L, Liu G, Li G, Hou Y, Tong Y, He J (2014) Current challenges in the bioinformatics of single cell genomics. Front Oncol 4:7

Picelli S (2017) Single-cell RNA-sequencing: the future of genome biology is now. RNA Biol 14(5):637–650. https://doi.org/10.1080/15476286.2016.1201618

Qi Y, Chen X, Zheng S, Wu T, Li Z, Cheng J, Yang X, Tao W, Huang Q, Gu J (2025) Single-cell and spatially resolved omics reveal transcriptional and metabolic signatures of ovarian endometriomas. Nat Commun 16:11539

Quatredeniers M, Serafin AS, Benmerah A, Rausell A, Saunier S, Viau A (2023) Meta-analysis of single-cell and single-nucleus transcriptomics reveals kidney cell type consensus signatures. Sci Data 10(1):361

Ramsköld D, Luo S, Wang Y-C, Li R, Deng Q, Faridani OR, Daniels GA, Khrebtukova I, Loring JF, Laurent LC (2012) Full-length mRNA-Seq from single-cell levels of RNA and individual circulating tumor cells. Nat Biotechnol 30(8):777–782

Reichard A, Asosingh K (2019) Best practices for preparing a single cell suspension from solid tissues for flow cytometry. Cytometry A 95(2):219–226. https://doi.org/10.1002/cyto.a.23690

Schmokel SS, Nordentoft I, Lindskrog SV, Lamy P, Knudsen M, Jensen JB, Dyrskjot L (2023) Improved protocol for single-nucleus RNA-sequencing of frozen human bladder tumor biopsies. Nucleus 14(1):2186686. https://doi.org/10.1080/19491034.2023.2186686

Schor SL, Schor AM (2001) Phenotypic and genetic alterations in mammary stroma: implications for tumour progression. Breast Cancer Res 3(6):373–379. https://doi.org/10.1186/bcr325

Skylaki S, Hilsenbeck O, Schroeder T (2016) Challenges in long-term imaging and quantification of single-cell dynamics. Nat Biotechnol 34(11):1137–1144

Smith J, Smith K, Libwa F, Kimmick M (2024) Ethical and privacy considerations in single-cell data sharing: addressing concerns in public data repositories using multi-scale computational approaches

Squair JW, Gautier M, Kathe C, Anderson MA, James ND, Hutson TH, Hudelle R, Qaiser T, Matson KJ, Barraud Q (2021) Confronting false discoveries in single-cell differential expression. Nat Commun 12(1):5692

Tang F, Barbacioru C, Wang Y, Nordman E, Lee C, Xu N, Wang X, Bodeau J, Tuch BB, Siddiqui A, Lao K, Surani MA (2009) mRNA-Seq whole-transcriptome analysis of a single cell. Nat Methods 6(5):377–382. https://doi.org/10.1038/nmeth.1315

Taniguchi K, Kajiyama T, Kambara H (2009) Quantitative analysis of gene expression in a single cell by qPCR. Nat Methods 6(7):503–506. https://doi.org/10.1038/nmeth.1338

Timp W, Timp G (2020) Beyond mass spectrometry, the next step in proteomics. Sci Adv 6(2):eaax8978

Volozonoka L, Miskova A, Gailite L (2022) Whole genome amplification in preimplantation genetic testing in the era of massively parallel sequencing. Int J Mol Sci 23(9):4819

Wang X, Liu Y, Liu H, Pan W, Ren J, Zheng X, Tan Y, Chen Z, Deng Y, He N (2022) Recent advances and application of whole genome amplification in molecular diagnosis and medicine. MedComm 3(1):e116

Wang X, Li N, Wang W, Liu B (2025) Single-cell sequencing: accurate disease detection. Clin Transl Oncol 28:1–20

Wankhede D, Grover S, Hofman P (2022) Circulating tumor cells as a predictive biomarker in resectable lung cancer: a systematic review and meta-analysis. Cancer 14(24):6112

Woo H, Eyun S-i (2025) Applications and techniques of single-cell RNA sequencing across diverse species. Brief Bioinform 26(4):bbaf354

Wu Y, Zhuang J, Song Y, Gao X, Chu J, Han S (2024) Advances in single-cell sequencing technology in microbiome research. Genes Dis 11(4):101129

Xu Y, Xie J, Chen R, Cao Y, Ping Y, Xu Q, Hu W, Wu D, Gu L, Zhou H (2016) Fluorescence-and magnetic-activated cell sorting strategies to separate spermatozoa involving plural contributors from biological mixtures for human identification. Sci Rep 6(1):36515

Yuan J, Sims PA (2016) An automated microwell platform for large-scale single cell RNA-Seq. Sci Rep 6:33883. https://doi.org/10.1038/srep33883

Yuan G-C, Cai L, Elowitz M, Enver T, Fan G, Guo G, Irizarry R, Kharchenko P, Kim J, Orkin S (2017) Challenges and emerging directions in single-cell analysis. Genome Biol 18(1):84

Zhu X, Li H-D, Guo L, Wu F-X, Wang J (2019) Analysis of single-cell RNA-seq data by clustering approaches. Curr Bioinforma 14(4):314–322

Zilionis R, Nainys J, Veres A, Savova V, Zemmour D, Klein AM, Mazutis L (2017) Single-cell barcoding and sequencing using droplet microfluidics. Nat Protoc 12(1):44–73. https://doi.org/10.1038/nprot.2016.154

Emerging Technologies in Single-Cell Analysis

2

Van-Anh Pham Thi, Khanh-Linh Nguyen, Thu-Hoai Nguyen Thi, and Dinh-Toi Chu

Abstract

Single-cell analysis (SCA) in recent years has emerged as a powerful framework for characterizing cellular heterogeneity and dynamic biological processes. The analytical potential of SCA has been enhanced by recent technological advances in microfluidics, next-generation sequencing, and high-sensitivity detection platforms. Complementary to single-cell RNA sequencing (scRNA-seq) and its integrative modalities such as CITE-seq and REAP-seq, which enable simultaneous profiling of the transcriptome and proteome, single-cell multiomics can incorporate genomic, epigenomic, and other omics data to construct an integrated cellular landscape. Third-generation long-read technology, such as nanopore sequencing, allow amplification-free analysis of DNA and RNA, enabling the identification of full-length isoforms, structural variations, and epigenetic modifications at the single-cell level. Additionally, advanced imaging methods such as stimulated emission depletion (STED) microscopy and quantum-enhanced single-cell nuclear magnetic resonance (NMR) spectroscopy provide noninvasive and spatially resolved molecular characterization. The emerging trends in SCA with novel approaches like combinatorial indexing, AI-based data integration, and spatial transcriptomics are transitioning SCA from a descriptive science to one that provides predictive insights and clinical implications. Despite ongoing limitations related to data quality, cost, and computational complexity of SCA

Van-Anh Pham Thi and Khanh-Linh Nguyen contributed equally with all other contributors.

V.-A. P. Thi · K.-L. Nguyen · T.-H. N. Thi
Faculty of Applied Sciences, International School, Vietnam National University, Hanoi, Vietnam

D.-T. Chu (✉)
Faculty of Applied Sciences, International School, Vietnam National University, Hanoi, Vietnam

The Interdisciplinary Research Group on Biomedicine and Health, International School, Vietnam National University, Hanoi, Vietnam
e-mail: toicd@vnu.edu.vn

V. Singh (ed.), *Advances in Single Cell Analysis*,
https://doi.org/10.1007/978-981-95-9789-5_2

and although data quality, high cost, and computational complexity are huge challenges of SCA, these emerging technologies collectively advance precision medicine, disease modeling, and drug discovery through improved understanding of cellular function and diversity.

Keywords

Single-cell analysis · scRNA-seq · Multiomics · Nanopore sequencing · Advanced imaging · AI-driven analysis

2.1 Introduction

Single-cell analysis (SCA) is one of the exciting breakthroughs in the field of omics and has become a common technique in molecular biology. SCA enables the separation and characterization of cellular heterogeneity as well as the identification of specific functional properties at a single-cell level (Galler and Brautigam 2014). With SCA, cellular details are resolved at the single-cell level, which is useful in oncology, immunology, and stem cell biology (Wang and Bodovitz 2010). The key is identifying heterogeneities at high resolution combined with the production of multiparametric datasets (Chattopadhyay and Winters 2019).

The origin of SCA was traced to the advent of flow cytometry and fluorescence-based detection in the 1960s, which enabled quantitative single-cell measurements for the first time (Chattopadhyay and Winters 2019). Next, microfluidic technologies and droplet-based systems have revolutionized the field, enabling the isolation and analysis of tens of thousands of individual cells within a single experiment (Galler and Brautigam 2014). Prominent platforms such as 10× Genomics Chromium, Drop-seq, and inDrops employ barcoded microfluidic beads to label mRNA molecules (Chattopadhyay and Winters 2019). SCA has been extended to various fields, including genomics, transcriptomics, proteomics, and metabolomics (Chattopadhyay and Winters 2019).

From the technological perspective, scRNA-seq has become a pillar of SCA (Chattopadhyay and Winters 2019; Cheow 2022). This approach allows the identification of cell-to-cell variability and the ability to profile gene expression from thousands of cells simultaneously. The development of integrative technologies such as REAP-seq (Peterson et al. 2017), CITE-seq (Stoeckius et al. 2017), and AbSeq (Shahi et al. 2017) has expanded the types of analyses that can be performed by allowing for the coassessment of proteomic and transcriptomic profiles at single-cell resolution, which offers a more complete and faithful view of biological states than provided by either factor alone (Chattopadhyay and Winters 2019). The implementation of Unique Molecular Identifiers (UMIs) reduces amplification bias and enables near-molecule counting, improving expression estimates across thousands of cells (Galler and Brautigam 2014; Shahi et al. 2017; Haile and Smith 2021).

Once scRNA-seq was sufficiently mature, researchers started to combine it with other data types, sparking the emergence of single-cell multiomics. On top of this, multiomics methodologies have been proposed that aim to combine different molecular levels (genomic, transcriptomic, proteomic, and epigenetic ones) in single cells. Such technologies offer a comprehensive view of cellular states and regulatory networks that are not fully measurable by single-modality assays (Ma et al. 2024).

There are many emerging technological trends that extend the potential areas for SCA. Microfluidics has isolated single cells by encapsulating cells into uniform picoliter droplets (Wang and Bodovitz 2010). At the same time, long-read sequencing technologies, particularly Oxford Nanopore platforms, have addressed the limitations of short-read methods by detecting full-length transcripts, alternative splicing, and RNA modifications, thereby improving the accuracy of transcriptome reconstruction at the single-cell level (Galler and Brautigam 2014; Wang and Bodovitz 2010; Fu et al. 2025).

SCA is quickly developing with various new technologies that bypass the bulk-measurement limitations of traditional methods and broaden single-cell biological studies. In the field of micro/nanofluidics, it has become feasible to integrate micro and nanofluidic systems with mass spectrometry (MS), ultra-miniaturized interfaces, and droplet-based encapsulation devices that lead to in situ proteomics and metabolomics with high sensitivity (Wang and Bodovitz 2010).

In addition to sequencing, recent developments of advanced imaging and spectroscopic tools, including super-resolution STED nanoscopy and single-cell nuclear magnetic resonance (NMR), have also extended the analytical scope of SCA from molecular profiling to spatial and structural analysis. These techniques have allowed real-time imaging of subcellular processes and dynamic biological processes at nanometer resolutions (Blom and Brismar 2014; Peng et al. 2024). Alongside these, automation and functional screening platforms, integrating impedance-based pipetting and nanovial systems with flow cytometry, are transforming SCA from a manual process into a high-throughput platform, enabling broad applications in biological research, medicine, and drug development (Mermans and Mattelin 2023).

2.2 Single-Cell RNA Sequencing (scRNA-seq)

Single-cell RNA sequencing (scRNA-seq) has emerged as a fundamental toolbox for transcriptome profiling by capturing cellular diversity. High-throughput systems, which enable tens of thousands of cells to be profiled simultaneously (e.g., microfluidics, microwell-based, and droplet-based systems such as 10× Genomics, BD Rhapsody), are now available. Automation tools, like DLP-scRNA-seq with cellenONE that automate single cells from hundreds to several thousand per run (Haile and Smith 2021; Desta and Birhanu 2025).

Improvements in full-length transcriptome sequencing compensate for the limitations of 3′/5′-biased methods, thus enabling identification of splicing variants and fusion transcripts (Haile and Smith 2021), and whole-RNA protocols (polyA+ and nonpolyadenylated) that incorporate unique molecular identifiers (UMIs) improve precision quantification and make it possible to analyze other RNA species such as lncRNAs and snRNAs (Shen and Zhao 2022). In microbiology, scRNA-seq offers single-cell level access to gene expression and translation data, bridging the gap between genotype information and phenotype/metabolism data, but background contamination noise and genome assembly issues remain (Haile and Smith 2021).

To restore the dimension of tissue context, scRNA-seq is commonly combined with spatial transcriptomics or in-situ assays; simultaneous RNA-protein analysis (CITE-seq/REAP-seq) connects transcription to phenotype. However, SCA and scRNA-seq also have their own disadvantages in terms of data quality and interpretation, despite recent technical advances. Technical noises and dropout are the two most critical challenges: transcripts originating from low-abundance genes are filtered out during the library preparation procedure, consequently increasing zero counts and possible underestimation of gene expression (Svensson et al. 2018). Similarly, amplification bias arising from PCR steps introduces artificial variation between cells that increases biological differences.

Another difficulty is batch effects, where cells from distinct experimental runs or sequencing lanes show systematic differences that are not of biological origin. These confounding factors are challenging for harmonizing datasets across platforms, labs, or time points and need sophisticated computational correction approaches (e.g., Harmony, Seurat integration) (Lähnemann et al. 2020).

Moreover, cell stress and dissociation artifacts can distort transcriptomic profiles: the enzymatic or mechanical separation of cells from tissue often induces stress-related gene expression that does not reflect in vivo state. This complicates the interpretation of cell-state transitions, particularly in sensitive tissues such as the brain or tumor microenvironments (Svensson et al. 2018; Lähnemann et al. 2020).

Finally, although scRNA-seq does profile gene expression, it typically does so without retaining spatial context—information about where cells are in tissue. To address this shortcoming and recover true tissue organization, integrating with spatial transcriptomics and multiomics technologies is required (Andrews and Hemberg 2018).

scRNA-seq remains subject to intrinsic limitations: loss of spatial information due to cell isolation, technical noise, and dropout causing false negatives, and amplification bias exaggerating intercellular variability. Many protocols focus solely on 3′/5′ ends and exclude non-polyadenylated RNAs (Haile and Smith 2021), limiting the ability to survey the full transcriptome. Moreover, high-throughput systems can lead to batch effects, which hinder the comparison between various biological samples. Such demands require streamlining of both experimental procedures and data analysis platforms to deliver accurate and reproducible results. The overall scRNA-seq workflow, from single-cell dissociation and library preparation to downstream multiomics and spatial integration, together with key technical limitations, is illustrated in Fig. 2.1.

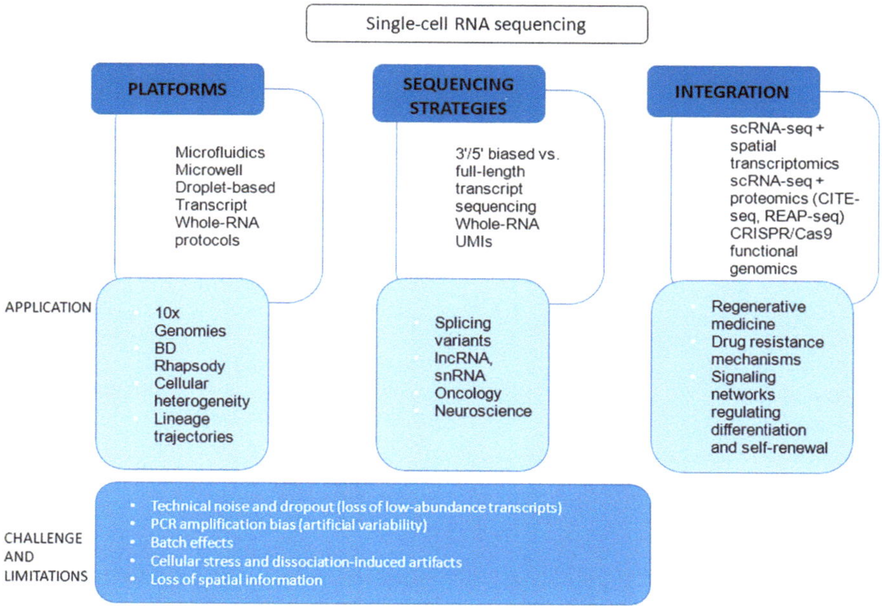

Fig. 2.1 The scRNA-seq pipeline from dissociation to multiomics/spatial integration, with key limitations

scRNA-seq is a key tool for analyzing cellular heterogeneity, identifying novel cell populations, mapping lineage trajectories, and detecting transitional cell states (Shen and Zhao 2022). In the context of regenerative medicine, it provides insight into signaling pathways that control differentiation and self-renewal. For oncology, SCA has accelerated the understanding of multicellular tumor ecosystems, opened a window to cancer stem cells, and brought to light transcriptomic variations. Neuroscience applications provide detailed information on neuronal communications (Desta and Birhanu 2025). Integration with CRISPR/Cas9 and multiomics platforms (e.g., REAP-seq, CITE-seq) extends the ability to simultaneously analyze gene and protein function, supporting drug development and precision medicine initiatives (Yuan et al. 2017). An overview of representative emerging technologies integrated with scRNA-seq and their major applications is summarized in Table 2.1, highlighting recent advances in throughput, automation, full-length transcript detection, and multimodal integration.

2.3 Single-Cell Multiomics

In recent years, single-cell multiomics technologies have made great strides and significantly expanded the ability to analyze at the single-cell level of biology (Ma et al. 2024). Unlike transcriptome, which is a method examining only one layer of

Table 2.1 Examples of emerging technologies for single-cell analysis integration and their applications

Technology	Single-cell analysis	Application	Reference
High-throughput platforms (Microfluidics, Microwell, Droplet-based: 10× Genomics, BD Rhapsody)	scRNA-seq	Large-scale profiling of tens of thousands of cells; analysis of cellular heterogeneity	Haile and Smith (2021) and Desta and Birhanu (2025)
Automation strategies (DLP-scRNA-seq, cellenONE)	scRNA-seq	Expansion of single-cell processing from hundreds to thousands per run	Haile and Smith (2021) and Desta and Birhanu (2025)
Full-length transcript sequencing	scRNA-seq	Detection of splicing variants and fusion transcripts; overcoming 3′/5′-end bias	Haile and Smith (2021)
Whole-RNA sequencing + UMIs (polyA+ and non-polyA)	scRNA-seq	Accurate quantification, detection of non-coding RNAs (lncRNA, snRNA)	Shen and Zhao (2022)
Microbiology application	scRNA-seq	Gene expression and translation at single-cell resolution; integration of genotypic, phenotypic and metabolic data	Haile and Smith (2021)
Integration with spatial transcriptomics and multiomics	scRNA-seq + spatial/multiomics	Reconstruction of tissue organization; improved biological interpretation	Andrews and Hemberg (2018)
CRISPR/Cas9 integration (e.g., REAP-seq, CITE-seq)	scRNA-seq + proteomics	Simultaneous gene–protein analysis; drug development; precision medicine	Yuan et al. (2017)

molecular information, the combination of multiple layers, including genome, proteome, transcriptome, epigenome, and metabolome, has allowed for a more comprehensive reconstruction of the functional states and regulatory mechanisms of single cells, which were described in Fig. 2.2 (Luo et al. 2024). Advanced methods simultaneously evaluate gene expression and surface proteins, such as scRNA-seq combined with simultaneous high-throughput assay for transposase-accessible chromatin (scATAC-seq), cellular indexing of transcriptomes and epitopes (CITE-seq) which have provided new insights into cell genesis and differentiation, cell heterogeneity, and microenvironment interactions (Flynn et al. 2023). At the same time, the advancements in sensitivity, resolution, and the capacity to analyze cells with large numbers have paved the way for broad application in fundamental research and precision medicine (Baysoy et al. 2023).

Although single-cell multiomics have many developments, these technologies still have many challenges and limitations. On the one hand, according to experimental criteria, the acquisition and processing of genetic material at the single-cell

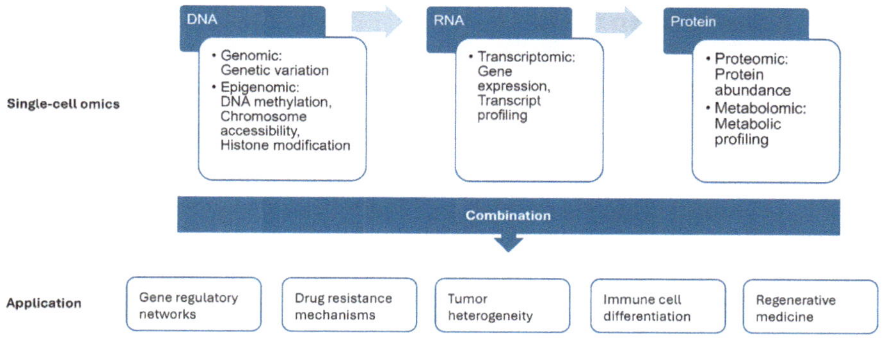

Fig. 2.2 The application of omics combination to analyze single cell

level are prone to data loss and quantification bias. On the other hand, computationally, the integration of multidimensional data layers requires complex analytical algorithms, large computational infrastructures, and reliable methods, while interlaboratory reproducibility is limited. Additionally, high implementation costs and the lack of international standardization procedures are major obstacles to widespread adoption in clinical practice (Baysoy et al. 2023).

Although the challenges have remained, the potential application of single-cell multiomics is significant (Fig. 2.2). In contrast to single-modality sequencing, which simply captures variants at a single level, multiomics approaches clarify not only that cells differ, but also the causes and mechanisms behind those changes. In regenerative medicine, decoding the gene regulatory and epigenetic systems at each stage of stem cell differentiation offers possibilities for optimizing cell therapy. Research combining scRNA-seq with scATAC-seq enables to map transcriptional states and their underlying regulatory landscapes, showing how identical genotypes can exhibit divergent phenotypes because of chromatin accessibility differences (Cao et al. 2018; Ma et al. 2020). This is further enhanced by spatial transcriptomics, which demonstrates how distinct subpopulations are handed out within the tissue microenvironment. Towards cancer research, single-cell multiomics supports to identify cell populations, elucidates drug resistance mechanisms, and directly links heterogeneity to the potential for malignant progression. In breast cancer, integrating spatial profiling and scRNA-seq found that invasive clones are preferentially localized at tumor margins, thereby connecting heterogeneity directly to metastatic potential (Wu et al. 2021a). In immunology, it introduces a resolution of the differentiation and activation state and signaling linkages of immune cell populations, contributing to mechanisms to account for changes in immune responses among individuals. For example, integrated scRNA-seq and scATAC-seq of T-cell differentiation revealed that regulatory heterogeneity in chromatin landscapes underpins divergent cell fates which provide mechanistic insights into variable immune responses among patients (Satpathy et al. 2019). Together, these studies highlight that single-cell multiomics goes beyond cell classification to expand our understanding of molecular mechanisms and microenvironments, thereby laying a solid foundation for precision medicine and treatment strategies (Fig. 2.2).

2.4 Nanopore Analysis

One of the first revolutions in genome exploration occurred in 1977 with the advent of Sanger sequencing, marking the beginning of complete genome decoding (Giani et al. 2020). Subsequently, next-generation sequencing (NGS) emerged, reducing costs and time commitment by a hundredfold in comparison with the first-generation method. However, it uses short-read sequences, including those produced by Illumina technology, so the techniques are limited in the quality of similarity analysis of repetitive regions, the identification of large structural variants, and accurate haplotype evidence (Giani et al. 2020; van Dijk et al. 2018). In contrast, long-read nanopore technology seems attractive due to the ability to directly sequence native DNA and RNA molecules with a theoretically unlimited read length and to detect base modifications without the necessity of amplification (Midha et al. 2019). Thus, considering these features, nanopore sequencing is regarded as the third-generation sequencing technology, opening deeper insights into the domain of genomic research and laying a critical foundation for single-cell analysis.

Fundamentally, the principle of nanopore sequencing is based on measuring changes in ionic current as a single DNA or RNA strand passes through nanoscale pores—either biological pores or solid-state nanopores—embedded in a high-resistance membrane. The applied voltage drives nucleic acid molecules from the cis to the trans chamber through the pore, transiently disrupting the ionic current and generating distinct electrical signals corresponding to each nucleotide type (Wang et al. 2021). Subsequently, these signals are being analyzed, a method that is known as nanopore analysis, and direct identification of nucleotide sequences (van Dijk et al. 2018; Wang et al. 2021). The fundamental principle of nanopore sequencing, based on ionic current disruptions during nucleic acid translocation through a nanopore, is schematically illustrated in Fig. 2.3.

In the early 1980s, the initial concept and pioneering ideas of nanopores were introduced, with the first nanopore was founded as a protein channel of

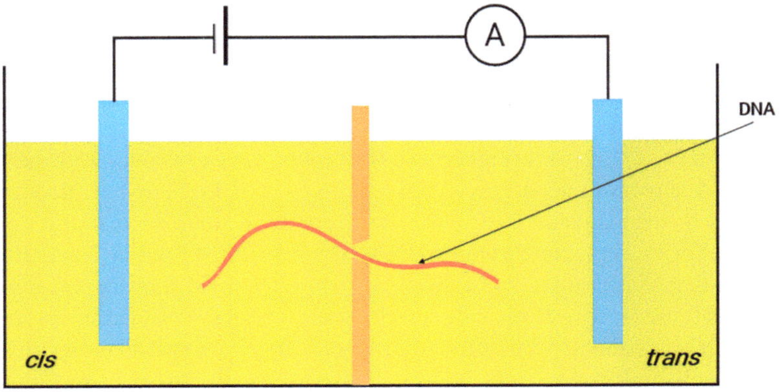

Fig. 2.3 The principle of nanopore sequencing

approximately 2 nm in diameter located on the membrane of *Staphylococcus aureus*, known as α-hemolysin (Wang et al. 2021; Chen et al. 2023). In the 1990s, Deamer, Branton, and Kasiannowicz observed the translocation of single-stranded DNA and RNA molecules through the α-hemolysin pore under an electric field. The result had detectable interruptions in ionic current across the membrane (Deamer et al. 2016; Wang et al. 2014). Further experiments found its ability to differentiate purine and pyrimidine bases and its dynamics of controlling DNA translocation through the pore, including signal analysis algorithms to improve base recognition accuracy (Deamer et al. 2016; Ashkenasy et al. 2005; Cockroft et al. 2008; Fologea et al. 2005; Bayley 2015).

In 2005, Professor Hagan Bayley established Oxford Nanopore Technologies (ONT) with the objective of developing and commercializing nanopore sequencing technology. In 2014, ONT launched the MinION—the first of many sequencer devices it has launched so far. This device acts as a compact portable sequencer having roughly 2048 nanopore proteins that have the ability to read DNA fragments up to 6000–48,000 bases (Deamer et al. 2016). And then, the company later introduced two increased-throughput systems, including the GridION in 2017 and PromethION in 2018. These systems increased the average read length to tens, hundreds, or even thousands of kilobases, compared to only several kilobases in the initial MinION release (Wang et al. 2021; Shafin et al. 2020; Payne et al. 2019). These developments enhanced precision and broadened the capabilities of the nanopore sequencer for genomic research, single-cell analysis, and clinical diagnostics.

Due to its ability to directly read long DNA molecules, nanopore sequencing can span large structural variants, minimizing assembly errors common to short-read sequencing that relies heavily on computational reconstruction. Mojarad et al. recently demonstrated that nanopore sequencing, even with low sequencing coverage (15×), can detect 987 structural variants—nearly twice as many as short-read sequencing (509 variants). In addition, they also identified 604 ncRNAs affected by the variants (Begum et al. 2021).

In SCA, nanopore sequencing enables the detection of rare isoforms, alternative splicing events, and single-cell transcriptomes (Wang et al. 2021; Leggett and Clark 2017). Individual B1a cells from mice were sequenced with the MinION platform, which identified 696 transcription start and end site variants and 354 genes with unreported alternative splicing (Byrne et al. 2017). In humans, using Oxford Nanopore long-read direct RNA sequencing (DRS) to analyze SH-SY5Y neuroblastoma cells during neuronal differentiation revealed 231 genes and 333 isoforms with differential expression compared to undifferentiated cells (Gleeson et al. 2022). Another DRS-based study on leukemia cell lines by Yerin Kim et al. detected *N6*-methyladenosine (m6A) modifications on mRNA and evaluated the relationship between m6A marks, poly(A) tail length, mRNA abundance, and stability, providing a more comprehensive view of gene expression regulation in humans (Kim et al. 2025).

In the future, it is believed that nanopore sequencing will revolutionize single-cell analysis by integrating long-read sequencing with single-cell barcodes and molecular identifiers, along with epigenetic modification profiling (Fu et al. 2025).

Furthermore, the long-read technology will allow mammalian cells to construct haplotype-resolved transcriptomes and epigenomes at a single-cell resolution, which remains a formidable task for the short-read technology (Kumari et al. 2024). Thus, nanopore sequencing is poised to become an excellent tool for understanding complex genomes and transcriptomes through single-cell analysis, and it will offer a comprehensive view of gene expression and its regulation through genetic and epigenetic means in various biological systems.

Despite its remarkable potential, nanopore sequencing still faces several limitations, the most prominent being its relatively high error rate of 5–20%, significantly greater than that of short-read platforms (Goodwin et al. 2016). Multiple studies found that high error rate is mainly attributed to ongoing difficulties in sequencing homopolymeric and repetitive regions. While low-GC regions exhibit fewer errors, the high-GC and homopolymer stretches are responsible for the significant amount of sequencing inaccuracies (Delahaye and Nicolas 2021). In some cases, the resulting data still requires complex bioinformatics pipelines and standardized input preparation, posing additional challenges for single-cell applications where sample quantities are inherently limited (Chen et al. 2023). Therefore, to expand the use of nanopore sequencing into high-precision application areas, such as single-nucleotide variant detection, clinical diagnostics, and single-cell analysis, transformations, and revolutions are also needed in nanopore chemistry, molecular translocation control, and bioinformatics algorithms for accuracy gains.

2.5 Advanced Imaging Techniques

The field of SCA has witnessed remarkable advances in optical and spectroscopic imaging technologies. These advances include the visualization of intracellular structures and activities at a resolution well below the diffraction limits of conventional light microscopies and real-time dynamic molecular information. One of the most significant milestones was the development of Stimulated Emission Depletion (STED) microscopy, proposed by Hell and Wichmann in the 1990s (Kner et al. 2020). Fundamentally, STED is an extension of confocal microscopy, in which a depletion beam is introduced alongside the excitation beam. When scanning the sample, the periphery of the focal spot is simultaneously irradiated with a second beam, typically in a so-called doughnut shape. This beam induces stimulated emission in the outer region and hence suppresses fluorescence emission by the molecules at the focal spot's periphery, drastically reducing the effective intensity point-spread function. This way, the effective point-spread function is smaller than the optical diffraction limit, and the method can achieve spatial resolutions in the nanometer range (Kner et al. 2020; Wu et al. 2021b).

Meanwhile, although conventional nuclear magnetic resonance (NMR) spectroscopy is a potent noninvasive analytical method in chemistry, its sensitivity is low due to physical limitations, and volumes with millions of cells or more are required for a detectable signal; this is impractical for single-cell analysis (Neuling et al. 2023). However, recent advances in quantum physics—particularly sensors based on nitrogen-vacancy (NV) centers in diamond—have enabled NMR signal detection within volumes as small as a few picoliters, comparable to that of a single eukaryotic cell. This breakthrough paves the way for noninvasive chemical analysis at the single-cell level (Allert et al. 2022).

Despite their potential in SCA, imaging-based techniques still face several challenges in practical applications. Specifically, in STED microscopy, the laser beams need to have a very high intensity to remain in a spatially confined state. High-intensity laser beams can cause damage to living cells (Blom and Brismar 2014). Additionally, prolonged illumination often leads to photobleaching, where fluorescent molecules permanently lose their ability to emit light, resulting in progressive image fading (Zhang et al. 2021; Göttfert et al. 2017). For single-cell NMR spectroscopy, the primary limitation lies in the inherently weak NMR signals obtained from such small sample volumes. Even with advanced quantum sensors like NV centers in diamond, signal strength remains low and highly susceptible to environmental noise (Peng et al. 2024; Neuling et al. 2023). Additionally, the diamond-based sensor fabrication is relatively cost-intensive. Furthermore, signal acquisition takes a lot of time. Thus, it is somewhat premature to speculate single-cell NMR within real expected application ranges (Peng et al. 2024).

While sequencing technologies can reveal which genes are being expressed, advanced imaging techniques such as STED and single-cell NMR spectroscopy provide spatial and dynamic insights—showing where genes are expressed and how their products interact. Their combination allows a more complete understanding of the functioning of a cell at the single-cell level. In cancer research, STED microscopy enabled the direct visualization of actin cytoskeleton remodeling and membrane receptor redistribution during the invasion and metastases of breast cancer cells, demonstrating the imminent prospects for such invasive mechanisms at the molecular level (Wu et al. 2021b). Meanwhile, single-cell NMR spectroscopy holds great promise for nondestructive metabolomic analysis. For example, the prior work of Neuling et al. demonstrated the ability of diamond NV sensors, single-cell NMR spectroscopy to record NMR signals in individual cells, and to reveal changes in enzyme activity and energy metabolism under physiological conditions (Neuling et al. 2023). This approach may help identify metabolic fingerprints characteristic of cancer cells, stem cells, or cells under oxidative stress, contributing to biomarker discovery for early diagnosis and personalized therapy (Neuling et al. 2023). In the future, as these technologies continue growing, these and similar technologies can enable simultaneous multiomics analysis of the genome, proteome, metabolome, and space in the genomic organization of molecules within a single cell. A comparative summary of STED microscopy and single-cell NMR spectroscopy, including their principles, advantages, limitations, and representative applications in single-cell analysis, is provided in Table 2.2.

Table 2.2 STED microscopy and single-cell NMR spectroscopy in single-cell analysis

Feature	STED Microscopy	Single-cell NMR
Principle	Super-resolution fluorescence imaging using an excitation and a donut-shaped depletion beam to confine emission below the diffraction limit (Wu et al. 2021b)	Quantum sensing of nuclear magnetic resonance signals via NV centers in diamond, allowing NMR from picoliter-scale volumes (Allert et al. 2022)
Resolution/ sensitivity	Spatial resolution: ~20 to 50 nm; limited by photobleaching and photon noise (Blom and Brismar 2014)	Detects NMR signals from single cells; limited by low signal-to-noise and long acquisition times (Neuling et al. 2023)
Sample type	Requires fluorescent labeling; semi-compatible with live cells but risk of phototoxicity (Blom and Brismar 2014)	Label-free and noninvasive; compatible with intact cells near the sensor surface (Allert et al. 2022)
Main advantages	High spatial and dynamic visualization of intracellular structures and molecular interactions (Wu et al. 2021b)	Nondestructive chemical and metabolic profiling at single-cell resolution (Allert et al. 2022)
Limitations	High laser power → cell damage; photobleaching limits long-term imaging (Blom and Brismar 2014)	Weak NMR signal, sensitive to environmental noise, expensive sensor fabrication (Neuling et al. 2023)
Example applications	Imaging actin remodeling and receptor distribution in cancer cell invasion (Wu et al. 2021b)	Monitoring enzyme activity and metabolism in individual cells (Neuling et al. 2023)

2.6 Other Emerging Technologies in Single-Cell Analysis

In single-cell analysis, beyond conventional sequencing and multiomics combination, several emerging technologies are becoming more developed due to the ability to scalability, resolution, and interpretability. Selected examples of technologies applied in single-cell multiomics integration, associated applications are shown in Table 2.3. One of the vital advancements is combinatorial indexing approaches such as Parse Evercode and Scale Biosciences. These methods enable high-throughput profiling of millions of cells without the need for physical isolation, leading to lower cost and reduction of technical complexity (Filippov et al. 2024). By barcoding cells or nuclei through many successive rounds, they significantly increase throughput which makes population-level research at single-cell resolution more feasible (Tran et al. 2022).

Another advancement of single-cell analysis is the exact manipulation of single cells in nanoliter volumes, that is made possible by microfluidics and advanced materials. By combining innovative hydrogel matrices, substrate droplet printing and/or nanomaterials, droplet-based microfluidics can improve the capture efficiency, minimize the sample loss, and preserve delicate cell types which are frequently lost or damaged in traditional workflows (Wang et al. 2024). When exposed cells to regulated mechanical or biochemical conditions, such material-based

Table 2.3 Emerging technologies instance for single-cell multiomics integration and their applications

Technology	Multiomics integration	Application	Reference
Combinatorial indexing (Parse Evercode, Scale Biosciences)	scRNA-seq	Developmental biology, large-scale cell atlas construction	Hatje et al. (2025) and De Simone et al. (2025)
Microfluidics and advanced materials	scRNA-seq, scProteomics	Functional assays, drug testing, personalized medicine	Wang et al. (2023) and Gebreyesus et al. (2022)
AI-driven data analysis	Cross-omics data integration scRNA-seq	Predictive modeling, precision medicine	Wu and Xie (2025) and Li et al. (2025)
Reference mapping	scRNA-seq, spatial transcriptomics	Meta-analysis	Tian et al. (2022)

development also makes it possible for dynamic perturbation tests, which spread out cell samples from millions to thousands in one single experiment (Datlinger et al. 2021).

The development of AI-driven data analysis is significantly transformative. In the past, computational workflows normally struggle with enormous dimensionality and noise included in single-cell datasets (Ge et al. 2025). In contrast, deep learning algorithms and graph-based models can improve clustering, trajectory inference, and cell type annotation with increased accuracy (Du et al. 2024; Brendel et al. 2022). In particular, the combination of reference mapping with AI allows new data collections to be aligned to high-quality cell atlases, consequently giving biological meaning and context in complex tissues. This makes meta-analysis across studies easier and accelerates the conversion of discoveries into clinically meanings (Lotfollahi et al. 2024).

Despite their noteworthy development, new technologies continue to encounter notable challenges that prevent their widespread acceptance. While combinatorial indexing approaches can face to lower sensitivity in detecting rare or low-abundance transcripts (Hatje et al. 2025) and reduce data quality in the case of application for heterogeneous tissues (Phan et al. 2021). Though powerful, microfluidic systems still require specialized equipment and often have limited compatibility with diverse tissue types, especially fragile or complex primary samples (Zhou et al. 2021). In the computational domain, AI-driven data analysis is predictably reliant on training datasets that raise issues with algorithmic bias, overfitting, and interpretability (Li et al. 2025; Chen et al. 2025). Furthermore, reference mapping is heavily dependent on the completeness and quality of existing cell atlases, which may lack of uncommon cell states. This could result in misclassification and limited applicability in novel disease contexts (Michielsen et al. 2023). All these limitations point to the necessity of continuous refinement, benchmarking, and improvement so that new single-cell technologies reach their full potential in both clinical application and research.

2.7 Conclusion

Single-cell analysis (SCA) has been reshaping biological research by uncovering cellular complexity rapidly beyond bulk-level research. Although current technologies face many issues with cost, sensitivity, and data integration, the continuous emergence in sequencing, imaging, and computational methods would attend to this weakness. The combination of multiomics, spatial, and AI-driven approaches is expected to improve diagnostic precision and accelerate the conversion of single-cell discoveries into clinical applications.

References

Allert RD, Briegel KD, Bucher DB (2022) Advances in nano- and microscale NMR spectroscopy using diamond quantum sensors. Chem Commun (Camb) 58(59):8165–8181

Andrews TS, Hemberg M (2018) False signals induced by single-cell imputation. F1000Res 7:1740

Ashkenasy N, Sánchez-Quesada J, Bayley H, Ghadiri MR (2005) Recognizing a single base in an individual DNA strand: a step toward DNA sequencing in nanopores. Angew Chem Int Ed Engl 44(9):1401–1404

Bayley H (2015) Nanopore sequencing: from imagination to reality. Clin Chem 61(1):25–31

Baysoy A, Bai Z, Satija R, Fan R (2023) The technological landscape and applications of single-cell multi-omics. Nat Rev Mol Cell Biol 24(10):695–713

Begum G et al (2021) Long-read sequencing improves the detection of structural variations impacting complex non-coding elements of the genome. Int J Mol Sci 22(4)

Blom H, Brismar H (2014) STED microscopy: increased resolution for medical research? J Intern Med 276(6):560–578

Brendel M et al (2022) Application of deep learning on single-cell RNA sequencing data analysis: a review. Genomics Proteomics Bioinformatics 20(5):814–835

Byrne A et al (2017) Nanopore long-read RNAseq reveals widespread transcriptional variation among the surface receptors of individual B cells. Nat Commun 8:16027

Cao J et al (2018) Joint profiling of chromatin accessibility and gene expression in thousands of single cells. Science 361(6409):1380–1385

Chattopadhyay PK, Winters AF (2019) High-parameter single-cell analysis. Annu Rev Anal Chem 12:411

Chen P et al (2023) Portable nanopore-sequencing technology: trends in development and applications. Front Microbiol 14:1043967

Chen X et al (2025) AI-driven transcriptome prediction in human pathology: from molecular insights to clinical applications. Biology 14. https://doi.org/10.3390/biology14060651

Cheow LF (2022) Single cell analysis technologies in biomedical research. SLAS Technol 27:107–108

Cockroft SL, Chu J, Amorin M, Ghadiri MR (2008) A single-molecule nanopore device detects DNA polymerase activity with single-nucleotide resolution. J Am Chem Soc 130(3):818–820

Datlinger P et al (2021) Ultra-high-throughput single-cell RNA sequencing and perturbation screening with combinatorial fluidic indexing. Nat Methods 18(6):635–642

De Simone M et al (2025) A comprehensive analysis framework for evaluating commercial single-cell RNA sequencing technologies. Nucleic Acids Res 53(2):gkae1186

Deamer D, Akeson M, Branton D (2016) Three decades of nanopore sequencing. Nat Biotechnol 34(5):518–524

Delahaye C, Nicolas J (2021) Sequencing DNA with nanopores: troubles and biases. PLoS One 16(10):e0257521

Desta GM, Birhanu AG (2025) Advancements in single-cell RNA sequencing and spatial transcriptomics: transforming biomedical research. Acta Biochim Pol 72:13922
Du JH, Chen T, Gao M, Wang J (2024) Joint trajectory inference for single-cell genomics using deep learning with a mixture prior. Proc Natl Acad Sci USA 121(37):e2316256121
Filippov I, Philip CS, Schauser L, Peterson P (2024) Comparative transcriptomic analyses of thymocytes using 10×genomics and parse scRNA-seq technologies. BMC Genomics 25(1):1069
Flynn E, Almonte-Loya A, Fragiadakis GK (2023) Single-cell multiomics. Annu Rev Biomed Data Sci 6:313–337
Fologea D et al (2005) Slowing DNA translocation in a solid-state nanopore. Nano Lett 5(9):1734–1737
Fu Y et al (2025) Single cell and spatial alternative splicing analysis with Nanopore long read sequencing. Nat Commun 16(1):6654
Galler K, Brautigam K (2014) Making a big thing of a small cell—recent advances in single cell analysis. Analyst 139:1237
Ge S et al (2025) Deep learning in single-cell and spatial transcriptomics data analysis: advances and challenges from a data science perspective. Brief Bioinform 26(2):bbaf136
Gebreyesus ST et al (2022) Streamlined single-cell proteomics by an integrated microfluidic chip and data-independent acquisition mass spectrometry. Nat Commun 13(1):37
Giani AM, Gallo GR, Gianfranceschi L, Formenti G (2020) Long walk to genomics: history and current approaches to genome sequencing and assembly. Comput Struct Biotechnol J 18:9–19
Gleeson J et al (2022) Accurate expression quantification from nanopore direct RNA sequencing with NanoCount. Nucleic Acids Res 50(4):e19
Goodwin S, McPherson JD, McCombie WR (2016) Coming of age: ten years of next-generation sequencing technologies. Nat Rev Genet 17(6):333–351
Göttfert F et al (2017) Strong signal increase in STED fluorescence microscopy by imaging regions of subdiffraction extent. Proc Natl Acad Sci USA 114(9):2125–2130
Haile S, Smith RD (2021) A scalable strand-specific protocol enabling full-length Total RNA sequencing from single cells. Front Genet 12:665888
Hatje K et al (2025) Comparison of single-cell RNA-seq methods to enable transcriptome profiling of neutrophils in clinical samples. Cell Rep Methods 5(9):101173
Kim Y et al (2025) Nanopore direct RNA sequencing of human transcriptomes reveals the complexity of mRNA modifications and crosstalk between regulatory features. Cell Genom 5(6):100872
Kner P, Manley S, Shechtman Y, Stallinga S (2020) 25(th) anniversary of STED microscopy and the 20(th) anniversary of SIM: feature introduction. Biomed Opt Express 11(3):1707–1711
Kumari P et al (2024) Advances in long-read single-cell transcriptomics. Hum Genet 143(9–10):1005–1020
Lähnemann D et al (2020) Eleven grand challenges in single-cell data science. Genome Biol 21(1):31
Leggett RM, Clark MD (2017) A world of opportunities with nanopore sequencing. J Exp Bot 68(20):5419–5429
Li N et al (2025) Artificial intelligence-driven integration of single-cell RNA sequencing and transcriptome analysis to decipher APOE's role in gastric cancer prognosis and therapy. Discov Oncol 16(1):314
Lotfollahi M, Yuhan H, Theis FJ, Satija R (2024) The future of rapid and automated single-cell data analysis using reference mapping. Cell 187(10):2343–2358
Luo Y, Zhao C, Chen F (2024) Multiomics research: principles and challenges in integrated analysis. BioDesign Research 6:0059
Ma S et al (2020) Chromatin potential identified by shared single-cell profiling of RNA and chromatin. Cell 183(4):1103–1116.e20
Ma J, Dong C, He A, Xiong H (2024) Single-cell multiomics: a new frontier in drug research and development. Front Drug Discov 4
Mermans F, Mattelin V (2023) Opportunities in optical and electrical single-cell technologies to study microbial ecosystems. Front Microbiol 14:1233705

Michielsen L et al (2023) Single-cell reference mapping to construct and extend cell-type hierarchies. NAR Genom Bioinform 5(3):lqad070

Midha MK, Wu M, Chiu KP (2019) Long-read sequencing in deciphering human genetics to a greater depth. Hum Genet 138(11–12):1201–1215

Neuling NR, Allert RD, Bucher DB (2023) Prospects of single-cell nuclear magnetic resonance spectroscopy with quantum sensors. Curr Opin Biotechnol 83:102975

Payne A, Holmes N, Rakyan V, Loose M (2019) BulkVis: a graphical viewer for Oxford nanopore bulk FAST5 files. Bioinformatics 35(13):2193–2198

Peng Y et al (2024) NMR spectroscopy for metabolomics in the living system: recent progress and future challenges. Anal Bioanal Chem 416(9):2319–2334

Peterson VM et al (2017) Multiplexed quantification of proteins and transcripts in single cells. Nat Biotechnol 35(10):936–939

Phan HV et al (2021) High-throughput RNA sequencing of paraformaldehyde-fixed single cells. Nat Commun 12(1):5636

Satpathy AT et al (2019) Massively parallel single-cell chromatin landscapes of human immune cell development and intratumoral T cell exhaustion. Nat Biotechnol 37(8):925–936

Shafin K et al (2020) Nanopore sequencing and the Shasta toolkit enable efficient de novo assembly of eleven human genomes. Nat Biotechnol 38(9):1044–1053

Shahi P et al (2017) Abseq: ultrahigh-throughput single cell protein profiling with droplet microfluidic barcoding. Sci Rep 7(1):44447

Shen X, Zhao Y (2022) Recent advances in high-throughput single-cell transcriptomics and spatial transcriptomics. Lab Chip 22(24)

Stoeckius M et al (2017) Simultaneous epitope and transcriptome measurement in single cells. Nat Methods 14(9):865–868

Svensson V, Vento-Tormo R, Teichmann SA (2018) Exponential scaling of single-cell RNA-seq in the past decade. Nat Protoc 13(4):599–604

Tian Y et al (2022) Single-cell immunology of SARS-CoV-2 infection. Nat Biotechnol 40(1):30–41

Tran V et al (2022) High sensitivity single cell RNA sequencing with split pool barcoding. bioRxiv: p. 2022.08.27.505512

van Dijk EL, Jaszczyszyn Y, Naquin D, Thermes C (2018) The third revolution in sequencing technology. Trends Genet 34(9):666–681

Wang D, Bodovitz S (2010) Single cell analysis: the new frontier in 'omics'. Trends Biotechnol 28:281

Wang Y, Yang Q, Wang Z (2014) The evolution of nanopore sequencing. Front Genet 5:449

Wang Y et al (2021) Nanopore sequencing technology, bioinformatics and applications. Nat Biotechnol 39(11):1348–1365

Wang C et al (2023) Dissecting order amidst chaos of programmed cell deaths: construction of a diagnostic model for KIRC using transcriptomic information in blood-derived exosomes and single-cell multi-omics data in tumor microenvironment. Front Immunol 14

Wang C et al (2024) Microfluidic biochips for single-cell isolation and single-cell analysis of multiomics and exosomes. Adv Sci (Weinh) 11(28):e2401263

Wu Y, Xie L (2025) AI-driven multi-omics integration for multi-scale predictive modeling of genotype-environment-phenotype relationships. Comput Struct Biotechnol J 27:265–277

Wu SZ et al (2021a) A single-cell and spatially resolved atlas of human breast cancers. Nat Genet 53(9):1334–1347

Wu Z, Xu X, Xi P (2021b) Stimulated emission depletion microscopy for biological imaging in four dimensions: a review. Microsc Res Tech 84(9):1947–1958

Yuan G-C, Cai L, Elowitz M (2017) Challenges and emerging directions in single-cell analysis. Genome Biol 18:84

Zhang J et al (2021) Low-power two-color stimulated emission depletion microscopy for live cell imaging. Biosensors (Basel) 11(9)

Zhou W-M et al (2021) Microfluidics applications for high-throughput single cell sequencing. J Nanobiotechnol 19:312

Advances in Single Plant Cell Genomics

3

Dhananjay K. Pandey and Bhupendra Chaudhary

Abstract

Single-cell and spatial omics technologies are rapidly reshaping plant science. These approaches allow researchers to examine individual plant cells and understand how they function within their native tissue environment. These advanced approachs, beyond bulk sequencing methods, reveal the true heterogeneity of plant tissues. This has enabled the annotation of rare cell types, reconstruction of developmental trajectories, and enhanced insight into plant responses to the environmental stimuli. Adoption of single-cell technologies to plant systems require advanced research and innovation due to plant's rigid cell walls, complex tissues, and diverse genomic structures. Recent advancement in protoplast isolation, microfluidic sorting, single-nucleus sequencing, and spatial transcriptomics have made it possible to profile thousands of plant cells with high accuracy. These tools have also expanded the study of specialized cell types which include callus cells, somatic embryos, fibers, trichomes, root hairs, and guard cells. Studies of such cell-types reveal their unique transcriptional programming and the associated regulatory networks. Single-cell multiomics approaches now integrate transcriptomic, epigenomic, proteomic, and metabolomic data at the cellular level, and are being increasingly adopted. Combined with improved computational tools, they offer a more complete view of gene regulation and cell-to-cell communication in plants. These insights are transforming our understanding of plant development, stress tolerance, pathogen interactions, and beneficial associations. These advancements also provide more precise strategies for crop improvement. This chapter highlights key advances, challenges, and future directions in single-cell and spatial genomics.

D. K. Pandey
Amity Institute of Biotechnology, Amity University, Ranchi, Jharkhand, India

B. Chaudhary (✉)
School of Life Sciences, Jawaharlal Nehru University, New Delhi, India
e-mail: bhupendra@jnu.ac.in

V. Singh (ed.), *Advances in Single Cell Analysis*,
https://doi.org/10.1007/978-981-95-9789-5_3

Keywords

Single-cell genomics · Spatial transcriptomics · Multiomics integration · Cellular heterogeneity · Gene regulatory networks · Plant stress responses · Crop improvement strategies

3.1 Introduction

Single-cell and spatial omics approaches have revolutionized biological studies by allowing detailed analysis of the individual cells and their spatial relationships within complex tissues (Nobori 2025). These advancements are particularly impactful in plant science, where multicellular organisms exhibit heterogeneous cell populations with distinct developmental programs and environmental responses (Nobori 2025). Integration of single-cell and spatial transcriptomics with spatial metabolomics enables a detailed, stage-specific view of gene expression, regulation and metabolic activity within defined plant tissues and cell types (Yu et al. 2022). This molecular level insights are important for dissecting the precise intercellular regulatory mechanisms that orchestrate plant development and their coordinated responses to diverse environmental stimuli (Arora et al. 2018; Hu et al. 2024; Pandey and Chaudhary 2016, 2017). The rigid nature of the plant cell wall presents major obstacles to isolating and handling individual cells, limiting the broader application of scRNA-seq in lignified and fully mature plant tissues (Zhang et al. 2023). Despite these challenges, advancements in enzymatic cell wall degradation, coupled with microfluidic and fluorescent activated cell sorting techniques, have facilitated the isolation of viable plant protoplasts and nuclei for downstream analyses (Giolai et al. 2019). This allows for detailed molecular profiling of plant cells/tissue at a resolution previously unattainable with bulk sequencing methods (Yu et al. 2022). These methodological innovations have paved the way for unprecedented insights into cell type-specific gene expression, developmental trajectories, and stress responses in various plant species (Cole et al. 2021). The emergence of nuclear RNA sequencing (snRNA-seq) has further eased the difficulties associated with protoplast isolation, particularly in woody plants with robust secondary cell walls, by enabling transcriptomic analysis directly from isolated nuclei (Zhang et al. 2023). However, despite these strides, the process of enzymatic digestion for protoplast preparation can introduce transcriptional changes due to stress responses, which might confound the interpretation of true cellular states (Pan et al. 2024). Furthermore, while scRNA-seq has significantly increased the resolution of transcriptomic studies, the process of tissue dissociation often leads to the loss of crucial spatial context, which is challenging to reconstruct for most of the plant organs (Depuydt et al. 2022). Spatially resolved transcriptomic approaches have arisen to tackle this limitation, permitting researchers to study gene expression patterns in their native tissue context (Giacomello et al. 2017). These techniques enable researchers to map gene activity to specific cell types and structures, providing a better understanding of plant development and physiological responses within their native tissue environment. The combination of single-cell and spatial omics creates new opportunities to unravel the intricate regulatory networks that drive plant biology, offering unparalleled resolution and broad potential for crop enhancement and sustainable farming. By characterizing all genes collectively within an organism, genomics illuminates their interrelationships and impacts, revealing detailed gene expression heterogeneity and spatial specificity that traditional bulk methods obscure (Lv et al. 2024). This augmented resolution confers a better comprehension of cellular functions amid biological processes, unveils rare and novel cell types, and accommodates the parallel assessment of heterogeneous cell populations across diverse conditions (Sang and Kong 2024).

3.1.1 The Evolution of Single-Cell Genomics

The rapid evolution of single-cell genomics, initially propelled by advancements in mammalian systems, has progressively extended its reach to plant biology, necessitating innovative adaptations to overcome unique plant-specific cellular and tissue characteristics. This expansion has played a crucial role in deepening our insight into plant development and environmental responses at an unmatched level of cellular detail. These adaptations address the distinct structural and biochemical properties of plant cells, such as the rigid cell wall and often polyploid genomes, which historically hindered single-cell isolation and analysis. Consequently, specialized protocols for protoplast generation, nuclear extraction, and high-throughput sequencing have been developed to enable the comprehensive investigation of the gene expression profiles from individual plant cells (Nobori et al. 2023; Zhang et al. 2023). Previous methodologies, such as single-cell-digital gene expression (1cell-DGE), utilized micromanipulation to extract living cell contents while preserving positional information, allowing for the detection of differentially expressed genes during developmental reprogramming (Zhang et al. 2023). Recent improvements in single-cell and spatial omics technologies have made it possible to examine the gene expression and metabolic profiles in the selected plant tissues and cell types with stage-specific resolution (Yu et al. 2022). These techniques have been particularly instrumental in deconstructing the cellular heterogeneity often masked by the bulk sequencing, thereby providing a detailed understanding of the intricate regulatory mechanisms within plant tissues (Zhang et al. 2023). For instance, differences in gene expression within meristematic tissues have been precisely determined and validated through confocal imaging, while developmental trajectories of various plant structures, including *Arabidopsis* female germline and rice floret, have been reconstructed using scRNA-seq (Zhang et al. 2023). These advancements have unveiled novel cell types and delineated the dynamic transcriptional changes underlying cell fate specification in different plant species, moving beyond the traditional model organisms (Zhang et al. 2023; Zhou et al. 2022). Despite these successes, substantial challenges persist, notably in obtaining sufficient numbers of viable single cells from diverse plant tissues and species, and in mitigating technical noise inherent in single-cell assays (Zenda et al. 2021). Furthermore, the diverse chemical composition of plant cell walls across species and tissues necessitates species-specific protoplast isolation protocols, posing a significant bottleneck for the universal applicability and high-throughput profiling (Cole et al. 2021). For example, while optimized enzymatic digestion protocols have improved the protoplast preparations for certain plant tissues, many cell types remain recalcitrant to such methods (Huo et al. 2024). Recent studies emphasize persistent challenges in plant single-cell research, particularly the demand for more efficient tissue dissociation and reliable single-cell isolation methods to extend the reach of single-cell genomics across the wide range of plant taxa (Minne et al. 2021). Additionally, the intricate nature of secondary metabolism in plants continues to restrict the development of a standardized extraction and analytical protocols for single-cell metabolomics (Zhang et al. 2023). Nevertheless, continuous advancements in scRNA-seq technology have significantly enhanced our understanding of plant tissue development beyond root systems, encompassing other organs such as leaves, flowers, pollen, and seeds (Cole et al. 2021). These developments have facilitated the discovery of previously unrecognized cell types and marker genes, thereby deepening our insight into the functional and structural diversity of plant cells (Zhang et al. 2023). Furthermore, applying scRNA-seq to complex tissue architectures, including the phloem poles, has illuminated the specialized roles of critical transport-associated cells like sieve elements and companion cells (Liu et al. 2022). This heightened level of resolution now enables the identification of cell type–specific regulatory and signaling networks that coordinate plant development, adaptation, and growth processes.

3.1.2 Technological Breakthroughs Enabling Single-Cell Analysis

Technological innovations, particularly in microfluidics and molecular barcoding, have been crucial in overcoming the technical hurdles associated with single-cell isolation and high-throughput transcriptomic profiling in plants, paving the way for comprehensive cell atlases. Although these advancements are very significant, but still have limitations in case of plant systems, compared to human and animal tissues for disease analysis and cell type identification (Bian et al. 2024). However, recent progress in the plant single-cell transcriptomics has demonstrated remarkable success in dissecting complex developmental processes and stress responses (Cho et al. 2025). For example, single-cell RNA sequencing (scRNA-seq) has been effectively utilized in *Arabidopsis* root cells to identify hundreds of cell-type-specific gene expression profiles, revealing both known and novel genes involve in different developmental stages (Zenda et al. 2021). Such studies have not only validated the existing knowledge of cell identities but also unveiled previously undefined cell subclasses (Cole et al. 2021). Furthermore, the ability to analyze single-nucleus RNA sequencing (snRNA-seq) data has provided insights into cell-type specific transcriptomic changes in complex plant tissues that are difficult to dissociate (Minne et al. 2021). These methodologies are essential for elucidating the intricate gene regulatory networks governing root development, as they yield high-resolution, cell type-specific expression profiles. (Wang et al. 2025a, b, c). This comprehensive understanding is crucial for deciphering the molecular mechanisms underlying plant plasticity and adaptation to the diverse environmental conditions, offering avenues for crop genetic improvement. Beyond *Arabidopsis*, single-cell transcriptomic atlases have also been developed for other economically important crops such as rice and maize, leading to the discoveries concerning unique cellular localization and mobility features, fungal response mechanisms, and cell-type-specific regulatory programs, including phytohormone biosynthesis and signaling (Zhang et al. 2023). The expanding range of studied plant species demonstrates the growing capacity of single-cell technologies to tackle key challenges in crop enhancement and plant biotechnology (Pan et al. 2024). This heightened level of resolution has enabled researchers to determine the conserved regulators involved in root formation and to map complex gene regulatory networks (Minne et al. 2021). Moving beyond conventional transcriptomic analysis, the combination of single-cell RNA sequencing (scRNA-seq) with complementary approaches, such as single-cell ATAC sequencing (scATAC-seq) and spatial transcriptomics, is expected to yield a deeper, integrative view of gene regulation and cellular activity across space and time (Minne et al. 2021). Such multiomics frameworks are vital for decoding how chromatin dynamics, transcriptional control, and spatial context interact within plant tissues, shedding light on the molecular mechanisms guiding cell differentiation and adaptation to environmental cues (Biswas et al. 2023; Minne et al. 2021). The application of spatial transcriptomics, in particular, allows for the precise localization of gene expression within tissues, bridging the gap between single-cell resolution and anatomical context, which is especially valuable for studying complex plant developmental processes like root meristem organization (Minne et al. 2021). Moreover, this technology is anticipated to accelerate the establishment of new molecular biology applications by enabling the study of crops and species that are difficult to transform (Zhang et al. 2023). The increasing spatial resolution of these technologies, including methods for three-dimensional image segmentation, further enables the deconstruction of tissue heterogeneity, making them highly suitable for the comprehensive gene regulatory network analysis in various plant systems (Cole et al. 2021; Minne et al. 2021). Such detailed analysis is critical for the identification of cell-specific regulators and cell types that can be targeted for agricultural crop improvement, especially in developing new promoters for tissue-specific CRISPR applications to avoid pleiotropic effects (Lindsay et al. 2023). Single-cell multiomics approaches, such as combining scRNA-seq and live-cell imaging, have further elucidated developmental patterns, revealing the role of GTPase signaling in phloem vasculature formation and identifying key transcription factors regulating root development (Zhang et al. 2023). These integrated approaches allow for tracking cellular dynamics over time and space, revealing how individual cells contribute to tissue-level functions and developmental trajectories (Drapek et al. 2017). Single-cell chromatin accessibility analyses have mapped over 8000 regulatory regions in *Arabidopsis* roots and revealed cell-type-specific *cis*-regulatory elements in maize, majority of which show enhancer-like activity within unmethylated LTR retrotransposons (Dorrity et al. 2021; Zhang et al. 2023). These insights into the epigenomic landscape at single-cell resolution are crucial for understanding cell-type-specific gene regulation and unraveling the intricate mechanisms of plant development and environmental responses (Thibivilliers and Libault 2021; Upadhyay et al. 2019). The continued advancement of these single-cell multiomics techniques promises to reveal the regulatory complexity, thereby enabling targeted genetic engineering. These advances are particularly promising for moving beyond model organisms like *Arabidopsis* to directly characterize complex systems such as plant-pathogen interactions or cell—communications in crop species, enabling rapid annotation of transcriptome atlases for less-studied plants (Minne et al. 2021). Furthermore, these methods facilitate the identification of cell-type-specific regulatory mechanisms in horticultural plants, thereby offering potential avenues for improving agronomic traits from early developmental stages (Zhang et al. 2023). The ability to dissect cellular heterogeneity through comparative single-cell transcriptomic analysis also enables the molecular characterization of mutant cell phenotypes, as seen in studies of root-hair defective and glabra2 mutants of *Arabidopsis* (Zhang et al. 2023). Such fine-resolution analyses are crucial for deciphering how individual genes influence cell fate and differentiation, thereby strengthening the foundation for targeted genetic manipulation. Complementary approaches, including single-cell DNase I hypersensitivity profiling and scATAC-seq, have further illuminated chromatin accessibility dynamics at the single-cell level (Zhang et al. 2023). This expanded understanding of the epigenome, combined with advanced methods like single-nucleus CUT&Tag for histone modification profiling, offers unprecedented opportunities to dissect epigenetic heterogeneity in individual plant cells (Zhang et al. 2023). This comprehensive epigenomic profiling, when coupled with advanced computational methods for network inference, can delineate the intricate regulatory circuits that govern plant development and stress responses at an unprecedented resolution (Huo et al. 2024). Integration of genomic and epigenomic data accelerate the identification of novel genetic targets for crop improvement, particularly for enhancing resilience to both biotic and abiotic stresses (Deep and Pandey 2024). Moreover, meta-analysis of existing omics data can identify candidate genes involved in stress tolerance, such as specific transcription factor families, which can then be further validated and characterized at single-cell resolution level (Zenda et al. 2021). Such detailed cellular atlases, including those generated from diverse tissues like rice root tips, poplar stems, and cotton ovules, are vital for unraveling cell-type-specific responses to the environmental perturbations and developmental cues (Li et al. 2023a, b, c). The application of plant single-cell ATAC-seq technology, for instance, is crucial for gaining a comprehensive preview of the plant cell epigenome and understanding gene regulation (Luo et al. 2020). Furthermore, advanced computational algorithms and machine learning models are being developed to interpret these complex multiomic datasets, offering deeper insights into gene regulatory networks and cell-to-cell communication that underpin the plant development and responses to environmental stimuli (Bhardwaj et al. 2022; Biswas et al. 2023; Pandey and Mishra 2024; Zhou et al. 2022). This integrative approach facilitates the comprehensive understanding of how plant cells function and interact, which provide new avenues for targeted genetic engineering and breeding strategies (Deshmukh et al. 2025). Achieving comprehensive models of gene regulatory dynamics during development will require larger datasets to encompass the full spectrum of cellular states (Dorrity et al. 2021). Moreover, integrating multiomics layers, such as single-cell transcriptomes and chromatin accessibility profiles offer a holistic understanding of the regulatory architecture, clearly demonstrating how chromatin accessibility shapes transcriptional activity (Depuydt et al. 2022; Li et al. 2023a, b, c). The advancement of artificial intelligence and machine learning algorithms will also be crucial in refining the inference and analysis of these complex regulatory networks (Bhardwaj et al. 2022; Biswas et al. 2023; Huo et al. 2024; Pandey and Mishra 2024; Zhou et al. 2022). This refined understanding will further facilitate the prediction of gene functions, identification of key regulatory elements, and engineering of crop plants with improved characteristics (Depuydt et al. 2022).

3.2 Fundamentals of Single Plant Cell Isolation and Processing

Accurate isolation of individual protoplasts or nuclei is essential for single-cell genomics, despite the constraints imposed by the rigid plant cell wall. Therefore, enzymatic digestion protocols require thorough optimization to maintain the cell integrity and RNA quality while effectively breaking down this protective barrier. After protoplast isolation, specialized microfluidic devices and robotic platforms enable high-throughput capture and encapsulation of individual cells. These cells are subsequently lysed to release their genetic material for the library preparation followed by RNA sequencing. Next, reverse transcription synthesizes complementary DNA (cDNA), which is then amplified and fragmented for transcript sequencing. Unique molecular identifiers are often used to reduce amplification biases. Thorough optimization of these protocols is essential to ensure high-quality data and accurately reflect cellular heterogeneity.

However, plant single-cell methodologies are less advanced than animal single-cell methods in terms of throughput, range of applications, and number of species studied, largely due to distinct biological and technical challenges (Cole et al. 2021). The robust plant cell wall requires specific dissociation protocols, and abundant secondary metabolites can interfere with subsequent molecular assays (Cole et al. 2021). This often leads to a lower cell viability and reduced RNA integrity, significantly affecting the quality and interpretation of single-cell omics data. To address these issues, ongoing research focuses on the development of moderate dissociation methods and improving current protocols, such as investigating new enzyme combinations and mechanical disruption techniques (Zhou et al. 2022). Even with optimized enzymatic digestion, natural variability in cell size and cell wall composition across plant cell types can introduce biases, potentially leading to the selective exclusion, or damage of larger or more resistant cells (Minne et al. 2021).

Therefore, researchers are increasingly exploring alternative approaches such as single-nucleus RNA sequencing (snRNA-seq), which avoids the need for protoplast isolation from the intact nuclei (D'Agostino et al. 2023). This method is particularly advantageous for tissues rich in secondary metabolites, or highly lignified cells, where protoplast isolation is challenging. Moreover, techniques such as laser capture microdissection offer precise isolation of specific cell types from complex tissues, though at lower throughput, ensuring the capture of rare, or spatially defined cell populations with high fidelity. Further innovations involve microfluidic platforms that integrate both dissociation and sorting capabilities, enabling a more streamlined workflow from tissue to single-cell libraries with minimal sample handling.

Beyond isolation, the subsequent analysis of single-cell RNA sequencing data from unique plant cells faces significant challenges, particularly in clustering analyses (Wang et al. 2025a, b, c). Large vacuoles can lead to RNA degradation during protoplast isolation process, while the rigid cell wall often necessitates harsh enzymatic treatments that can induce stress responses, thereby altering gene expression profiles (Ren and Wang 2023; Xin et al. 2023). These artifacts complicate cell type identification and pseudotime trajectory inference, often requiring specialized computational algorithms adapted for the plant single-cell data (Cole et al. 2021). Consequently, methods for plant protoplast isolation need further optimization and standardization to preserve cell viability and integrity, minimizing stress-related gene expression changes during dissociation (Kim et al. 2023). This is particularly evident in highly lignified tissues or those with suberized cell walls, where harsh digestion conditions impede efficient protoplast release, or significantly alter transcriptional profiles (Chen et al. 2021). Recent advancements in dissociation techniques, such as microfluidic-assisted cell sorting and improved enzymatic cocktails, are therefore, crucial for capturing the true transcriptional landscape of diverse plant cell types. Furthermore, the nonuniform size and robust cell walls of plant cells present significant hurdles for high-throughput microfluidic encapsulation techniques, limiting the broad applicability of these advanced platforms (Shulse et al. 2019). The persistent challenge of dissociating rigid plant cell walls while maintaining cellular viability and RNA integrity remains a critical bottleneck, hindering the widespread adoption of single-cell technologies in plant biology (Cole et al. 2021; Marchant et al. 2021).

For instance, developing efficient protoplast isolation protocols for recalcitrant species, such as *Chirita pumila*, using consecutive enzymatic digestions can significantly broaden the scope of single-cell RNA sequencing applications in plants (Wang et al. 2022). The use of a dead cell removal kit can also enhance the yield of viable protoplasts, improving the quality of downstream single-cell analyses (Kim et al. 2023). Despite these advancements, certain cell types, like trichomes, remain difficult to isolate *via* viable protoplasts due to their cutin and wax coverings, underscoring the need for alternative approach such as snRNA-seq (Zhou et al. 2022). While snRNA-seq offers a promising avenue for such challenging cell types, it may not fully capture the cytoplasmic RNA pool, which can contain important localized transcripts (Madrid et al. 2025). Therefore, developing methods that efficiently capture both nuclear and cytoplasmic RNA from intact single plant cells without inducing stress, or compromising the RNA integrity remains a critical frontier. Further research into microfluidic platforms modified as per the unique biophysical properties of plant cells, such as their diverse sizes and robust cell walls, could bridge this knowledge gap. Therefore, the refinement of both dissociation protocols and analytical techniques is crucial to fully harness single-cell genomics for uncovering the intricate cellular landscapes and regulatory networks governing plant development and stress responses. The emerging power of droplet-based single-cell RNA sequencing offers a transformative approach to dissecting cell-to-cell heterogeneity, addressing fundamental questions related to developmental trajectories, tissue regeneration, and stress responses at an unprecedented resolution (Denyer and Timmermans 2021). This technology has been successfully applied to numerous biological systems, comprehensively profiling gene expression in thousands of individual cells simultaneously (Cole et al. 2021). However, adapting droplet-based single-cell methods for plant cells presents unique challenges due to the rigid cell wall and often large cellular sizes, necessitating specialized enzymatic digestion and careful optimization of the microfluidic parameters for successful encapsulation. Despite these difficulties, advancements in snRNA-seq are overcoming the protoplast preparation limitations, offering a precise view into gene expression regulation, and are increasingly adopted in horticultural research (Zhang et al. 2023). This method allows researchers to bypass the often-damaging protoplasting step, which can introduce artifacts and stress responses in gene expression profiles (Sunaga et al. 2020). Moreover, recent innovations in capturing full-length transcripts from single cells, including methods like SMART-seq, offer the potential to identify cell type-specific mRNA isoforms, providing a deeper functional genomic annotation of transcriptome data (Cole et al. 2021; Zhang et al. 2023).

3.3 Single-Cell Sequencing Technologies in Plants

Single-cell RNA sequencing (scRNA-seq) has become a transformative approach for resolving gene expression patterns at high resolution, enabling precise characterization of cellular heterogeneity within complex plant tissues (Zheng et al. 2023) (Fig. 3.1). This approach allows researchers to distinguish discrete cell populations, trace developmental pathways, and reveal previously unknown regulatory mechanisms that shape plant growth and environmental responses (Yu et al. 2022). It has been widely utilized to characterize various cell types across different plant tissues and organs, including roots, shoots, leaves, and inflorescences in plant species (Chau et al. 2025).

Furthermore, single-cell transcriptomic analyses help infer gene function by resolving transcript abundance within precise spatial and subcellular contexts, enabling the identification of tissue-domain-specific genes and their associated pathways (Depuydt et al. 2022). Such fine analysis reveals how different cell types respond to various perturbations and provides a more complete understanding of cell population function, adaptive properties, and molecular toolkits for biotechnological engineering (Cole et al. 2021). For example, single-cell RNA sequencing studies have extensively characterized root development in *Arabidopsis*, enabling

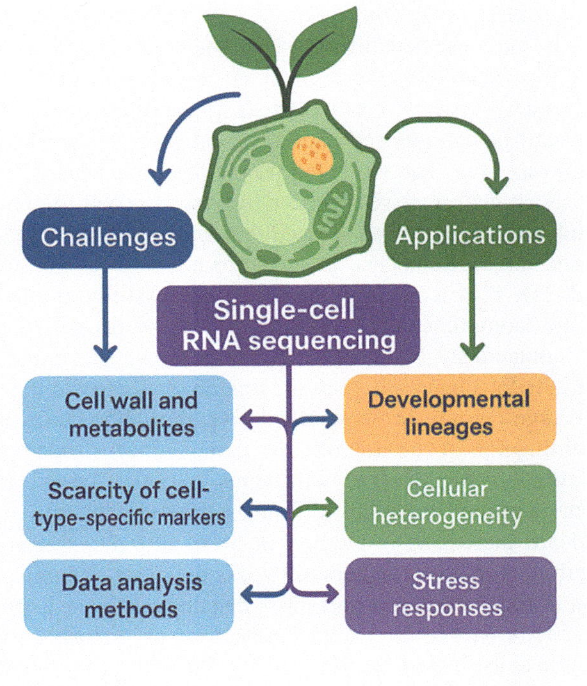

Fig. 3.1 Overview of Single-cell RNA sequencing and their applications in resolving plant cell identity, development, and stress responses

the identification of all major root cell types and elucidating the developmental trajectories underlying root growth (Cole et al. 2021). Such detailed cellular maps are crucial for understanding complex developmental processes and engineering plants for improved traits. Moreover, developmental trajectory analysis, applied to scRNA-seq data, has been instrumental in reconstructing differentiation processes, such as fiber cell development in *Bombax ceiba* and root cell fate determination in Moso bamboo, revealing species-specific mechanisms and key regulatory genes like *PheWOX13* (Zhang et al. 2023). The high-throughput nature of scRNA-seq enables the rapid detection of rare or previously unknown cell types while capturing a broad spectrum of cellular states and conditions, thereby providing a more complete understanding of their contributions to biological processes (Sang and Kong 2024). For example, scRNA-seq has been effectively employed to analyze the transcriptome atlas during cotton somatic embryogenesis, partitioning calli into distinct cell populations and clusters (Pan et al. 2024). This allows for the identification of specific gene expression patterns associated with different developmental stages and cell fates during plant regeneration. Similarly, single-cell analysis has been applied to characterize early cotyledon development in cotton, revealing cell heterogeneity and the differentiation trajectory of palisade and spongy cells (Pan et al. 2024). Beyond developmental insights, single-cell approaches have been instrumental in unraveling plant responses to environmental stressors, such as salt stress, by enabling the construction of high-resolution transcriptional maps of root tips and the identification of genes involved in salt tolerance within specialized cells such as epidermal bladder cells (Liu et al. 2024). This comprehensive approach also broadens the understanding of anther development and its response to high temperatures, further illustrating the versatility of scRNA-seq in uncovering intricate regulatory mechanisms (Pan et al. 2024). These single-cell studies have provided unprecedented insights into the transcriptional landscapes of various plant tissues, from vegetative organs to reproductive structures, under both normal and stress conditions (Liu et al. 2022; Pan et al. 2024). Further, the application of single-cell RNA sequencing (scRNA-seq) has also revolutionized cotton transcriptome analysis and gene functional investigations, enabling the distinction of transcriptomes even in the adjacent cell types (Pan et al. 2024). This refined resolution is essential for deciphering the precise molecular mechanisms underpinning cellular differentiation and specialization within plant tissues. Specifically, scRNA-seq has been used to identify genes controlling fiber cell initiation and rhythmic growth in cotton, as well as to investigate the molecular mechanisms of gland morphogenesis in cotyledons (Pan et al. 2024). Moreover, the technology has been pivotal in uncovering hierarchical transcriptional regulatory networks, such as those governing terpenoid biosynthesis in cotton secretory glandular cells, by identifying key transcription factors such as HSF4a and NAC42 that directly impact biosynthetic gene expression (Pan et al. 2024). Advanced imaging techniques further confirmed this depth of scRNA-seq analysis to elucidate gene expression differences among meristematic domains (Zhang et al. 2023).

3.3.1 Advancement in Single-Cell RNA Sequencing (scRNA-seq)

Advancements in single-cell methodologies, such as nuclear isolation strategies and improved enzymatic cocktails, are broadening the applicability of scRNA-seq to a wider array of plant species and tissue types (Zenda et al. 2021). For instance, optimized protocols have been developed to successfully generate comprehensive single-cell transcriptional landscapes from highly lignified tissues, such as poplar stems, which were previously considered challenging (Chen et al. 2021). This progress enables researchers to delve into the intricate cell type diversity and differentiation trajectories within woody plants, where traditional bulk RNA sequencing approaches fall short in resolving cellular heterogeneity (Chen et al. 2021). Moreover, single-nucleus RNA sequencing (snRNA-seq) offers a promising alternative, circumventing the need for protoplast isolation, which is particularly advantageous for plant tissues with robust cell walls and for perennial woody plants where protoplast isolation is difficult (Zhang et al. 2023). The ongoing developments that overcome the need for protoplast isolation, such as single-nucleus sequencing, is thus critical for expanding single-cell analyses to a broader range of plant species and tissues, thereby overcoming a major bottleneck in plant single-cell genomics (Minne et al. 2021). These innovative approaches allow for the investigation of gene regulatory networks in cell-type-specific contexts, offering an advanced understanding of how complex genetic interactions govern various biological processes in plants (Huo et al. 2024; Cole et al. 2021). This is particularly significant given that conventional bulk RNA-seq methods often obscure cell-type-specific gene expression due to the averaging of signals across heterogeneous cell populations (Cho et al. 2025). Furthermore, the integration of single-cell ATAC-seq with scRNA-seq datasets provides a powerful framework for elucidating gene regulatory networks and classifying cell types with unprecedented resolution by simultaneously profiling chromatin accessibility and gene expression (Cole et al. 2021). Such combined approaches have already demonstrated their utility in uncovering how chromatin architecture influences cell type identity and developmental trajectories in plant root cells (Zenda et al. 2021). The ability to recognize new cell types and infer gene regulatory networks has greatly advanced our understanding of developmental dynamics within complex biological systems (Cole et al. 2021). In *Populus* stems, for instance, single-cell RNA sequencing (scRNA-seq) has revealed 20 distinct cellular clusters, encompassing various vascular tissue cell types, and enabled the reconstruction of differentiation pathways for phloem and xylem cells, demonstrating the method's remarkable capacity to capture cellular diversity often overlooked by traditional bulk analyses (Chen et al. 2021). Similarly, the use of single-cell ATAC sequencing (scATAC-seq) in maize has produced a detailed *cis*-regulatory atlas across several organs, offering crucial insights into gene regulation at individual-cell resolution (Huo et al. 2024). Collectively, these investigations exemplify the growing sophistication of single-cell research tools and their transformative impact on exploring the molecular foundations of plant growth and adaptive responses (Bian et al. 2024; Chen et al. 2021). With improved throughput and sensitivity, current single-cell transcriptomic methods are achieving enhanced precision in defining cell identities, thereby demanding more advanced analytical strategies to interpret the resulting high-dimensional data (Nobori et al. 2023). To make sense of these complex datasets, the continual refinement of computational algorithms for trajectory mapping and cell type annotation is essential, particularly when examining dynamic developmental transitions or plant responses to environmental changes (Bhardwaj et al. 2022; Biswas et al. 2023). Progress in both experimental and computational fronts of single-cell genomics is fueling a shift toward a comprehensive system-level view of plant biology, exceeding the limitations of population-averaged measurements.

The integration of multiomics approaches spanning genomics, transcriptomics, epigenomics, and proteomics promises to deliver an all-encompassing understanding of cellular interactions and functions within complex plant tissues (Depuydt et al. 2022). Notably, detailed single-cell atlases, especially for root systems, are shedding light on cell lineage specification and developmental trajectories (Wang et al. 2025a, b, c). These atlases have already uncovered previously unknown cell types, such as protophloem and metaphloem sieve elements in *Arabidopsis* roots, alongside new marker genes and pivotal regulatory elements that broaden our grasp of plant cellular diversity (Zhang et al. 2023). Such detailed cellular and molecular mapping is instrumental for deciphering the mechanisms underlying root development and adaptation to diverse environmental conditions. Indeed, pairing single-cell RNA sequencing (scRNA-seq) with single-cell Assay for Transposase-Accessible Chromatin using sequencing (scATAC-seq) across various growth stages or environmental perturbations can further elucidate the regulatory networks governing root development and response (Minne et al. 2021). The application of single-cell multiomics approaches in diverse plant species, including rice and maize, has led to significant discoveries, such as cell-type-specific regulatory programs related to phytohormone biosynthesis and responses, and the identification of stem cell clusters in the maize shoot apical meristem (Zhang et al. 2023). However, despite advancements, some specialized cell types, such as those found within the phloem poles, remain challenging to fully characterize due to their low abundance and complex heterogeneity, necessitating targeted enrichment strategies (Liu et al. 2022). This challenge highlights the need for advanced cell isolation techniques and computational methods capable of resolving subtle molecular distinctions within highly similar cell populations (Zhang et al. 2023).

3.3.2 Advancement in Single-Cell DNA Sequencing (scDNA-seq)

While scRNA-seq offers profound insights into transcriptomic dynamics, scDNA-seq complements it by exposing genome-level variations, including copy number alterations and ploidy changes, at single-cell resolution in plants. This technology is particularly vital for plant systems, where somatic mutations and polyploidy are common, offering unprecedented resolution into clonal diversity and genomic heterogeneity that bulk sequencing methods cannot provide. Advances in single-cell DNA sequencing have made it possible to trace cellular lineages and monitor chromosomal rearrangements that occur during plant growth or under environmental stress, thereby illuminating the evolutionary processes shaping plant genomes. Moreover, applying single-cell DNA sequencing can shed light on the key developmental events, such as the divergence of protophloem into metaphloem sieve elements and procambial cells by detecting lineage-specific genetic modifications and regulatory components (Zhang et al. 2023). When combined with spatial transcriptomics, this approach enables the mapping of genomic variation to defined cell types and their physical arrangement within tissues, ultimately delivering an integrated perspective on genotype-phenotype relationships with an exceptional spatial and molecular precision (Yu et al. 2022). This capability is particularly beneficial for dissecting complex developmental processes, such as lateral root initiation, where specific cell types and their regulatory genes can be precisely identified (Zhang et al. 2023). Beyond genomic variation, single-cell DNA sequencing also facilitates the identification of novel promoters through the precise mapping of *cis*-regulatory elements, which can then be harnessed for targeted genetic modifications of crop plants (Lindsay et al. 2023). Moreover, the ability to correlate high-throughput phenotyping with single-cell omics data enables the identification of genes and pathways essential for specific product biosynthesis, thereby optimizing production yields in bioengineering applications (Cole et al. 2021). The advent of single-cell multiomic assays, combining DNA and RNA sequencing from the same cell, further extends this capability by directly linking genomic alterations to transcriptomic programs at a single-cell resolution (Edrisi et al. 2023). Such integrated approaches are instrumental for dissecting complex regulatory networks governing cellular identity and function (Cole et al. 2021). Such approach holds significant promise for advancing plant synthetic biology, particularly in optimizing metabolic pathways for enhanced biomaterial and natural product biosynthesis (Cole et al. 2021).

3.3.3 Single-Cell Multiomics Approaches

By integrating genomic, transcriptomic, proteomic, and epigenomic data at the level of individual cells, single-cell multiomics provides a unified and detailed understanding of cellular behavior and function (Zhang et al. 2022). This integrative framework is vital for decoding the complex regulatory circuits that orchestrate plant development, metabolic processes, and stress adaptation, which are often obscured in bulk tissue analyses (Cole et al. 2021). For instance, combining single-cell RNA sequencing with single-cell epigenomics enables the correlation of gene expression profiles with chromatin accessibility, or methylation patterns, thereby identifying the epigenetic mechanisms that control cell fate decisions and plasticity. These combined datasets enable the identification of regulatory elements and transcription factors that drive specific cellular processes, offering deeper insights into how plant cells differentiate and respond to the environmental cues. Furthermore, the application of single-cell metabolomics and proteomics, though lagging behind transcriptomics in plants due to detection sensitivities, is a beginning to bridge critical gaps in understanding functional outputs at the cellular level (Yang et al. 2021a, b). These nascent technologies hold significant promise for characterizing complex cellular states and elucidating metabolic pathways with unprecedented details (Cole et al. 2021). The combined use of snRNA-seq and scATAC-seq has enabled the direct linking of chromatin accessibility with gene expression in individual cells across various plant tissues, including those of *Arabidopsis* and maize (Zhang et al. 2023). These advancements are paving the way for a comprehensive understanding of the interplay between genetic, epigenetic, and transcriptional regulatory networks within individual plant cells, enabling the identification of novel targets for crop improvement (Liu et al. 2023a, b). Therefore, holistic understanding encompassing various omic layers, is essential for unraveling the complexities of plant cellular heterogeneity and development (Zhang et al. 2023).

3.4 Applications of Single Cell Genomics in Plants

Single plant cell genomics has diverse applications, such as unraveling complex cellular organization and accurately tracing how cells develop across various plant organs and tissues (Bian et al. 2024).

3.4.1 Unraveling Plant Development and Cell Lineage

Single-cell technologies have become indispensable tools for revealing plant cellular diversity and developmental logic. These methods enable the discovery of previously unrecognized cell types, the delineation of their specialized functions, and the reconstruction of developmental trajectories with exceptional resolution (Chau et al. 2025). By analyzing individual cells rather than bulk populations, researchers can track how transcriptional programs shift during differentiation and morphogenesis, thereby uncovering the molecular mechanisms that guide plant development (Nobori 2025). Trajectory inference from scRNA-seq datasets can depict the pace and direction of cell state transitions, highlighting the developmental branch points, and also identify new regulators involved in these processes. This is especially valuable in species with limited prior characterization, where conventional tissue classifications may not adequately reflect the true functional units of plant organization (Minne et al. 2021). Such high-resolution analyses also make it possible to detect rare or transient cell states that are typically obscured in bulk tissue profiling, significantly enriching our understanding of plant cellular complexity (Cho et al. 2025; Nobori 2025).

Furthermore, single-cell genomics has also transformed the study of intricate plant structures, exemplified by detailed analyses of the *Arabidopsis* root phloem vasculature from initial cell formation to enucleation (Zhang et al. 2023). These efforts have made it possible to isolate and study phloem pole cell types that were previously difficult to access, such as protophloem and metaphloem sieve elements, phloem pole pericycles, and companion cells, key players in the plant's long-distance transport system (Liu et al. 2022). Access to these rare populations enables detailed characterization of their transcriptional identities and functional contributions to nutrient allocation (Zhang et al. 2023). Beyond cataloguing cell identities, lineage-tracing approaches at single-cell resolution allow researchers to follow the progression from progenitor cells to fully differentiated states. This capability is particularly advantageous for dissecting complex developmental processes such as lateral root formation and early specification events (Minne et al. 2021). The highly ordered structure of *Arabidopsis* root meristem, for example, is particularly amenable to such lineage tracing and trajectory analysis (Minne et al. 2021). Moreover, these techniques are very helpful in dissecting the molecular mechanisms underlying phenotypic variations in mutants, as evidenced by studies characterizing root-hair defective and *glabra2* mutants, which illuminate how genetic perturbations affect cell fate specification and differentiation pathways (Zhang et al. 2023). Collectively, these advanced single-cell analyses not only reconstruct developmental pathways with unprecedented precision but also enable the assembly of detailed gene regulatory networks. Such networks are essential for identifying conserved regulatory modules across plant species (Minne et al. 2021). scRNA-seq has proven powerful especially in mapping developmental trajectories within root tissues, offering granular insights into how cells acquire identity and progress through maturation (Drapek et al. 2017; Minne et al. 2021). This level of resolution provides a foundational understanding of the transcriptional circuits governing root development and reveals how these programs evolve across developmental transitions (Dorrity et al. 2021; Roszak et al. 2021).

3.4.2 Investigating Plant Responses to Stress and Environment

Single-cell genomics provides a unique opportunity to investigate how individual plant cells respond to biotic and abiotic stresses, uncovering cell-specific transcriptional changes crucial for adaptation and resilience (Liu et al. 2022; Shulse et al. 2019). For instance, single-cell RNA sequencing has revealed stress-induced accumulation of abscisic acid and jasmonoyl isoleucine in individual cells of water- and wound-stressed leaves, demonstrating how these phytohormones mediate cellular stress responses (Zhang et al. 2023). This approach allows for detailed analysis of diverse cellular responses to environmental changes, identifying specific cell populations that are primary responders or highly susceptible to stress (Zhang et al. 2023). Furthermore, single-cell analyses have revealed crucial intercellular signaling pathways involved in plant cell differentiation and interactions with environmental stimuli and pathogens, uncovering noncell-autonomous phytohormone responses and developmental regulation (Zhang et al. 2023). For instance, under phosphate deficiency, TMO5/LHW-mediated cytokinin biosynthesis becomes activated in plant vasculature, and the transported cytokinin promotes transition to root hair cell by inducing TMO5/LHW target genes in epidermal cells (Zhang et al. 2023). These detailed insights into cellular responses to environmental cues, including dynamic changes in chromatin accessibility under heat stress, are crucial for understanding the complex regulatory networks underlying plant plasticity and adaptation (Zhang et al. 2023). These high-resolution analyses aid in identifying key regulatory genes and pathways that allow plants to manage adverse conditions, thereby guiding targeted genetic engineering for crop resilience. The application of spatial transcriptomics further enhances this understanding by providing spatially resolved gene expression profiles within tissues, enabling the deconstruction of tissue heterogeneity and the analysis of complex gene regulatory networks that coordinate growth and development in changing environments (Minne et al. 2021; Zhou et al. 2022). This comprehensive approach moves beyond bulk tissue analysis, allowing researchers to pinpoint how specific cell types contribute to the plant's overall stress response and adaptation mechanisms (Berrío et al. 2025; Zenda et al. 2021).

3.4.3 Exploring Plant Immunity and Pathogen Interactions

The adoption of cutting-edge single-cell and spatial profiling methods in plant–pathogen research is unveiling intricate details of how plants respond to infection across individual cells and tissue landscapes (Zhu et al. 2023). The single cell profiling is crucial because only a small number of plant cells directly engage with pathogens, but these localized interactions significantly influence how severe a disease becomes (Cole et al. 2021). These tools can identify the complex communication between infected and uninfected cells, showing local defense strategies and how immune signals spread throughout the plant (Thibivilliers and Libault 2021). This allows researchers to pinpoint which cell types start immune responses and observe when and where defense genes become active in the infected areas. For example, single-cell RNA sequencing has been used to study root infections, identifying how specific cell types respond and revealing the defense strategies different cell populations use against microbes (Zhu et al. 2023). This method is especially important for pathogens such as arbuscular mycorrhizal fungi, which only attack certain cortical root cells. Standard bulk RNA sequencing would not capture the specific host responses from these targeted interactions due to dilution (Cole et al. 2021). Therefore, because pathogens often target specific cells, single-cell methods are essential to accurately understand the molecular communication and defense strategies at the infection site (Giolai et al. 2019).

Moreover, combining spatial transcriptomics with scRNA-seq enables researchers to map pathogen-triggered gene activity within target tissues. This provides insights into how immune responses are spatially organized and how cells communicate during an infection (Cole et al. 2021; Giolai et al. 2019). This combined approach can identify uncommon cell populations that play key roles in immunity and show where gene transcripts are located within infected tissues. This is vital for understanding how plants defend themselves against different pathogens (Cho et al. 2025; Giolai et al. 2019; Sang and Kong 2024). For instance, by integrating single-cell and spatial transcriptomics, studies have revealed a three-dimensional response in rice leaf cells to *Magnaporthe oryzae* infection. This showed increased expression of immune receptors in vascular tissues and higher activity of genes involved in diterpene phytoalexin production in procambium cells (Wang et al. 2025a, b, c). This detailed spatial analysis also helps in understanding how plants interact with microbes beyond just pathogens, clarifying the complex relationships between plants and their various microbial communities (Saarenpää et al. 2023). Specifically, these high-resolution methods can map out local and cell type-specific pathways that control interactions with the root microbiome. This shows how plants either encourage beneficial microbes, or protect themselves from harmful ones (Yang et al. 2025).

3.4.4 Understanding Plant Symbioses

The application of single-cell and spatial technologies has also transformed the study of plant symbioses, moving beyond the generalized understanding provided by bulk analyses to reveal the intricate cellular and molecular communications underpinning these relationships (Hu et al. 2024). These techniques offer high resolution to examine the exact cellular processes of nutrient exchange, signaling, and host integration within symbiotic structures like nodules or mycorrhizal associations. For instance, spatial transcriptomics can pinpoint the specific cells engaged in nutrient transfer during mycorrhizal symbioses, detailing gene expression patterns that support phosphate uptake and carbon allocation (Yu et al. 2022). Similarly, single-cell analysis of nitrogen-fixing root nodules can identify cell types that house bacteria and facilitate metabolic exchange, revealing the genetic mechanisms behind these crucial interactions (Minne et al. 2021). These detailed analyses help identify critical regulatory elements and transporters needed for beneficial plant–microbe relationships, marking a significant step forward for agricultural and environmental biotechnology (Cole et al. 2021; Yu et al. 2022). Additionally, these methods are crucial for understanding the dynamic gene expression changes within individual cells as symbiotic interfaces form and are maintained, offering key insights into the molecular basis of mutualistic relationships. This detailed molecular insight is vital for explaining the mechanisms of nutrient exchange and immune suppression, which are hallmarks of successful symbiotic relationships (Bhimani et al. 2024).

3.5 Single-Cell Omics in Key Plant Cell Types

3.5.1 Callus Cells

Callus cells are a heterogeneous population of dedifferentiated plant cells exhibiting totipotency, enabling their regeneration into complete plants. They form the basis for plant tissue culture, genetic transformation, and somatic embryogenesis (Long et al. 2022). Molecularly, callus formation entails a significant reprogramming of cell identity, triggered by phytohormone-induced dedifferentiation and the activation of pluripotency-associated regulatory networks (Ikeuchi et al. 2013). The balance between auxin and cytokinin critically determines callus fate, directing it towards shoot organogenesis, root regeneration, or embryogenesis. Single-cell RNA sequencing (scRNA-seq) studies have further elucidated the molecular mechanisms underpinning callus formation and regeneration, revealing dynamic changes in gene expression and cell-type-specific responses during dedifferentiation and redifferentiation (Liu et al. 2025). These transcriptomic studies have demonstrated that callus cells are heterogeneous, comprising subpopulations with distinct developmental trajectories and stress-response states. For instance, auxin signaling components and cytokinin response regulators are enriched in separate subclusters, indicating hormonal domain partitioning (Liu et al. 2025). Transient cell states marked by high expression of wound-response genes, ROS-scavenging enzymes, and chromatin remodeling factors appear early in callus induction before lineage commitment. Epigenomic profiling has further uncovered global chromatin decondensation and dynamic DNA methylation changes, suggesting that epigenetic plasticity is central to callus totipotency (Pírek et al. 2023).

However, a significant analytical challenge lies in distinguishing true developmental intermediates from stress-induced transcriptional artifacts caused by explant excision, enzymatic digestion, and prolonged in vitro culture. Callus cells can also develop somaclonal variation, which complicates the fidelity of subsequent regeneration. Furthermore, batch effects related to genotype and explant origin frequently obscure conserved transcriptional modules. Despite these challenges, single-cell genomics has been instrumental in identifying crucial transcription factors and signaling pathways that regulate callus formation and differentiation, including the negative regulation of SAM formation from callus in *Arabidopsis* by WUSCHEL-RELATED HOMEOBOX 13 (Zhang et al. 2023). Understanding the single-cell landscape of callus formation provides actionable strategies for enhancing plant biotechnology. Identifying molecular markers of regenerative competence could enable the early selection of highly totipotent cell populations, thus improving transformation efficiency in recalcitrant species. Moreover, precise adjustment of hormone regimes to selectively promote desired subpopulations could accelerate regeneration timelines. Ultimately, these insights can reduce somaclonal variations and enhance the predictability of in vitro plant propagation systems (Lardon and Geelen 2020).

3.5.2 Somatic Embryos

Somatic embryogenesis is a process where specialized plant cells dedifferentiate and then redifferentiate to form a complete embryo without sexual reproduction (Pandey et al. 2012; Pandey and Chaudhary 2014a, b). This technique is a powerful tool for clonal propagation and genetic engineering in many plant species, offering advantages such as producing large numbers of genetically identical plants and overcoming sexual reproduction barriers, making it valuable for crop improvement (Park et al. 2023). Single-cell genomic approaches are now providing detailed insights into the molecular events governing somatic embryogenesis, revealing the intricate regulatory networks that drive cellular reprogramming and developmental progression (Tang et al. 2025). For instance, exposure to 2,4-dichlorophenoxyacetic acid during somatic embryogenesis triggers extensive chromatin remodeling, which subsequently influences the activity of auxin-responsive genes and pivotal developmental regulators, including LEAFY COTYLEDON1 and LEC2 (Deep and Pandey 2025; Firdous et al. 2025; Lindsay et al. 2023; Pandey and Chaudhary 2025). Further research on callus development in *Arabidopsis* has used single-cell RNA sequencing to identify distinct cellular developmental paths and transcription factor networks during callus initiation, proliferation, and greening (Liu et al. 2025). These studies demonstrate the dynamic shifts in gene expression and cellular states that precede the formation of organized structures, emphasizing the role of transcriptional reprogramming in developmental plasticity (Ibáñez et al. 2020; Islam et al. 2023).

Furthermore, single-cell analyses have revealed distinct cell subpopulations within the embryogenic callus that possess different developmental potentials, ranging from highly proliferative undifferentiated cells to those prepared for embryonic development (Pais 2019). This cellular diversity is crucial for successful regeneration, as specific cell types are more capable of differentiating into somatic embryos (Salaün et al. 2021). Therefore, understanding the particular gene regulatory networks and signaling pathways active in these embryogenic cell populations is essential for optimizing somatic embryogenesis protocols (Salaün et al. 2021). For instance, recent transcriptomic analyses of cotton have identified key transcription factors, such as LEAFY COTYLEDON 1, BABY BOOM, FUSCA, and AGAMOUS-LIKE15, which are significantly upregulated in embryogenic tissues, highlighting their roles as master regulators (Li et al. 2020). Additionally, the interplay between key plant growth regulators, particularly auxins and cytokinins, plays a vital role in triggering the initial phases of somatic embryogenesis, shaping both the timing and spatial patterns of gene expression that drive early developmental transitions (De-la-Peña et al. 2015).

The molecular mechanisms underlying somatic embryogenesis involve complex gene regulatory networks and chromatin remodeling events that enable cellular totipotency reprogramming (Altamura et al. 2015; Deep and Pandey 2025; Gulzar et al. 2020; Pandey and Chaudhary 2015, 2025). Specifically, genes like *SERK* and *BBM*, along with the *LEAFY COTYLEDON* family, are highly expressed during embryogenic induction and play central roles in transitioning nonembryogenic cells to an embryogenic state (Gulzar et al. 2020). For example, SERK1 has been shown to inhibit LEC1, LEC2, and FUS3, indicating complex regulatory interactions among these crucial factors (Yavuz et al. 2020). The interaction between SERK1 and BBM is also critical, with SERK1 directly activating BBM and indirectly influencing the BBM pathway through the activation of *AGL15* and *WUS* genes (Yavuz et al. 2020). Many genes, including SOMATIC EMBRYOGENESIS LIKE RECEPTOR KINASE, BABYBOOM, WUSCHEL, and CLAVATA, are crucial in regulating somatic embryogenesis (Gulzar et al. 2020; Guo et al. 2021; Pandey and Chaudhary 2014a, b). The synchronized function of *BBM*, *FUS3*, and *AGL15* genes in conjunction with auxin pathway is essential for triggering somatic embryogenesis, as shown in cotton transcriptomic studies (Yavuz et al. 2020). Beyond these, other transcription factors like WOX5, WUSHEL, WRKY2, GRD/RKD4, BBM, LEC1, FUS3, ABI3, and AGL15 have been identified as crucial for successful somatic embryogenesis across various plant species (Li et al. 2020). Moreover, the intricate regulation of auxin metabolism and signaling pathways, involving genes like ARF and IAA, is crucial during somatic embryogenesis, with different expression patterns observed in various developmental stages and species, such as *Arabidopsis* and cotton (Yavuz et al. 2020). These diverse genes and pathways work together in a spatiotemporally regulated manner, with their precise coordination being essential for the successful transition from somatic cells to a fully developed embryo (Yavuz et al. 2020). Studies have also revealed that a complex interplay of chromatin remodeling, gene activation and deactivation, and intricate transcription networks, including those involving POLYCOMB REPRESSIVE COMPLEX 1 and REVERSIBLY GLYCOSYLATED POLYPEPTIDE 1, are intrinsically linked with both somatic and zygotic embryogenesis processes (Gulzar et al. 2020). For example, the *RWP-RK DOMAIN-CONTAINING 4* gene has been identified as a critical regulator, promoting embryogenesis by stimulating root cell proliferation and causing developmental arrest in its absence (Gulzar et al. 2020). Furthermore, the coordinated action of transcription factors such as BABYBOOM, EMBRYOMAKER, LEAFY COTYLEDON, FUSCA3, and ABSCISIC ACID INSENSITIVE 3, often forming complex regulatory networks with auxin signaling pathways, is critical for orchestrating seed maturation and suppressing premature germination (Gulzar et al. 2020; Ranade and Egertsdotter 2021).

These molecular insights into somatic embryogenesis, particularly those gained through single-cell genomic approaches, are vital for advancing agricultural applications. They enable the development of more efficient plant regeneration protocols for difficult-to-regenerate species and facilitate genetic engineering strategies for crop improvement through precise manipulation of embryogenesis-related genes (Li et al. 2023a, b, c; Yavuz et al. 2020). Moreover, understanding these complex regulatory networks can lead to the development of new approaches for in vitro plant propagation, enhancing biomass production and disease resistance in agriculturally important crops. Furthermore, identifying key regulatory genes like AUX22, ABI3, ARF3, ARF5, AIL1, AIL5, and AGL15 provides a foundation for targeted genetic manipulation to enhance embryogenic competence and improve regeneration efficiency across diverse plant species (Qi et al. 2021). The understanding of genes like *WOX9*, *LEC2*, *FUS3*, and *ABI3*, which are crucial for embryogenesis and seed development, facilitates the targeted modification of these pathways to improve agricultural yields and resilience (Islam et al. 2023). This detailed molecular knowledge also provides a foundation for the application of advanced biotechnological tools, such as CRISPR-Cas9 genome editing, to precisely modify regulatory elements or coding sequences, thereby fine-tuning embryogenic responses and enhancing desirable agricultural traits.

3.5.3 Cotton Fibers

Understanding the genetic makeup of cotton fibers requires comprehending the intricate regulatory networks that control their development (Pandey et al. 2022; Pandey and Chaudhary 2019, 2020, 2021, 2024). Single-cell genomics significantly enhances this understanding. Single-cell RNA sequencing (scRNA-seq) has proven to be a powerful method for analyzing cellular diversity and the complex regulatory processes involved in cotton fiber initiation, growth, and the formation of secondary cell walls (Pan et al. 2024). This technique allows the identification of previously unknown genes and transcription factors specific to certain cell types, which direct different stages of fiber development. It provides an exceptionally detailed view of the molecular events critical to this important agricultural trait. Key genes in cotton fiber development include those responsible for transcription factors, enzymes for cellulose production, and proteins that modify cell walls; their expression patterns can be precisely determined using single-cell analyses (Yavuz et al. 2020). This detailed analysis of gene expression offers improved opportunities to discover new regulatory mechanisms controlling fiber initiation and growth with a precision not previously possible (Li et al. 2020; Pan et al. 2024).

Advanced single-cell methods are demonstrating the gene activity programs specific to each developmental stage, cell wall production, metabolic changes, and how fibers respond to environmental cues during cotton fiber development (Pan et al. 2024). For example, scRNA-seq has been used to study the formation of glands in wild cotton, revealing new genes and transcription factors involved in the pigment gland development (Pan et al. 2024). This has identified specific genes and pathways for the genetic engineering of cotton to enhance fiber quality and yield (Pan et al. 2024). Furthermore, combining single-cell genomics with population genetics, such as large-scale genome-wide association studies, can identify additional genetic regions and DNA variations linked to fiber quality traits, particularly in the cotton D-subgenome, which shows a stronger association (Liu et al. 2023a, b; Shudi et al. 2020; Yasir et al. 2022). This comprehensive approach is essential for finding candidate genes and understanding the molecular processes involved in the complex mechanism of cotton fiber development.

Identifying genes such as alpha-expansins, which are vital for cell wall loosening, or those involved in polysaccharide synthesis and the dynamics of the internal cellular scaffold (actin cytoskeleton), can guide efforts to improve fiber elongation and strength (Khan et al. 2018). Further research into highly expressed genes, such as xyloglucan endotransglucosylase/hydrolase and expansins, at specific developmental times (e.g., 0 and 10 days post-anthesis), can uncover crucial control points in fiber elongation and secondary cell wall development (Hande et al. 2017). This precise molecular mapping of gene expression provides a basis for identifying new targets for the genetic modification to improve fiber quality and yield in cotton (Grover et al. 2024; Shudi et al. 2020). Moreover, such comprehensive gene expression maps allow researchers to analyze how specific cell types respond to environmental stresses, laying the groundwork for breeding cotton varieties with increased resilience (Hu et al. 2024; Li et al. 2023a, b, c).

Single-cell technologies have also enabled the characterization of 1258 genes within 36 genetic modules that coordinate fiber development across stages. This has revealed genetic components that control how duplicated genes in different subgenomes are expressed (You et al. 2023). This in-depth understanding of gene expression during fiber development, achieved through scRNA-seq, allows for the precise identification of candidate genes that can be utilized for cotton improvement (Pan et al. 2024). For instance, combining meta-quantitative trait loci analysis with genome-wide association studies has identified 297 candidate genes related to cotton fiber quality, with 20 of these showing high expression specifically in developing fibers (Shudi et al. 2020). Confirming the functions of these candidate genes, such as Gh_D11G1626 (involved in cellulose biosynthesis) and Gh_A10G2036 (involved in polarized cell growth), is critical for validating their roles in improving fiber quality (Shudi et al. 2020). Additionally, detailed expression analysis has highlighted genes like Gh_A11G2663 and Gh_D11G2059 as highly expressed during fiber secondary wall biosynthesis and elongation, respectively, providing specific targets for genetic manipulation to enhance fiber properties (Shudi et al. 2020). For example, increasing the expression of specific actin genes, such as *GhFIM2*, has been shown to accelerate fiber elongation and increase mature fiber length, while genes like *GhEXPA8*, an expansin family member, can improve both fiber length and tenacity (Yang et al. 2021a, b). These advancements demonstrate the transformative potential of single-cell genomics in clarifying the complexities of cotton fiber biology and opening new paths for targeted genetic improvement strategies

(Grover et al. 2024; Pan et al. 2024). The ability to examine gene activity at the single-cell level has elaborated the molecular mechanisms involved in fiber development, thus creating new opportunities for engineering superior fiber traits (Pan et al. 2024). This detailed insight into gene expression allows for precise genetic modifications, such as introducing beneficial gene variants identified through genomic studies, to enhance fiber length, strength, and fineness (You et al. 2023).

3.5.4 Root Hair Cells

Beyond fiber development, single-cell genomics is increasingly used to understand the details of root hair cell differentiation and function, uncovering new regulatory networks that control nutrient uptake and how roots interact with their environment. This method offers a detailed view of gene activity in the individual root cells, identifying specific gene expression patterns linked to the different stages of root hair development and responses to stress conditions (Pan et al. 2024). For instance, developmental analysis has been used to map how root cells develop into specific types in *Moso bamboo*, with single-cell precision (Zhang et al. 2023). This approach also helps identify all genes and transcription factors that control root hair growth and how roots adapt to stress conditions (Liu et al. 2024). Furthermore, comparing single-cell gene activity in wild-type and mutant root cells (e.g., those without hair cells) describe the differences among cells and also identify genes essential for the cell identity (Zhang et al. 2023). Such studies have also provided a clearer understanding of how cells influence each other, showing how plant hormones like brassinosteroids and cytokinins coordinate cellular differentiation and cell expansion in root tissues (Zhang et al. 2023). This detailed view explains the molecular ways roots respond to environmental signals, such as nutrient availability or pathogenic infections (Cole et al. 2021). It has also allowed researchers to pinpoint specific cell types (such as pericycle, cortex, endodermis, and root hair cells) that react differently to rhizobium infection, providing insight into early symbiotic relationships (Zhang et al. 2023). Moreover, single-cell gene activity analyses have been essential for mapping the developmental paths of root hair cells and proving that high-throughput single-cell sequencing can be successfully used in plants (Liu et al. 2024).

3.5.5 Trichomes

Similarly, scRNA-seq has become a crucial tool for understanding the complex development and specialized roles of trichomes, which are epidermal outgrowths involved in defense and environmental interactions. These studies have revealed the distinct patterns of gene activity (transcriptional landscapes) of different trichome types and developmental stages, offering in-depth knowledge into the gene regulatory networks that govern their formation and specialized metabolism (Zhou et al. 2022). Applying single-cell RNA sequencing to *Nicotiana tenuifolia* leaf tissue has produced a detailed single-cell gene expression profile, uncovering cellular diversity and offering a foundational dataset for studying trichome formation and gene functionality (Zhou et al. 2022). This granular approach has been particularly effective in identifying key transcription factors and metabolic pathways active in glandular trichomes, which are responsible for synthesizing and secreting a variety of specialized metabolites, many of which have medicinal or industrial value. This high-resolution transcriptomic profiling facilitates the identification of novel genes and regulatory elements involved in trichome development and specialized metabolism, thereby offering targets for engineering enhanced production of valuable compounds (Zhou et al. 2022). Integrating spatial transcriptomics with scRNA-seq further refines our understanding by mapping gene expression back to specific cellular locations within the trichome, thus providing a holistic view of cellular heterogeneity and spatial organization (Sang and Kong 2024). Applications of single-cell genomics in plants have broadened our ability to identify previously unrecognized cell types, reconstruct gene regulatory networks, and also to dissect developmental pathways within highly heterogeneous tissues (Cole et al. 2021). In particular, scRNA-seq has revolutionized the characterization of cell identities and their core functions during plant growth and differentiation (Zhou et al. 2022). This high-resolution approach enables researchers to unravel the intricate interactions among molecular regulators and pinpoint active cellular processes with true single-cell precision, including within highly specialized structures such as glandular trichomes (Hurgobin et al. 2020).

3.5.6 Guard Cells and Stomatal Lineage

Single-cell genomics has greatly enhanced our understanding of guard cell development and stomatal lineage, providing unparalleled detail into the gene expression changes that drive their formation and specialized functions. These techniques clarify the complex gene regulatory networks controlling the differentiation of stomatal precursor cells into mature guard cells, emphasizing the interaction of signaling pathways and transcription factors (Zhang et al. 2023). Such insights are crucial for understanding how plants regulate gas exchange and water loss, offering potential avenues for engineering crops with improved water use efficiency (Yu et al. 2022). Tracking cell lineage trajectories with scRNA-seq provides a powerful means to decipher the developmental progression of guard cells and other stomatal lineage cells, revealing the stepwise gene expression programs that guide their maturation (Minne et al. 2021; Zhou et al. 2022). This level of resolution also enables the identification of cell-type-specific regulatory modules and expression signatures, offering valuable insights for improving stomatal performance and enhancing plant resilience to environmental stressors (Han et al. 2023). Trajectory analyses have specifically highlighted the ordered activation of the transcription factors SPEECHLESS, MUTE, and FAMA, which collectively drive the transition of protodermal cells into mature guard cells (Han et al. 2023). Such progress emphasizes

Table 3.1 Applications of single-cell genomics across key plant cell types

Cell type/system	Major biological insights enabled by single-cell or spatial omics	References
Callus / dedifferentiated cells	Reveals subpopulations, trajectories of dedifferentiation, early embryogenic competence	Guo et al. (2024)
Somatic embryos	Identifies regulators of embryogenic transition, developmental modules, SE-specific transcription factors	Ge et al. (2025)
Cotton fibers	Maps elongation and secondary wall trajectories; identifies stage-specific marker genes	Pan et al. (2024)
Cotton roots (stress response)	Cell-type-specific salt response; ion transporters; stress-responsive cell clusters	Li et al. (2023a, b, c)
Root hair cells	Developmental trajectories, root epidermal differentiation, cell-identity mutants	Denyer et al. (2019)
Trichomes (glandular and nonglandular)	Specialized metabolism, trichome lineage differentiation, biosynthetic pathways	Zhou et al. (2022)
Guard cells/stomatal lineage	Identifies stomatal precursor states	Lopez-Anido et al. (2021)
Light-responsive leaf cells (incl. stomatal)	Time-series atlas of guard cell maturation under light cues	Han et al. (2023)

the value of single-cell genomic approaches for dissecting cellular diversity in complex plant systems and pinpointing rare cell populations with critical biological functions (Sang and Kong 2024) (Table 3.1).

3.6 Challenges and Future Perspectives

While single-cell genomics has brought major progress in the field of plant biology, several important challenges still remain. One main difficulty is getting good single-cell suspensions. Plant cells vary greatly in their sizes, wall makeup, and chemical contents. These differences can damage RNA, or create false stress signals during the process of isolating protoplasts (plant cells without cell walls) (Ren and Wang 2023). While single-nucleus methods can mitigate some of these challenges, they inherently provide only a partial view of the transcriptome, potentially missing critical regulatory processes occurring outside the nucleus in the cytoplasm (Conde et al. 2021; Xin et al. 2023). From a computational standpoint, plant genomes are also complex due to multiple sets of chromosomes (polyploidy), presence of duplicated genes, and large noncoding regions, which makes it hard to align genetic sequences and measure gene activity. Therefore, more specialized analysis tools are required for plant research (Sun et al. 2021; Upton et al. 2023; Voshall and Moriyama 2019). These biological and computational difficulties lead to less complete data, which makes it harder to understand small differences in gene expression between cell types.

However, the future of single-cell research in plants looks very promising. A major breakthrough will come from combining single-cell data with observable traits (phenotypes) (Cho et al. 2025; Cole et al. 2021; Depuydt et al. 2022; Nobori 2025). Understanding how and what happens inside individual cells connects to traits like root growth, fiber length, leaf shape, or resistance to drought and salt, will greatly change plant science. Combining this information creates a crucial link between how cells act and how the whole plant performs. It uncovers insights that neither molecular nor phenotypic data alone can fully reveal. New methods for rapid phenotyping, such as automated imaging, 3D modeling, and field sensors, now allow us to measure traits with extreme accuracy (Hamidon and Ahamed 2024; Jin et al. 2020; Pandey and Mishra 2024). When used with single-cell transcriptomics, these tools will help researchers find which cellular processes are most important for agricultural results.

Machine learning and statistical models will also be crucial for putting all this information together. Algorithms that can analyze different cell types data will help uncover the complex connections between gene activity, how DNA is packaged (chromatin states), where cells are located, and observable traits (Bhardwaj et al. 2022; Pandey and Mishra 2024). Importantly, this approach not only identifies cell types but also allows researchers to investigate how plants integrate and coordinate functions across tissues, how local signals affect the overall shape of the plant, and which regulatory systems make plants strong or weak in changing environments.

Spatial transcriptomics (mapping gene expression in space) and CRISPR tools for single cells will further improve location information and facilitate the functional experiments (Aldridge and Teichmann 2020; Lindsay et al. 2023; Rao et al. 2021). These methods will also help find the key regulators of complex traits and confirm predictions from computer models. Eventually, combining diverse "omics" data with observations of physical traits will accelerate the identification of target genes for future research (Mishra and Pandey 2021; Wang et al. 2024). This is especially true for crops where important developmental changes or cell types that respond to stress are not well known (Abbas et al. 2024; Depuydt et al. 2022; Huo et al. 2024).

So, combining single-cell genomics, spatial technologies, advanced phenotyping, and computational learning will provide a thorough understanding of plant life. It will help explain how individual cells lead to the overall actions of tissues and organs, how plants adapt to their environment, and how specific genetic variations affect performance in the field. Such knowledge will support targeted approaches for crop improvement, for a changing world.

3.7 Conclusion

By surpassing the limits of traditional bulk techniques, single-cell and spatial omics are revolutionizing how plant tissues are studied at an unprecedented resolution. These technologies reveal the diversity of cellular states, track developmental transitions, and clarify how individual cells respond to the environmental cues. Despite this progress, important challenges persist. Many plant tissues remain difficult to dissociate into high-quality single-cell suspensions, and enzymatic digestion can alter transcriptional profiles. In addition, the size, complexity, and duplicated nature of many plant genomes continue to create analytical difficulties, underscoring the need for improved plant-focused protocols and computational tools. Developments in single-nucleus sequencing, microfluidic platforms, and gentler dissociation strategies are expanding the range of species and tissues amenable to single-cell study. Spatial transcriptomics is gaining finer resolution and broader coverage, while multiomics integration at single-cell resolution is offering comprehensive understanding of gene regulation. Machine-learning approaches are further enhancing these advances by identifying patterns and regulatory relationships that conventional analyses may overlook. High-resolution cellular maps can identify genes and pathways underlying stress tolerance, nutrient uptake, pathogen resistance, and traits related to yield and quality aspects. These discoveries open new opportunities for the targeted breeding, gene editing, and synthetic biology. As data resources expand and techniques become more widely available, single-cell and spatial omics are expected to drive innovation in sustainable crop production and the creation of resilient, high-performing plant varieties.

References

Abbas Q, Wilhelm M, Küster B, Poppenberger B, Frishman D (2024) Exploring crop genomes: assembly features, gene prediction accuracy, and implications for proteomics studies. BMC Genomics 25(1). https://doi.org/10.1186/s12864-024-10521-w

Aldridge S, Teichmann SA (2020) Single cell transcriptomics comes of age. Nat Commun 11(1). Nature Portfolio. https://doi.org/10.1038/s41467-020-18158-5

Altamura MM, Rovere FD, Fattorini L, D'Angeli S, Falasca G (2015) Recent advances on genetic and physiological bases of in vitro somatic embryo formation. In: Methods in molecular biology, vol 47. Springer Science+Business Media. https://doi.org/10.1007/978-1-4939-3061-6_3

Arora S, Pandey DK, Chaudhary B (2018) Target-mimicry based diminution of miRNA167 reinforced flowering-time phenotypes in tobacco via spatial-transcriptional biases of flowering-associated miRNAs. Gene 682:67. https://doi.org/10.1016/j.gene.2018.10.008

Berrío RT, Verhelst E, Eekhout T, Seyfferth C, Veylder LD, Rybel BD, Dubois M (2025) Dual and spatially resolved drought responses in the Arabidopsis leaf mesophyll revealed by single-cell transcriptomics. New Phytol. https://doi.org/10.1111/nph.20446

Bhardwaj A, Kishore S, Pandey DK (2022) Artificial intelligence in biological sciences. Life 12(9):1430. https://doi.org/10.3390/life12091430

Bhimani P, Mahavar P, Rajguru B, Bhatt VD, Nathani NM, Shri M (2024) Unveiling the green dialogue: advancements in omics technologies for deciphering plant–microbe interactions in soil. Discover Plants 1(1). https://doi.org/10.1007/s44372-024-00004-3

Bian J, Zhuang Z, Ji X, Tang R, Li J, Chen J, Li Z, Peng Y (2024) Research Progress of single-cell transcriptome sequencing technology in plants. Agronomy 14(11):2530. https://doi.org/10.3390/agronomy14112530

Biswas A, Kumari A, Gaikwad DS, Pandey DK (2023) Revolutionizing biological science: the synergy of genomics in health, bioinformatics, agriculture, and artificial intelligence. OMICS 27(12):550. https://doi.org/10.1089/omi.2023.0197

Chau TN, Timilsena PR, Bathala SP, Kundu S, Bargmann BOR, Li S (2025) Orthologous marker groups reveal broad cell identity conservation across plant single-cell transcriptomes. Nat Commun 16(1). https://doi.org/10.1038/s41467-024-55755-0

Chen Y, Tong S, Jiang Y, Ai F, Feng Y, Zhang J, Gong J, Qin J, Zhang Y, Zhu Y, Liu J, Ma T (2021) Transcriptional landscape of highly lignified poplar stems at single-cell resolution. Genome Biol 22(1). https://doi.org/10.1186/s13059-021-02537-2

Cho Y, Kadam US, Park B, Amariillis S, Nguyen K, Can M-HT, Lee KO, Park SJ, Chung WS, Hong JC (2025) Recent progress in single-cell transcriptomic studies in plants. In: Plant biotechnology reports. https://doi.org/10.1007/s11816-025-00967-z

Cole B, Bergmann DC, Blaby-Haas CE, Blaby IK, Bouchard KE, Brady SM, Ciobanu D, Coleman-Derr D, Leiboff S, Mortimer JC, Nobori T, Rhee SY, Schmutz J, Simmons BA, Singh AK, Sinha N, Vogel JP, O'Malley RC, Visel A, Dickel DE (2021) Plant single-cell solutions for energy and the environment. Commun Biol 4(1). Nature Portfolio. https://doi.org/10.1038/s42003-021-02477-4

Conde D, Triozzi PM, Balmant KM, Doty A, Miranda MA, Boullosa A, Schmidt HW, Pereira WJ, Dervinis C, Kirst M (2021) A robust method of nuclei isolation for single-cell RNA sequencing of solid tissues from the plant genus Populus. PLoS One 16(5). https://doi.org/10.1371/journal.pone.0251149

D'Agostino L, Yong-Villalobos L, Herrera-Estrella L, Patil G (2023) Development of high-quality nuclei isolation to study plant root–microbe interaction for single-nuclei transcriptomic sequencing in soybean. Plants 12(13):2466. https://doi.org/10.3390/plants12132466

Deep A, Pandey DK (2024) Genome-wide analysis of VILLIN gene family associated with stress responses in cotton (Gossypium spp.). Curr Issues Mol Biol 46(3):2278. https://doi.org/10.3390/cimb46030146

Deep A, Pandey DK (2025) Cellular and molecular mechanisms of plant regeneration. In Regeneration Potential in Plants. CRC Press https://doi.org/10.1201/9781032702391-2

De-la-Peña C, Nic-Can GI, Galáz-Ávalos RM, Avilez-Montalvo RN, Loyola-Vargas VM (2015) The role of chromatin modifications in somatic embryogenesis in plants. Frontiers in Plant Science, 6:635 https://doi.org/10.3389/fpls.2015.00635

Denyer T, Timmermans MCP (2021) Crafting a blueprint for single-cell RNA sequencing. Trends Plant Sci 27(1):92. Elsevier BV. https://doi.org/10.1016/j.tplants.2021.08.016

Denyer T, Ma X, Klesen S, Scacchi E, Nieselt K, Timmermans MCP (2019) Spatiotemporal developmental trajectories in the Arabidopsis root revealed using high-throughput single-cell RNA sequencing. Dev Cell 48(6):840. https://doi.org/10.1016/j.devcel.2019.02.022

Depuydt T, Rybel BD, Vandepoele K (2022) Charting plant gene functions in the multi-omics and single-cell era. Trends Plant Sci 28(3):283. Elsevier BV. https://doi.org/10.1016/j.tplants.2022.09.008

Deshmukh R, Gupta SK, Patil G (2025) Integrated omics approaches for biotic and abiotic stress tolerance in plants. J Plant Biochem Biotechnol. https://doi.org/10.1007/s13562-025-00969-6

Dorrity MW, Alexandre CM, Hamm MO, Vigil A-LBG, Fields S, Queitsch C, Cuperus JT (2021) The regulatory landscape of Arabidopsis thaliana roots at single-cell resolution. Nat Commun 12(1):3334. https://doi.org/10.1038/s41467-021-23675-y

Drapek C, Sparks EE, Benfey PN (2017) Uncovering gene regulatory networks controlling plant cell differentiation. Trends Genet 33(8):529. Elsevier BV. https://doi.org/10.1016/j.tig.2017.05.002

Edrisi M, Huang X, Ogilvie HA, Nakhleh L (2023) Accurate integration of single-cell DNA and RNA for analyzing intratumor heterogeneity using MaCroDNA. Nat Commun 14(1). https://doi.org/10.1038/s41467-023-44014-3

Firdous, M., Gupta, A., Yadav, P. K., & Pandey, D. K. (2025). Somatic embryogenesis: A key pathway of regeneration. In: Pandey, D. K. & Chaudhary, B. (Eds.), Regeneration Potential in Plants. CRC Press (Taylor & Francis), pp. 59–80. https://doi.org/10.1201/9781032702391-4

Ge X, Yu X, Liu Z, Yuan J, Qin A, Wang Y, Chen Y, Qin W, Liu Y, Liu X, Zhou Y, Wang P, Yang J, Liu H, Zhao Z, Hu M, Zhang Y, Sun S, Herrera-Estrella L et al (2025) Spatiotemporal transcriptome and metabolome landscapes of cotton somatic embryos. Nat Commun 16(1):859. https://doi.org/10.1038/s41467-025-55870-6

Giacomello S, Salmén F, Terebieniec BK, Vicković S, Navarro JF, Alexeyenko A, Reimegård J, McKee LS, Mannapperuma C, Bulone V, Ståhl PL, Sundström JF, Street NR, Lundeberg J (2017) Spatially resolved transcriptome profiling in model plant species. Nat Plants 3(6). https://doi.org/10.1038/nplants.2017.61

Giolai M, Verweij W, Lister A, Heavens D, Macaulay IC, Clark MD (2019) Spatially resolved transcriptomics reveals plant host responses to pathogens. Plant Methods 15(1). https://doi.org/10.1186/s13007-019-0498-5

Grover CE, Jareczek JJ, Swaminathan S, Lee Y, Howell AH, Rani H, Arick MA, Leach AG, Miller ER, Yang P, Hu G, Xiong X, Mallery EL, Peterson DG, Xie J, Haigler CH, Zabotina OA, Szymanski DB, Wendel JF (2024) A high-resolution model of gene expression during Gossypium hirsutum (cotton) fiber development. BMC Genomics. https://doi.org/10.21203/rs.3.rs-5291809/v1

Gulzar B, Mujib A, Malik MQ, Sayeed R, Mamgain J, Ejaz B (2020) Genes, proteins and other networks regulating somatic embryogenesis in plants. J Genet Eng Biotechnol 18(1):31. Elsevier BV. https://doi.org/10.1186/s43141-020-00047-5

Guo B, Zeng S, Yin Y, Li L, Ma G, Wu K, Fang L (2021) Characterization of phytohormone and transcriptome profiles during protocorm-like bodies development of Paphiopedilum. BMC Genomics 22(1). https://doi.org/10.1186/s12864-021-08087-y

Guo H, Zhang LX, Guo H, Cui X, Fan Y, Li T, Qi X, Yan T, Chen A, Shi F, Zeng F (2024) Single-cell transcriptome atlas reveals somatic cell embryogenic differentiation features during regeneration. Plant Physiol 195(2):1414. https://doi.org/10.1093/plphys/kiae107

Hamidon MH, & Ahamed T (2024) High-throughput plant phenotyping techniques in controlled environments. In: Ahamed, T. (Ed.), IoT and AI in Agriculture: Smart Automation Systems for Increasing Agricultural Productivity to Achieve SDGs and Society 5.0. Springer, Singapore, pp. 159–181. https://doi.org/10.1007/978-981-97-1263-2_11

Han X, Zhang Y, Lou Z, Li J, Wang Z, Gao C, Liu Y, Ren Z, Liu W, Li B, Pan W, Zhang H, Sang Q, Wan M, He H, Deng XW (2023) Time series single-cell transcriptional atlases reveal cell fate differentiation driven by light in Arabidopsis seedlings. Nat Plants 9(12):2095. https://doi.org/10.1038/s41477-023-01544-4

Hande AS, Katageri IS, Jadhav MP, Adiger S, Gamanagatti S, Padmalatha K, Gurusamy D, Mogilicherla K, Kumar PA, Reddy VS (2017) Transcript profiling of genes expressed during fibre development in diploid cotton (Gossypium arboreum L.). BMC Genomics 18(1). https://doi.org/10.1186/s12864-017-4066-y

Hu Y, Dash L, May GD, Sardesai N, Deschamps S (2024) Harnessing single-cell and spatial transcriptomics for crop improvement. Plants 13(24):3476 https://doi.org/10.3390/plants13243476

Huo Q, Song R, Ma Z (2024) Recent advances in exploring transcriptional regulatory landscape of crops. The Crop Journal, 12(1): 1–13. https://doi.org/10.3389/fpls.2024.1421503

Hurgobin B, Tamiru-Oli M, Welling MT, Doblin MS, Bacic A, Whelan J, Lewsey MG (2020) Recent advances in cannabis sativa genomics research. New Phytol 230(1):73. Wiley. https://doi.org/10.1111/nph.17140

Ibáñez S, Carneros E, Testillano PS, Pérez-Pérez JM (2020) Advances in plant regeneration: shake, rattle and roll. Plants 9(7):897 https://doi.org/10.3390/plants9070897

Ikeuchi M, Sugimoto K, Iwase A (2013) Plant callus: mechanisms of induction and repression. Plant Cell 25(9):3159 https://doi.org/10.1105/tpc.113.116053

Islam MK, Mummadi ST, Liu S, Wei H (2023) Regulation of regeneration in Arabidopsis thaliana. aBIOTECH 4(4):332. https://doi.org/10.1007/s42994-023-00121-9

Jin X, Zarco-Tejada PJ, Schmidhalter U, Reynolds M, Hawkesford MJ, Varshney RK, Yang T, Nie C, Li Z, Ming B, Xiao Y, Xie Y, Li S (2020) High-throughput estimation of crop traits: a review of ground and aerial phenotyping platforms. IEEE Geosci Remote Sens Mag 9(1):200 https://doi.org/10.1109/mgrs.2020.2998816

Khan A, Pan X, Ullah N, Tan DKY, Fahad S, Zahoor R, Luo H (2018) Coping with drought: stress and adaptive mechanisms, and management through cultural and molecular alternatives in cotton as vital constituents for plant stress resilience and fitness. Biol Res 51(1) https://doi.org/10.1186/s40659-018-0198-z

Kim M, Lee S-W, Kim K, Kim Y, Hwang H, Hinterdorfer P, Choi D, Ko K (2023) Single-cell transcriptome sequencing of plant leaf expressing anti-HER2 VHH–FcK cancer therapeutic protein. Sci Data 10(1). https://doi.org/10.1038/s41597-023-02833-5

Lardon R, Geelen D (2020) Natural variation in plant pluripotency and regeneration. Plants 9(10):1261. https://doi.org/10.3390/plants9101261

Li W, Li W, Parris S, West M, Lawson JW, Smathers M, Li Z, Jones DC, Jin S, Saski C (2020) Transcriptomic profiles of non-embryogenic and embryogenic callus cells in a highly regenerative upland cotton line (Gossypium hirsutum L.). BMC Dev Biol 20(1). https://doi.org/10.1186/s12861-020-00230-4

Li L, Sun X, Yu W, Gui M, Qiu Y, Tang M, Hai T, Liang G (2023a) Comparative transcriptome analysis of high- and low-embryogenic Hevea brasiliensis genotypes reveals involvement of phytohormones in somatic embryogenesis. BMC Plant Biol 23(1). https://doi.org/10.1186/s12870-023-04432-3

Li P, Liu Q, Wei Y, Xing C, Xu Z, Ding F, Liu Y, Lú Q, Hu N, Wang T, Zhu X, Cheng S, Li Z, Zhao Z, Li Y, Han J, Cai X, Zhou Z, Wang K et al (2023b) Transcriptional landscape of cotton roots in response to salt stress at single-cell resolution. Plant Commun 5(2):100740. https://doi.org/10.1016/j.xplc.2023.100740

Li Y, Ma H, Wu Y, Ma Y, Yang J, Li Y, Yue D, Zhang R, Kong J, Lindsey K, Zhang X, Min L (2023c) Single-cell transcriptome atlas and regulatory dynamics in developing cotton anthers. Adv Sci 11(3). https://doi.org/10.1002/advs.202304017

Lindsay PL, Swentowsky KW, Jackson D (2023) Cultivating potential: harnessing plant stem cells for agricultural crop improvement. Mol Plant 17(1):50. https://doi.org/10.1016/j.molp.2023.12.014

Liu J, Majeed A, Mukhtar MS (2022) Rooting through single-cell sequencing in phloem pole cells. Commun Biol 5(1). https://doi.org/10.1038/s42003-022-04180-4

Liu S, Zenda T, Tian Z, Huang Z (2023a) Metabolic pathways engineering for drought or/and heat tolerance in cereals. Plant Physiol and Biochem, 194: 1–14. https://doi.org/10.3389/fpls.2023.1111875

Liu Z, Sun Z, Ke H, Chen B, Gu Q, Zhang M, Wu N, Chen L, Li Y, Meng C, Wang G, Wu L, Zhang G, Ma Z, Zhang Y, Wang X (2023b) Transcriptome, ectopic expression and genetic population analysis identify candidate genes for fiber quality improvement in cotton. Int J Mol Sci 24(9):8293. https://doi.org/10.3390/ijms24098293

Liu H, Liu Z, Zhou Y, Qin A, Li C, Liu Y, Gao P, Zhao Q, Song X, Li M, Kong L, Xie Y, Yan L, Guo E, Sun X (2024) Single-cell transcriptomic analysis reveals the developmental trajectory and transcriptional regulatory networks of quinoa salt bladders. Stress Biol 4(1). https://doi.org/10.1007/s44154-024-00189-3

Liu Z, Zhang Y, Zhao Q, Liu H, Zhou Y, Qin A, Li C, Yan L, Li M, Gao P, Song X, Xie Y, Guo E, Kong L, Guan L, An G, Sun X (2025) Single-cell RNA sequencing reveals developmental trajectories and environmental regulation of callus formation in Arabidopsis. Stress Biol 5(1). https://doi.org/10.1007/s44154-025-00255-4

Long Y, Yang Y, Pan G, Shen Y (2022) New Insights into tissue culture plant-regeneration mechanisms. Front Plant Sci 13: 926752. https://doi.org/10.3389/fpls.2022.926752

Lopez-Anido C, Vatén A, Smoot NK, Sharma N, Guo V, Gong Y, Gil MXA, Weimer AK, Bergmann DC (2021) Single-cell resolution of lineage trajectories in the Arabidopsis stomatal lineage and developing leaf. Dev Cell 56(7):1043. https://doi.org/10.1016/j.devcel.2021.03.014

Lv Z, Jiang S, Kong S, Zhang X, Yue J, Zhao W, Li L, Lin S (2024) Advances in single-cell transcriptome sequencing and spatial transcriptome sequencing in plants. Plants 13(12):1679. https://doi.org/10.3390/plants13121679

Madrid G, Raimundo GAS, Reyes F, de Picoli EA, T, Resende MFR, Balmant KM (2025) An improved nuclei isolation protocol from leaf tissue for single-cell transcriptomics. PLoS One 20(9). https://doi.org/10.1371/journal.pone.0302118

Marchant DB, Nelms B, Walbot V (2021) FX-cell: quantitative cell release from fixed plant tissues for single-cell genomics. bioRxiv (Cold Spring Harbor Laboratory). https://doi.org/10.1101/2021.10.11.463960

Minne M, Ke Y, Saura-Sánchez M, Rybel BD (2021) Advancing root developmental research through single-cell technologies. Curr Opin Plant Biol 65:102113. https://doi.org/10.1016/j.pbi.2021.102113

Mishra R, Pandey DK (2021) Agri/Bioinformatics: shaping next-generation agriculture 111. In: Bioinformatics for agriculture: high-throughput approaches. Springer. https://doi.org/10.1007/978-981-33-4791-5_7

Luo M, Ríos G, Sarnowski TJ, Zhang S, Mantri N, Charron J-B and Libault M (2020) Editorial: New Insights Into Mechanisms of Epigenetic Modifiers in Plant Growth and Development. Front. Plant Sci. 10:1661. doi: 10.3389/fpls.2019.01661

Nobori T (2025) Exploring the untapped potential of single-cell and spatial omics in plant biology. Curr Opinion in Plant Biol, 82: 102820. https://doi.org/10.1111/nph.70220

Nobori T, Oliva M, Lister R, Ecker JR (2023) Multiplexed single-cell 3D spatial gene expression analysis in plant tissue using PHYTOMap. Nat Plants 9(7):1026. https://doi.org/10.1038/s41477-023-01439-4

Pais MS (2019) Somatic embryogenesis induction in Woody species: the future after OMICs data assessment. Front Plant Sci 10:240. https://doi.org/10.3389/fpls.2019.00240

Pan X, Peng R, Zhang B (2024) Single-cell RNA sequencing opens a new era for cotton genomic research and gene functional analysis. J Cotton Res 7(1). https://doi.org/10.1186/s42397-024-00181-2

Pandey DK, Chaudhary B (2014a) Oxidative stress responsive *SERK1* gene directs the progression of somatic embryogenesis in cotton (*Gossypium hirsutum* L. cv. Coker 310). Am J Plant Sci 5(1):80. https://doi.org/10.4236/ajps.2014.51012

Pandey DK, Chaudhary B (2014b) Role of plant somatic embryogenesis receptor kinases (SERKs) in cell-to-embryo transitional activity: key at novel assorted structural subunits. Am J Plant Sci 5(21):3177. https://doi.org/10.4236/ajps.2014.521334

Pandey DK, and Chaudhary B (2015) Genes and Trans Factors Underlying Embryogenic Transition in Plant Soma Cell, In Advances in the Understanding of Biological Sciences Using Next Generation Sequencing (NGS) Approaches. (Eds.) Sablok G, Kumar S, Ueno S, Kuo J, Varotto C, (Eds.) (155-178), Publisher: Springer International Publishing. https://doi.org/10.1007/978-3-319-17157-9_10

Pandey DK, Chaudhary B (2016) Domestication-driven Gossypium profilin 1 (GhPRF1) gene transduces early flowering phenotype in tobacco by spatial alteration of apical/floral-meristem related gene expression. BMC Plant Biol 16(1):112. https://doi.org/10.1186/s12870-016-0798-0

Pandey DK, Chaudhary B (2017) Evolutionary expansion and structural functionalism of the ancient family of profilin proteins. Gene 626:70. https://doi.org/10.1016/j.gene.2017.05.024

Pandey DK, Chaudhary B (2019) Synchronous transcription of cytoskeleton-associated genes is critical to cotton fiber elongation. J Plant Growth Regul 38(3):1037. https://doi.org/10.1007/s00344-019-09913-0

Pandey DK, Chaudhary B (2020) Evolution of functional diversity among actin-binding profilin genes in land plants. Front Cell Dev Biol 8:588689. https://doi.org/10.3389/fcell.2020.588689

Pandey DK, Chaudhary B (2021) Transcriptional loss of domestication-driven cytoskeletal GhPRF1 gene causes defective floral and fiber development in cotton (Gossypium). Plant Mol Biol 107(6):519. https://doi.org/10.1007/s11103-021-01200-5

Pandey DK, Chaudhary B (2024) Ectopic overexpression of domestication-driven cytoskeletal profilin (GhPRF1) gene improves flowering and trichome development in cotton (Gossypium hirsutum L.). J Plant Growth Regul 43(10):3543. https://doi.org/10.1007/s00344-024-11329-4

Pandey DK, Chaudhary B (2025) Regeneration potential in plants. CRC Press. https://doi.org/10.1201/9781032702391

Pandey DK, Mishra R (2024) Towards sustainable agriculture: harnessing AI for global food security. Artif Intell Agric 12:72. https://doi.org/10.1016/j.aiia.2024.04.003

Pandey DK, Singh AK, Chaudhary B (2012) Boron-mediated plant somatic embryogenesis: a provocative model. J Bot 2012:1. https://doi.org/10.1155/2012/375829

Pandey DK, Kumar V, Chaudhary B (2022) Concomitant expression evolution of cell wall cytoskeletal geneic triad(s) controls floral organ shape and fiber emergence in cotton (Gossypium). Front Plant Sci 13:900521. https://doi.org/10.3389/fpls.2022.900521

Park J, Choi Y, Jeong M, Jeong Y-I, Han J-H, Choi H (2023) Uncovering transcriptional reprogramming during callus development in soybean: insights and implications. Front Plant Sci 14. https://doi.org/10.3389/fpls.2023.1239917
Pírek P, Kryštofová K, Kováčová I, Kromerová A, Zachová D, Helia O, Panzarová K, Fajkus J, Zdráhal Z, Lochmanová G, Fojtová M (2023) Unraveling epigenetic changes in A. thaliana Calli: impact of HDAC inhibitors. Plants 12(24):4177. https://doi.org/10.3390/plants12244177
Qi S, Zhao R, Yan J, Fan Y, Huang C, Li H, Chen S, Zhang T, Kong L, Zhao J, Zhang J (2021) Global transcriptome and coexpression network analyses reveal new insights into somatic embryogenesis in hybrid sweetgum (Liquidambar styraciflua × Liquidambar formosana). Front Plant Sci 12. https://doi.org/10.3389/fpls.2021.751866
Ranade SS, Egertsdotter U (2021) In silico characterization of putative gene homologues involved in somatic embryogenesis suggests that some conifer species may lack LEC2, one of the key regulators of initiation of the process. BMC Genomics 22(1). https://doi.org/10.1186/s12864-021-07718-8
Rao A, Barkley D, França GS, Yanai I (2021) Exploring tissue architecture using spatial transcriptomics. Nature 596(7871):211. https://doi.org/10.1038/s41586-021-03634-9
Ren S, Wang Y (2023) Protoplast isolation for plant single-cell RNA-seq. Methods Mol Biol:301. https://doi.org/10.1007/978-1-0716-3299-4_14
Roszak P, Heo J, Blob B, Toyokura K, Sugiyama Y, Balaguer MA d L, Lau W, Hamey F, Cirrone J, Madej E, Bouatta AM, Wang X, Guichard M, Ursache R, Tavares H, Verstaen K, Wendrich JR, Melnyk CW, Oda Y et al (2021) Cell-by-cell dissection of phloem development links a maturation gradient to cell specialization. Science 374(6575). https://doi.org/10.1126/science.aba5531
Saarenpää S, Shalev O, Ashkenazy H, Carlos V, Lundberg DS, Weigel D, Giacomello S (2023) Spatial metatranscriptomics resolves host–bacteria–fungi interactomes. Nat Biotechnol 42(9):1384. https://doi.org/10.1038/s41587-023-01979-2
Salaün C, Lepiniec L, Dubreucq B (2021) Genetic and molecular control of somatic embryogenesis. Plants 10(7):1467. https://doi.org/10.3390/plants10071467
Sang Q, Kong F (2024) Applications for single-cell and spatial transcriptomics in plant research. New Crops 1:100025. https://doi.org/10.1016/j.ncrops.2024.100025
Shudi X, Pan Z, Yin F, Yang Q, Lin Z, Wen T, Zhu L, Zhang D, Nie X (2020) Identification of candidate genes controlling fiber quality traits in upland cotton through integration of meta-QTL, significant SNP and transcriptomic data. J Cotton Res 3(1). https://doi.org/10.1186/s42397-020-00075-z
Shulse CN, Cole B, Ciobanu D, Lin J, Yoshinaga Y, Gouran M, Turco G, Zhu Y, O'Malley RC, Brady SM, Dickel DE (2019) High-throughput single-cell transcriptome profiling of plant cell types. Cell Rep 27(7):2241. https://doi.org/10.1016/j.celrep.2019.04.054
Sun Y, Shang L, Zhu Q, Fan L, Guo L (2021) Twenty years of plant genome sequencing: achievements and challenges. Trends Plant Sci 27(4):391. https://doi.org/10.1016/j.tplants.2021.10.006
Sunaga DY, Muiño JM, Braeuning C, Xu X, Zong M, Smaczniak C, Yan W, Fischer C, Vidal R, Kliem M, Kaufmann K, Sauer S (2020) Single-nuclei RNA-sequencing of plant tissues. bioRxiv (Cold Spring Harbor Laboratory). https://doi.org/10.1101/2020.11.14.382812
Tang LP, Zhai LM, Li J, Gao Y, Li Q, Li R, Liu Q, Zhang W, Yao WJ, Mu B, Qin C, Tian X, Shaw R, Xia K, Xu J, Su YH, Zhang XS (2025) Time-resolved reprogramming of single somatic cells into totipotent states during plant regeneration. Cell 188:6923-6938. https://doi.org/10.1016/j.cell.2025.08.031
Thibivilliers S, Libault M (2021) Enhancing our understanding of plant cell-to-cell interactions using single-cell omics. Front Plant Sci 12. https://doi.org/10.3389/fpls.2021.696811
Upadhyay AK, Arora S, Pandey DK, Chaudhary B (2019) Interspersed 5′cis-regulatory elements ascertain the spatio-temporal transcription of cytoskeletal profilin gene family in Arabidopsis. Comput Biol Chem 80:177. https://doi.org/10.1016/j.compbiolchem.2019.03.023

Upton RN, Correr FH, Lile J, Reynolds G, Falaschi K, Cook JP, Lachowiec J (2023) Design, execution, and interpretation of plant RNA-seq analyses. Front Plant Sci 14. https://doi.org/10.3389/fpls.2023.1135455

Voshall A, Moriyama EN (2019) Next-generation transcriptome assembly and analysis: impact of ploidy. Methods 176:14. https://doi.org/10.1016/j.ymeth.2019.06.001

Wang J, Wang Y, Lü T, Yang X, Liu J, Dong Y, Wang Y (2022) An efficient and universal protoplast isolation protocol suitable for transient gene expression analysis and single-cell RNA sequencing. Int J Mol Sci 23(7):3419. https://doi.org/10.3390/ijms23073419

Wang P, Lehti-Shiu MD, Lotreck S, Abá KS, Krysan PJ, Shiu S (2024) Prediction of plant complex traits via integration of multi-omics data. Nat Commun 15(1):6856. https://doi.org/10.1038/s41467-024-50701-6

Wang H, Fu X, Liu L, Wang Y, Hong J, Pan B, Cao Y, Chen Y, Cao Y, Ma X, Fang W, Yan S (2025a) PhytoCluster: a generative deep learning model for clustering plant single-cell RNA-seq data. aBIOTECH. https://doi.org/10.1007/s42994-025-00196-6

Wang T, Wang F, Deng S, Wang K, Feng D, Xu F, Guo W, Yu J, Wu Y, Wuriyanghan H, Li S, Gu X, Le L, Pu L (2025b) Single-cell transcriptomes reveal spatiotemporal heat stress response in maize roots. Nat Commun 16(1). https://doi.org/10.1038/s41467-024-55485-3

Wang W, Zhang X, Zhang Y, Zhang Z, Yang C, Cao W, Liang Y, Zhou Q, Hu Q, Zhang Y, Wang Y, Xing Y, Qian W, Yao N, Xu N, Liu J (2025c) Single-cell and spatial transcriptomics reveals a stereoscopic response of rice leaf cells to Magnaporthe oryzae infection. Adv Sci. https://doi.org/10.1002/advs.202416846

Xin X, Du F, Jiao Y (2023) Plant nuclei isolation for single-nucleus RNA sequencing. Methods Mol Biol:307. https://doi.org/10.1007/978-1-0716-3299-4_15

Yang J, Gao L, Liu X, Zhang X, Wang X, Wang Z (2021a) Comparative transcriptome analysis of fiber and nonfiber tissues to identify the genes preferentially expressed in fiber development in Gossypium hirsutum. Sci Rep 11(1). https://doi.org/10.1038/s41598-021-01829-8

Yang X, Liu D, Lu H, Weston DJ, Chen J, Muchero W, Martin S, Liu Y, Hassan MM, Yuan G, Kalluri UC, Tschaplinski TJ, Mitchell JC, Wullschleger SD, Tuskan GA (2021b) Biological parts for plant biodesign to enhance land-based carbon dioxide removal. Biodes Res 2021. American Association for the Advancement of Science. https://doi.org/10.34133/2021/9798714

Yang Q, Li Z, Guan K, Wang Z, Tang X, Hong Y, Liu Z, Zhai J, Huang AC, Long Y, Song Y (2025) Comparative single-nucleus RNA-seq analysis revealed localized and cell type-specific pathways governing root-microbiome interactions. Nat Commun 16(1). https://doi.org/10.1038/s41467-025-58395-0

Yasir M, Kanwal HH, Hussain Q, Riaz MW, Sajjad M, Rong J, Jiang Y (2022) Status and prospects of genome-wide association studies in cotton. Front Plant Sci 13. https://doi.org/10.3389/fpls.2022.1019347

Yavuz C, Tillaboeva S, Bakhsh A (2020) Apprehending the potential of BABY BOOM transcription factors to mitigate cotton regeneration and transformation. J Cotton Res 3(1). https://doi.org/10.1186/s42397-020-00071-3

You J, Liu Z, Qi Z, Ma Y, Sun M, Su L, Niu H, Peng Y, Luo X, Zhu M, Huang Y, Chang X, Hu X, Zhang Y, Pi R, Liu Y, Meng Q, Li J, Zhang Q et al (2023) Regulatory controls of duplicated gene expression during fiber development in allotetraploid cotton. Nat Genet 55(11):1987. https://doi.org/10.1038/s41588-023-01530-8

Yu X, Liu Z, Sun X (2022) Single-cell and spatial multi-omics in the plant sciences: technical advances, applications, and perspectives. Plant Commun 4(3):100508. Elsevier BV. https://doi.org/10.1016/j.xplc.2022.100508

Zenda T, Liu S, Dong A, Li J, Wang Y, Liu X, Wang N, Duan H (2021) Omics-facilitated crop improvement for climate resilience and superior nutritive value. Front Plant Sci 12. https://doi.org/10.3389/fpls.2021.774994

Zhang R, Zhang C, Yu C, Dong J, Hu J (2022) Integration of multi-omics technologies for crop improvement: status and prospects. Front Bioinf 2. https://doi.org/10.3389/fbinf.2022.1027457

Zhang J, Ahmad M, Gao H (2023) Application of single-cell multi-omics approaches in horticulture research. Mol Hortic 3(1). https://doi.org/10.1186/s43897-023-00067-y

Zheng D, Xu J, Lu Y, Chen H, Chu Q, Fan L (2023) Recent progresses in plant single-cell transcriptomics. Crop Des 2(2):100041. https://doi.org/10.1016/j.cropd.2023.100041

Zhou P, Chen H, Dang J, Shi Z, Shao Y, Liu C, Fan L, Wu Q (2022) Single-cell transcriptome of Nepeta tenuifolia leaves reveal differentiation trajectories in glandular trichomes. Front Plant Sci 13. https://doi.org/10.3389/fpls.2022.988594

Zhu J, Moreno-Pérez A, Coaker G (2023) Understanding plant pathogen interactions using spatial and single-cell technologies. Commun Biol 6(1). Nature Portfolio. https://doi.org/10.1038/s42003-023-05156-8

Single-Cell RNA Sequencing

4

Jhalak Aggarwal, Eeba, and Nisheeth Agarwal

Abstract

The advent of single-cell RNA sequencing (scRNA-seq) has transformed our understanding of the biological world by providing unprecedented insights into the intricate realm of cellular individuality. The scRNA-seq technique has been extremely useful in unraveling rare and complex cell populations and dynamic cellular states, thereby elegantly capturing the nuances of heterogeneity that otherwise remain masked by bulk RNA sequencing. This capacity to delineate cellular diversity at the resolution of a single cell has not only revolutionized fundamental biology but also unveiled transformative avenues in precision medicine. The art of dimensionality reduction, clustering and lineage inference together transform vast datasets into structured biological narratives. The narrative further explores emerging frontiers, including multimodal single-cell profiling, spatial transcriptomics and integrative frameworks that elegantly combine transcriptomes, proteomes, and epigenomes. Beyond its technological sophistication, scRNA-seq has emerged as a potent lens for investigating complex biological processes of immunity, metabolism, developmental biology, neurobiology, or cancer biology.

This chapter provides an overview of the scRNA-seq highlighting major applications of this technique in various biological domains. It describes the workings of scRNA-seq from the meticulous isolation of individual cells and the development of sophisticated library preparation techniques to the advanced sequencing methods and computational pipelines that transform the raw data into valuable insights. Overall aim of the chapter is to showcase the potential of the scRNA-seq technique, which continues to advance the boundaries of biological discoveries by providing not only the huge datasets but also an atlas of cellular world at the unparalleled resolution.

J. Aggarwal · Eeba · N. Agarwal (✉)
Translational Health Science and Technology Institute, NCR Biotech Science Cluster, Faridabad, Haryana, India
e-mail: nisheeth@thsti.res.in

V. Singh (ed.), *Advances in Single Cell Analysis*,
https://doi.org/10.1007/978-981-95-9789-5_4

Keywords

Transcription · Gene expression · Transcriptome profiling · Transcriptomics · RNA sequencing · Single-cell RNA sequencing

4.1 Introduction

The journey of transcriptomics chronicles a plethora of transformative advancements that have progressively unveiled the intricate cellular complexities. From meticulously switching from painstaking gene-by-gene analysis to contemporary high-throughput single-cell atlases, over the years the discipline has undergone remarkable transformations profoundly reinvigorating our understanding of gene expression and cellular identity. Initially transcriptomics employed low-throughput methods such as Northern blotting, reverse transcription polymerase chain reaction (RT-PCR), or in situ hybridization. Although these pioneering techniques established a foundational framework for deciphering the gene expression, these were incapable to comprehensively study the global transcriptome in its entirety (Alwine et al. 1977). In the mid-1990s, the advent of microarray technology revolutionized transcriptomics, ushering it into an era of large-scale, high-dimensional analysis. The microarray technology enabled the simultaneous quantification of thousands of transcripts through sequence-specific hybridization (Schena et al. 1995). This marked the first substantive leap from single-gene analysis to genome-wide profiling of transcripts leading to the delineation of intricate regulatory networks and discovery of coordinated gene expression programs. However, this platform remains intrinsically constrained by its reliance on prior sequence information and a restricted dynamic range. These limitations restrict the wide applicability of microarray in studying rare variants, novel transcripts, or isoforms (Wang et al. 2009). The advent of next-generation sequencing (NGS) in the 2000s marked the emergence of bulk RNA sequencing (RNA-seq) (Mortazavi et al. 2008). The RNA-seq approach circumvented the inherent issues with the microarray technique due to its ability of high-resolution analysis of the global transcriptional changes with significantly enhanced sensitivity and dynamic range. As RNA-seq is a generalized approach which does not require prior knowledge of target sequences, this technique aided in identification of several novel transcripts including noncoding RNAs as well as alternative splicing events.

Biological systems are inherently heterogenous at the cellular level, as exhibited in various tissues and organs comprising a plethora of distinct cell types, unique subpopulations, and their transient intermediates. Although the RNA-seq has gained popularity among biologists, a major limitation of bulk transcriptomic approach is that it overlooks the cellular diversity by averaging signals across vast cell populations. As a consequence, the subtle yet critical differences specific to individual cells that drive cellular homeostasis remain unknown (Stark et al. 2019). It was ~15 years ago, when single-cell RNA-seq (scRNA-seq) profoundly redefined transcriptomics by shifting the unit of analysis from bulk populations to individual cells

(Tang et al. 2009). Early protocols such as SMART-seq, Tang-seq, and CEL-seq demonstrated that entire transcriptomes could be captured from individual cells. The field underwent subsequent technological advancements in droplet microfluidics and combinatorial indexing having dramatically increased throughput, thereby enabling parallel profiling of tens of thousands of individual cells (Svensson et al. 2018). This substantial advancement in resolution unveiled the extensive cellular heterogeneity within tissues, resulting in the identification of uncommon or rare cell types, transient transcriptional states, and developmental trajectories that were previously imperceptible in comprehensive measurements. Spatial transcriptomics, proteomics, and epigenomics are now being integrated with scRNA-seq to create multimodal, spatially resolved cellular atlases (Stuart and Satija 2019). These integrative approaches are revolutionizing our understanding of every domain of biology such as cancer biology, immunology, developmental biology, neuroscience, etc., thereby enabling a deeper view into the intricate choreography of gene expression at a resolution of an individual cell.

This chapter introduces the concept of single-cell transcriptomics in the biological research. Beginning with major advancements in the transcriptomics, the chapter delves into the relevance of cellular heterogeneity. It further highlights the impact of scRNA-seq on various biological domains justifying the requirement of transcript analysis at the individual cell level, followed by a brief overview of the work flow of scRNA-seq and data analysis. The chapter ends with potential limitations and future perspectives of single-cell RNA sequencing technology.

4.2 Chronological Advances in Transcriptome Profiling

Whole-cell transcriptome profiling has long been a preferred approach for researchers to comprehend the genetic expression patterns. The transcriptome is defined as the fraction of the genetic code that is transcribed into RNA, predominantly represented by the total messenger RNA (mRNA) in a cell. It elucidates cellular processes and disease mechanisms, while uncovering novel pathways for biomarker identification and drug discovery. Understanding of the transcriptome began in the late 1970s with the invention of the Northern blotting, the first technique used to detect specific RNA molecules (Alwine et al. 1977). This hybridization-based method screens for RNA of interest by resolving total RNAs by gel electrophoresis, transferring the RNA samples to nylon membrane, followed by detection using labeled probes that are complementary to the target RNA. This strategy was considered ideal for recognition of microRNAs (miRNAs) and precursor miRNAs. However, the Northern blotting is a labor-intensive and low throughput technique with reduced sensitivity, which relies on prior information of the target sequence for probe generation, thus is unable to detect unknown transcripts (Krumlauf 1994). Advent of quantitative RT-PCR (qRT-PCR) revolutionized transcriptomics (Higuchi et al. 1993). The qRT-PCR can not only detect RNAs but also measure relative expression of the transcripts via complementary DNA (cDNA) synthesis followed by PCR-based amplification. A fluorescent dye is added to the reaction which precisely detects double-stranded DNA, enabling real-time quantification. This technique offers high sensitivity and is a cost-effective and reproducible alternative of blotting technique. Because of simplicity, the qRT-PCR has emerged as the gold standard in various diagnostic applications (Bustin and Mueller 2005). Since the qRT-PCR is restricted to known sequences, it remains a low-throughput targeted approach for detection and quantification of specific transcripts.

Transcriptome profiling saw a breakthrough in the mid-1990s with the discovery of two important techniques, namely microarray and Serial Analysis of Gene Expression (SAGE). A microarray is a state-of-the-art technology capable of detecting and quantitating thousands of transcripts synchronously (Schena et al. 1995). In a dual-color microarray technique, mRNA is converted to cDNA followed by differential dye-labeling. For instance, the test cDNA sample is labeled with a red fluorescent dye, while the control sample is labeled with a green fluorescent dye or vice versa. These labeled cDNAs are then loaded on microscopic slides that are precoated with complementary DNA probes, also known as gene chips. The labeled cDNAs from both the control and the test samples are mixed together to compete for hybridization with the probe on the array. According to the relative abundance of genes, different fluorescent intensities are emitted that are detected by the array scanner. In contrast to the dual-color microarray, a single-color microarray involves labeling of both the cDNA using single color fluorescent dyes, followed by hybridization on two arrays. Fluorescent intensities of each cDNA in both the samples are subsequently detected and quantitated by the array scanner. Microarray allows simultaneous detection of thousands of transcripts and has been applied in diverse fields, including developmental biology, cancer research, medicine, and biomarker discovery. With the innovation of Sanger sequencing, researchers tried to study the "expressed segment" of the genome via expressed sequence tags (ESTs). These are generated by cloning cDNA obtained from reverse transcription of mRNA into a suitable vector and randomly sequencing these clones. EST libraries are then created by aligning obtained reads, which can be further used for the generation of microarray chips (Lowe et al. 2017). Although microarray is a high-throughput technology, it cannot be used against unknown transcripts as array requires probes with known sequences. Moreover, the technique involves multiple steps and relies on incorporation of photosensitive dyes that often yield inconsistent data among replicative experiments. SAGE is a global profiling method, evolved from EST that allows users to detect RNA species without having any prior sequence information. Briefly, a short tag (10–12 bp) library of mRNA is constructed using reverse transcription and restriction enzymes. Subsequently, these oligonucleotide tags are concatenated and sequenced using Sanger sequencing with each tag's occurrence corresponding to the abundance of the particular mRNA (Velculescu et al. 1995). This technique offers quantitative analysis of transcripts with the added advantage of identification of unknown transcripts.

The emergence of NGS redefined the fields of genomics and transcriptomics, facilitating rapid sequencing of large datasets. RNA-seq provides a deeper understanding of differentially expressed genes, offering a more holistic picture of

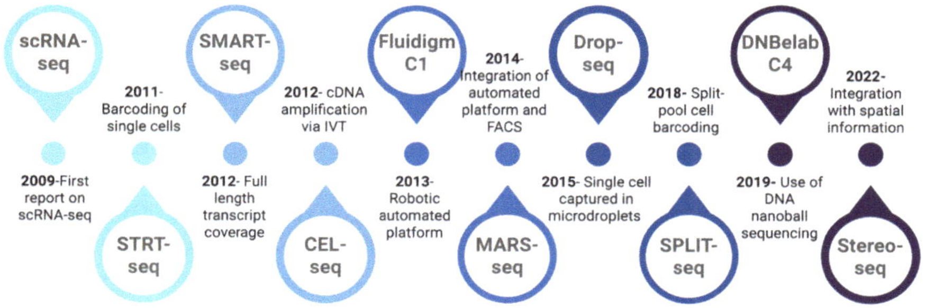

Fig. 4.1 Chronology of the technological advancements in single cell RNA sequencing. Shown is the timeline of major technological advancements in scRNA-seq. Since the first report of scRNA-seq by Tang et al. in 2009, sequencing of single-cell mRNA has witnessed significant technological advancements such as bar coding of single cells, automation and integration of automated platform with FACS, capturing of single cells in microdroplets, split-pool barcoding of cells, use of DNA nanoball sequencing and the recently-developed approach of integration of transcriptomic with spatial information. (The illustration was created in BioRender. https://BioRender.com/yewyeb6)

mutations, splice variants, and newly identified RNA species. RNA-seq can also identify single nucleotide changes, that are not feasible with hybridization-based methods. Nowadays, Bulk RNA-seq is routinely applied in clinical research for biomarker discovery, understanding disease mechanisms, and identifying novel therapeutic targets. Bulk RNA-seq measures average gene expression across the pool of cells in a sample and is considered an ideal strategy for assessing homogeneous samples (Nagalakshmi et al. 2008), whereas scRNA-seq measures gene expression in individual cells of a sample. scRNA-seq was first employed to profile mouse blastomere (Tang et al. 2009), and has proven advantageous for the identification of rare cell populations in heterogeneous samples (Li and Wang 2021). The scRNA-seq has witnessed major advancements since its inception as discussed in subsequent sections of this chapter, and summarized in Fig. 4.1. The latest technology, currently in its infancy stage, is spatial RNA sequencing which integrates RNA-seq data with the spatial information of cells. It helps align the gene expression data with the histological details of tissue, unlocking its potential applications in a variety of disciplines (Salmen et al. 2018).

4.3 Why scRNA-seq

Multicellular eukaryotic systems are heterogeneous that exhibit molecular and phenotypic variations among individual cells within a population that may otherwise appear genetically or morphologically similar. Heterogeneity at the cellular level does not merely arise as a consequence of genetic variations but also as a complex interplay of intrinsic molecular fluctuations and various epigenetic mechanisms. Cellular heterogeneity is a characteristic fundamental to multicellular systems. Cells of the same tissue may vary due to histone modifications, chromatin accessibility, and transcriptional noise. Heterogeneity in signaling cascades and cellular metabolism significantly influences cell behavior. It facilitates the development of stochastic survival strategies and also underpins developmental trajectories, immune responses, tissue homeostasis, and disease progression. Hence, it is of paramount importance to understand key cellular events in their entirety at the single-cell level. In the present time, single-cell metabolomics and scRNA-seq are commonly in practice to elucidate these phenomena (Evers et al. 2019; Trapnell 2015). Analysis of molecular events in the individual cells has enabled researchers to comprehend transcriptional noise and stochastic gene expression during embryogenesis that drive lineage specification, thereby allowing cells to explore various developmental fates (Carter and Zhao 2021). Single-cell analysis led to the identification of various subpopulations of stem cells exhibiting varied differentiation and proliferative potentials, ultimately safeguarding tissue renewal under injury or stress (Klein and Treutlein 2019). Tremendous heterogeneity has been observed even at the level of signaling pathways. Variability in SMAD/TGF-β signaling is one such example which results in diverse responses in otherwise identical cells upon exposure to the same stimuli. This has profound implications for differentiation and therapeutic responses (Sarma et al. 2020). One of the most striking examples of functional heterogeneity is illustrated by the immune system. Over time, single-cell transcriptomics has revealed remarkable diversity among myeloid lineages, B cells, and T cells. Even within the canonical subsets such as Th2, Th1, and Th17, hybrid and transient states have been identified, challenging conventional definitions (Villani et al. 2017). This biological heterogeneity enables the immune system to generate context-specific and nuanced responses to pathogens. Simultaneously, it monitors for autoimmunity and excessive inflammation that could arise as a consequence of excessive proinflammatory responses. Like every other cell type, heterogeneity is also found in cancerous cells, making intratumoral heterogeneity a key feature of malignancies which contributes to immune evasion, therapy resistance, and cancer dissemination (Meacham and Morrison 2013). A tumor, in itself, constitutes a dynamic ecosystem comprising both benign and malignant cells with distinct transcriptional regulations, infiltrating immune cells and stromal components, all of which interact in a temporally and spatially heterogeneous manner. The scRNA-seq technique has been instrumental in segregating these ecosystems, thereby facilitating the identification of rare malignant cell populations and drug-resistant persister cells that were previously imperceptible in bulk profiles (Tirosh et al. 2016).

Heterogeneity also plays a pivotal role in host–pathogen interactions and microbial biology. For instance, bacterial populations exhibit phenotypic variability that provides survival advantages under diverse stresses, including antibiotic exposure and host immune assaults. Similarly, in the context of host–pathogen interactions, single-cell transcriptomics has revealed diverse transcriptional states within infected cells, implying that not all, but a subset of cells within a population contribute to infection dynamics (Avraham et al. 2015). Cellular heterogeneity holds profound implications for both translational medicine and fundamental science, solidifying its position as a cornerstone of modern biology.

Population-averaged assays, such as bulk RNA-seq, can obscure the intricate diversity within cell populations. By reporting average expression values across millions of cells, these assays may conceal transient or rare subpopulations that may hold significant physiological roles (Trapnell 2015). By unraveling the concealed layers of cellular diversity, single-cell analysis offers a potent lens to delineate the fundamental principles of functional and cellular organization and plasticity. A comprehensive synthesis of this heterogeneity is paramount for the development of precision therapeutics that specifically target subpopulations rather than treating the entire cell populations as homogeneous entities. Cellular heterogeneity stands as a pivotal axis in physiology, development and pathology. scRNA-seq therefore serves as a dual lens, both indispensable and exquisite, capturing the concealed choreography between diversity and uniformity.

4.4 Impact of scRNA-seq on the Biological Research

Bulk RNA-seq has been a cornerstone of gene expression studies, providing comprehensive summaries of transcriptomic profiles across diverse conditions and tissues. Nevertheless, the bulk RNA-seq data rely on averages, condensing the intricacies of heterogeneous populations into a homogenized signal. In the process, the individuality of cells is ignored, rendering the dynamic cellular states and rare subpopulations invisible that often drive the most biologically significant processes. In contrast, scRNA-seq visualizes gene expression at the resolution of a single cell, thereby unveiling hierarchies and substructures that otherwise remain obscured by bulk methods (Fig. 4.2). Seemingly uniform tissue, once believed to consist solely of homogeneous population, is found to comprise distinct populations upon examination by single-cell analysis. Some of these may be differentiated, while others may be proliferative or may represent unique transitional or stem-like states. This resolution offers a comprehensive framework for comprehending the cellular ecosystem with unparalleled granularity having a profound biological impact.

Unique cell populations such as tissue resident immune subsets, tumor cells, or early progenitor cells may only constitute a minor fraction of total cells. The scRNA-seq has been useful in resolving their distinctive molecular signatures, thereby unfolding their roles in therapeutic resistance, disease progression or tissue homeostasis (Tang et al. 2009; Svensson et al. 2018). Remarkably, the scRNA-seq captures transitional and dynamic states, the liminal stages of activation, differentiation, or stress response that bulk sequencing often averages out. Because of the above-mentioned promising features, the scRNA-seq has emerged as one of the most sought-after technologies of the present time (Tang et al. 2009; Trapnell 2015). By harnessing advancements in molecular barcoding, cell isolation, and high-throughput sequencing, this technique has undergone significant advancements enabling simultaneous transcriptional profiling of thousands of cells composed of both rare and abundant populations with exceptional sensitivity (Macosko et al. 2015). This capacity has profoundly transformed the biological research. For instance, the scRNA-seq can reconstruct lineage trajectories, thereby unveiling

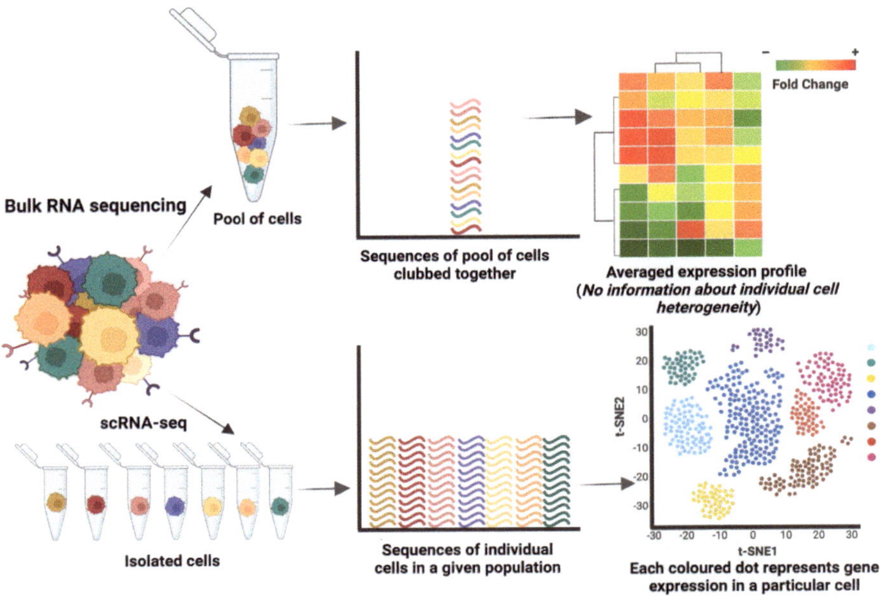

Fig. 4.2 Comparative analysis of bulk RNA sequencing and single cell RNA sequencing. The illustration depicts comparison of the bulk RNA sequencing and scRNA-seq approaches. Bulk RNA sequencing measures average gene expression across the heterogenous cell population, thus masking cellular heterogeneity. In contrast, scRNA-seq takes into account the transcriptomes of individual cells, revealing distinct cell types and rare cell populations. (The illustration was created in BioRender. https://BioRender.com/bvdkqsz)

transitional states that serve as guides for cell fate decisions (Klein and Treutlein 2019). It has added to our understanding of the complexity of the immune systems, uncovering novel monocyte and dendritic cell subsets and redefining classical T helper cell categories (Villani et al. 2017). The power of scRNA-seq extends to the field of oncology, enabling the unraveling of transcriptional heterogeneity within tumors which helps in the identification of drug-resistant persister cells, diverse stem cell populations, and tumor–stromal interactions that contribute to therapeutic resistance (Tirosh et al. 2016; Patel et al. 2014). Additionally, it has significantly illuminated the field of infection biology by portraying how the transcriptional variability across cells drives immune responses and pathogens' survival (Avraham et al. 2015). It is imperative to recognize that scRNA-seq does not render the classical bulk RNA-seq methods obsolete. Bulk sequencing continues to be the most cost-effective approach, exhibiting enhanced sensitivity for genes that are naturally less expressed with easier scalability across extensive sample cohorts. However, the scRNA-seq is revolutionary breakthrough that has fundamentally reimagined cellular identity not as a static entity but as a dynamic spectrum. This paradigm shift has propelled fundamental discoveries and holds immense promise for precision medicine (Fig. 4.3).

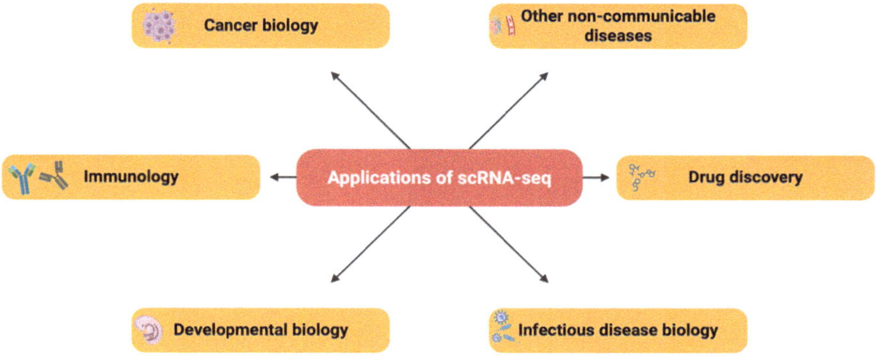

Fig. 4.3 Impact of scRNA-seq on the biological research. The illustration highlights some of the major applications of scRNA-seq technique in the biological research. As scRNA-seq facilitates high resolution profiling of gene expression at the granularity of a single cell, it holds multiple applications ranging from developmental lineage tracing, immune cell profiling, tumor heterogeneity analysis and assessment of cellular responses to environmental cues, microbial infections and drug exposure. (The illustration was created in BioRender. https://BioRender.com/j7tcrzx)

4.5 Workflow of scRNA-seq

RNA plays a crucial link in transferring the information encoded in DNA to functional molecules within cells, precisely proteins, thus underscoring the vitality of studying the transcriptome profile of cells. To discern the similarities and differences at the transcriptome level among the population of different cells, there are a number of steps undertaken to generate a scRNA-seq dataset. Currently, there are numerous protocols available such as STRT-seq, SMART-seq, CEL-seq, Fluidigm C1, Cyto-seq, Drop-seq, 10X genomics, SPLiT-seq, Smart-seq2, Seq-well, MATQ-seq, DNBelabC4, inDrop-seq, etc. which employ the work-flow in different orders. However, the basic framework of scRNA-seq involves the key steps that include: (i) collection of samples and isolation of viable single cells, (ii) extraction of intact RNA from these single cells, (iii) barcoding of mRNA and conversion into cDNA, (iv) preparation of cDNA library, (v) sequencing using various NGS platforms, and (vi) in silico analysis of data quality and quantification of read counts (Fig. 4.4a, b).

4.5.1 Sample Collection and Single-Cell Isolation

The initial yet challenging step of preparation of a single-cell suspension includes obtaining the sample, processing, and preserving its transcriptome, followed by the acquisition of healthy single cells. Blood samples are generally easier to process, while tissue samples must be dissociated first using a few enzymes like trypsin or collagenase. Moreover, cell–cell communication is critical for many cellular functions, and disrupting cellular crosstalk can drastically alter the transcriptome profile (Kolodziejczyk et al. 2015). A group of researchers attempted to compare

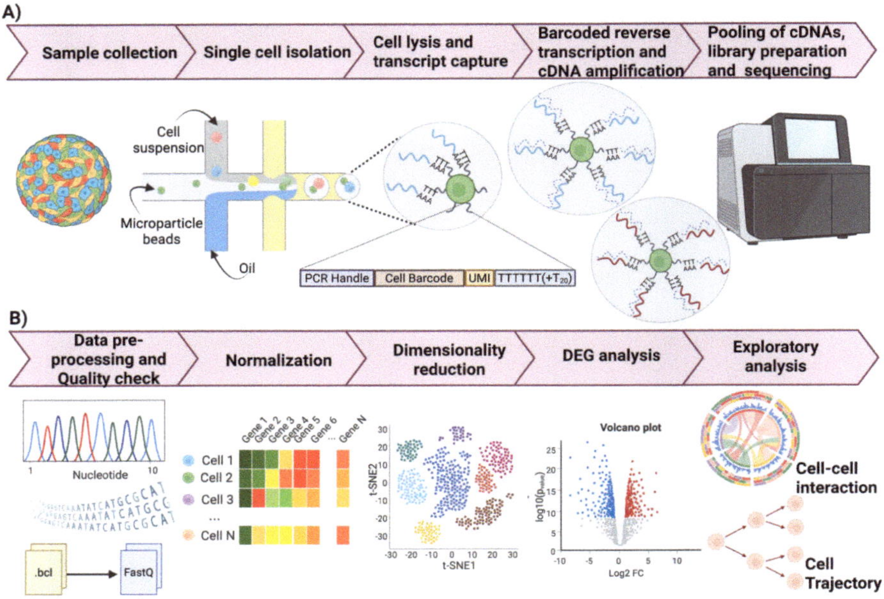

Fig. 4.4 Comprehensive roadmap of the single cell RNA sequencing workflow. (**a**) Schematic of different steps of sample preparation for scRNA-seq. Shown is the representative workflow of the Drop-seq method, which involves heterogeneous sample collection and encapsulation of single cells with beads using the droplet method. Subsequently, the cells are lysed, and their transcripts are captured and reverse transcribed. Finally, cDNAs from all beads are pooled and subjected to library preparation followed by deep sequencing. (**b**) Schematic of the workflow of in silico data analysis. The data analysis begins with pre-processing and quality check followed by conversion of data to *FASTQ* format. The formatted data are normalized and subsequently subjected to dimensionality reduction for better visualization and clustering of cells based on the analysis of differentially expressed genes. Finally, exploratory analysis of data is performed to understand relationship between cells. (The illustration was created in BioRender. https://BioRender.com/cqfs85x)

transcriptomes obtained from single nuclei RNA-seq and scRNA-seq to overcome the challenges associated with the isolation of single cells, and found comparative profiles (Zeng et al. 2016; Lacar et al. 2016). Earlier studies used to involve freshly isolated samples for the preparation of single-cell suspension, however, later on protocols were refined to successfully use cryopreserved (Guillaumet-Adkins et al. 2017) or fixed (Thomsen et al. 2016) samples in scRNA-seq studies.

Micromanipulation was one of the primitive and manual technique used in which single cells were captured with the help of micropipettes. However, it cannot be applied for isolating cells from complex tissues and demands expertise. To overcome this, researchers developed another technique known as laser capture microdissection which utilizes laser beams to capture single cell from tissues of interest (Frumkin et al. 2008). Both techniques were dependent on microscope-based viewing of cells to capture the cell of interest, and were time consuming. To isolate single cells in a high-throughput fashion, fluorescence-activated cell sorting (FACS) was adopted to segregate a larger number of cells on the basis of cell size and markers. FACS introduced a method to store data regarding cell phase and size and was termed as "index sorting" (Hayashi et al. 2010). With time, the efficiency and fidelity of single-cell isolation was improved with microfluidics and microdroplet-based methods, where reactions are scaled up in nanoliter volume, thus reducing the reagents' cost. The microfluidics platform implements a chip along with a fluidic circuit containing 96 wells, and each well can hold one cell. A commonly used robotic platform is Fluidigm C1, which captures cells and can subsequently integrate with reverse transcription. However, one major limitation of this system is the lower capture efficiency of cells (Trapnell et al. 2014). The microdroplet-based platform is a newer approach which compartmentalizes each cell in a droplet with the reagents capable of performing downstream steps (Mazutis et al. 2013). Drop-seq protocol utilizes this methodology, where a flow of microparticles or beads and a flow of single-cell suspension are combined, resulting in encapsulation of single cells in tiny oil droplets. Each bead has a PCR handle, cell barcode, unique molecular identifiers (UMIs), and oligo dT sequence to which mRNA binds, culminating in the formation of single-cell transcriptomes attached to microparticles after reverse transcription (Macosko et al. 2015). Another method namely Split-pool ligation-based single-cell transcriptome (SPLit-seq) assigns unique barcodes to single cells by splitting and pooling of cells in multiwells having barcodes. The SPLit-seq approach offers an added advantage to simultaneously process large number of samples without the need of a costly microfluidic device (Kuijpers et al. 2024).

4.5.2 Barcoded Reverse Transcription and Library Preparation

Once the single cells are captured, these are subjected to lysis for RNA isolation. It is a crucial step as effective cell lysis and good quality RNA are necessary to minimize technical variability. To distinguish the RNA of different cells in a pooled sequencing sample, barcoding strategy is adopted in nearly all methods. Barcoding can be performed either at the 5′end (STRT-seq) or the 3′ end (CEL-seq) during reverse transcription, or while library preparation (SMART-seq). Given that RNA is sensitive to degradation, and also sequencing platforms are compatible with DNA, RNA is converted to cDNA by reverse transcription. In the protocol described by Tang et al., first-strand synthesis of cDNA is conducted using Poly (T) primers to specifically reverse transcribe mRNA followed by attachment of poly A tail to the 3′ end of the first-strand using the terminal transferase enzyme. Subsequently, primers are enzymatically removed and added afresh to initiate the second-strand synthesis (Tang et al. 2009). However, this method suffers with potential bias as it does not account for incomplete reverse transcription of the first-strand. In contrast, scRNA-seq involving methods such as "switching mechanism at 5′ end" of RNA template (SMART-seq) (Ramskold et al. 2012) and single-cell tagged reverse transcription (STRT-Seq) (Natarajan 2019) employ a "template switching" mechanism with the help of Moloney murine leukemia virus reverse transcriptase (MMLV-RT). The MMLV-RT possesses an intrinsic property to add a few cytosine bases at the 3′

end of the first-strand of cDNA. Then, template switching oligonucleotides, consisting of PCR adaptors at the 5′ end and guanine bases at the 3′ end, are added which base pair with the first-strand of cDNA having cytosine residues at the 3′ end. The enzyme then transitions its template from RNA to DNA and completes the process of reverse transcription.

The resulting cDNA is amplified using either PCR or in vitro transcription (IVT). PCR is applied in the majority of the protocols despite the fact that it is a nonlinear process in comparison to IVT, whereas latter is utilized in the "cell expression by linear amplification and sequencing" (CEL-seq) (Hashimshony et al. 2012) and "massively parallel RNA single-cell sequencing" (MARS-seq) techniques (Jaitin et al. 2014). A major drawback with IVT is that it requires additional rounds of reverse transcription, contributing to the 3′ coverage bias. UMIs are used to tag individual RNA molecules, eliminating the coverage bias arising from cDNA amplification (Kivioja et al. 2011). These UMIs are implemented in all protocols where sequencing is done by 5′ or 3′ end counting. In the budding phase of this technology, only either the 5′ or the 3′ end used to be sequenced with unique barcodes, which was insufficient to decipher isoforms or splice variants of RNA. Later on, reading of full-length transcripts was followed which can efficiently detect all RNA, however, it increased the running cost of the experiment.

It is routine to add predefined amount of spike-ins, the commercially available external RNAs, in the reactions to decrease technical variability. The number of obtained reads is compared to the amount of spike-in RNA added, which allows for normalization using computational tools. The external RNA control consortium or ERCC has developed a set of RNA spike-in controls that help correct the batch effects between samples (External RNA Controls Consortium 2005).

4.5.3 Deep Sequencing

scRNA-seq employs various sequencing platforms, including Illumina, Oxford Nanopore, DNBSeq, and Pacific Biosciences to analyze the library. Each platform differs in its working principle, thereby having various advantages and disadvantages. For instance, the Illumina platform is a high-throughput and cost-effective option which works on the principle of short-read sequencing. In contrast, Oxford Nanopore uses a long-read sequencing principle allowing for read lengths that can exceed 100 kb, which makes it effective for achieving full-length transcript coverage. Pacific Biosciences is a third-generation sequencing technology which operates on "single molecule real-time" reading of DNA molecules. This method offers longer read lengths, which makes it an ideal technique for isoform characterization albeit at the substantially high operational cost. DNBSeq is another strategy which is based on DNA nanoball sequencing. It utilizes modified version of rolling circle replication with the help of Phi29 polymerase. This method generates circular single-stranded DNA molecules which are sequenced with short read-lengths (Drmanac et al. 2010) (Fig. 4.4a).

4.5.4 In-Silico Data Analysis

After the wet-lab processing of samples and sequencing, a massive amount of data is generated, which needs computational processing. The in silico data analysis is therefore considered the second half of the scRNA-seq protocol. Bioinformaticians follow a pipeline to analyze the extensive datasets as presented in Fig. 4.4b, and briefly summarized below:

4.5.4.1 Quality Check and Preprocessing

The first step in the process of analyzing an extensive dataset is to check the quality of reads. The output sequencing files are generally available in two formats, *FASTQ* and *BCL*. Since, majority of the Quality check (QC) tools are compatible with *FASTQ* formats, the *BCL* files are also converted to *FASTQ* format before data processing (Jovic et al. 2022). QC is vital to differentiate between the obtained true reads from any technical artifacts, and reads from "low quality" cells, which are dead or stressed cells and can mislead the data interpretation (Ilicic et al. 2016). Read quality can be checked using different tools such as FastQC, Seurat, Scanpy, or PopsicleR that deploy different algorithms and programming languages.

The next objective is to remove adaptors from the reads by various tools such as Trim Galore or Trimmomatic. Quality assessment for cells of eukaryotic origin is done based on three parameters: (i) count depth (UMI count), (ii) number of transcripts from mitochondria, and (iii) number of genes identified. UMI count is a crucial filter applied during QC, since low count depth means poor quality cells, whereas higher depth may indicate the presence of doublets. Cell ranger is commonly used for assessing the count depth of the library and also mapping the reads to a reference. Another potential barrier usually encountered by bioinformaticians is an abrupt high number of reads having the same cell barcode, indicating cell doublets or multiplets. Therefore, it is necessary to remove such cells from the analysis which is achieved by different in silico approaches such as Doublet Finder, Scrublet, Solo, Doublet Decon, etc. The fraction of mitochondrial gene count is an additional measure in cell QC as a high ratio can be indicative of dying cells. It is usually performed by using miQC, a data-driven QC metric (Hippen et al. 2021). Free-floating RNA species represent another source of confounding factor that need to be filtered out. It is usually achieved by using tools such as Cell Bender, SoupX, and DecontX (Young and Behjati 2020; Yang et al. 2020). Sometimes QC has to be performed at the gene level to further improve the quality of the dataset. It is required particularly when a few transcripts are unable to be detected either due to low expression or stochastic sampling, known as "dropout." Such issues are solved by carefully adjusting the threshold value in the analysis software. It is necessary to address such events so that "true zero" i.e., no or negligible expression levels can be distinguished from "nonbiological zeros" which arise due to technical errors such as absence of reads or UMIs of specific RNA in a cell in scRNA-seq data despite its presence in other cells of same type (Jiang et al. 2022). Various imputation methods have been developed to address such problems such as data smoothing which adjusts the technical variability by comparing all expression values recorded from a cell to those observed in similar cells, or model-based methods which evaluate the datasets by comparison with previously available scRNA-seq datasets. A few online platforms also utilize deep learning techniques wherein data are corrected by the construction of an expression matrix. IGSimpute, Deep Impute, SciGAN, and ScHinter are a few examples of available imputation tools that are commonly used in scRNA-seq data analysis (Ye et al. 2020; Arya et al. 2025).

4.5.4.2 Normalization

The number of filtered reads obtained in the scRNA-seq experiment varies among different cells within or between samples, and need to be normalized to omit variations occurring due to sampling factors or processing anomalies. Sequencing depth in a cell is normalized by either transcripts per kilobase million, fragments per kilobase of transcript per million reads, or reads per kilobase of transcript per million reads. Spike-ins play a crucial role at this stage, wherein variations obtained in the observed spike-ins expression from the expected amounts are used for normalization. Data scientists derive a scaling factor through this process which is subsequently used to correct the technical noise. Several online tools are available such as Seurat and Scran which normalizes scRNA-seq data using the R language or Scanpy, a Python-based tool, which performs normalization using multiple methods. In addition, few pipelines such as DESeq2 have their own normalization algorithm which work with raw count data and do not accept prenormalized data as input for the differential expression analysis (McDavid et al. 2013).

4.5.4.3 Feature Selection and Dimensionality Reduction

The next stage in the bioinformatic analysis of scRNA-seq data comprises of feature selection and dimensionality reduction to aid the visualization of results. Feature selection is applied to detect the highly variable genes (genes having varied expression across cells), and to segregate biological variance from technical variance. Such genes are crucial to analyze as these are important for cellular heterogeneity determination and rare cell identification. GiniClust is one such tool which is used to detect rare cell-types from large-scale scRNA-seq data (Jiang et al. 2016).

The scRNA-seq count matrices contain thousands of genes and cells, each of which contributes to the dimension of the data. However, it is challenging to analyze data with the high dimensionality and sparsity. Hence, it becomes necessary to reduce their dimensionality so that these can be easily comprehended. Here, experts use two approaches, i.e., linear and nonlinear. Principal component analysis (PCA) is a linear approach, which is also widely used in bulk RNA-seq data analysis. PCA can handle large datasets while preserving the genetic makeup. Since it is a low-throughput approach, it is to be followed by nonlinear approaches. The t-distributed stochastic neighbor embedding and "Diffusion Maps" algorithms are nonlinear techniques that are used for reduction of the dimension of the data and visualization of high-dimensional data (Rostom et al. 2017). Uniform manifold approximation and projection is a recent dimensionality reduction approach which conserves both local and long-range patterns. Though these methods effectively reduce the dimensions to two or three, one needs to be careful as these may also generate technological noise (Becht et al. 2018).

4.5.4.4 Cell Clustering and Differential Expression Analysis

Cell clustering is one of the vital components of in silico data analysis, which discriminates scRNA-seq from bulk RNA-seq. Based on expression profiles, different cells are grouped in an unbiased approach which can be achieved by using five different methods. The k-means clustering was the first approach of clustering cells which works by recurrently finding the centroids of clusters in order to decrease the distance between each data point and cluster. Though this method can be efficiently used with high-throughput data, it is not suitable for identifying rare cell clusters because it iteratively works on predefined clusters. The second method is hierarchical clustering, in which cell clusters are arranged in a hierarchical structure or dendrogram, and is therefore useful in rare cell-type determination. The third approach is density-based clustering, where grouping is done based on cell density, and low-density regions are labeled as outliers. Apart from these, graph-based and deep learning are the other two methods that are employed in cell clustering. There are various tools such as SC3, SAIC, Gini Clust, GRACE, BackSPIN, scDeepCluster, etc., that employ either of the above-mentioned approaches. Each of these tools has its own strengths and challenges, and are used as per the requirement (He et al. 2022).

Understanding the differential expression of genes (DEGs) is the ultimate goal of any RNA sequencing experiment, as it helps in comprehending the biological differences between the control and the test samples. In a scRNA-seq study, distinct cell types are identified on the basis of these DEGs and DESeq2 is widely used to analyze scRNA-seq data. It implements a generalized linear model to quantify the fold-change difference for each gene and diverse statistical tests for estimating the significance of the same. Advanced computational tools such as multidisciplinary-design adaptation and sensitivity toolkit, or MAST, have been developed such that they can take into account the dropout events. In this advancing field, newer tools are constructed very frequently having the potential to perform multiple tasks. For instance, Monocle can locate DEGs along with pseudo-time analysis. EdgeR, D3E, GREIN, and Scotty are few other examples that are employed to execute the same task (Lu et al. 2023).

4.5.4.5 Trajectory Analysis and Other Optional Analysis

The advanced analysis of scRNA-seq data involves trajectory inference of cell clusters, wherein bioinformaticians try to map the developmental stages of cells using pseudo-time analysis. Monocle is a widely accepted algorithm among researchers for analyzing cell trajectories (Trapnell et al. 2014). Besides, harmony, wishbone, PAGA, and RNA velocity are other computational tools that are used in scRNA-seq analysis to understand various developmental stages of cells and gene regulation. These can offer insights into diverse biological networks and reveal rare subpopulation of cells.

Cell–cell communication is other kind of analysis, which is mainly driven by ligand–receptor interactions and gene network analysis that may provide some information about disease states and aid researchers with therapeutic discoveries (Efremova et al. 2020). Moreover, pathway enrichment analysis and metabolic state analysis, along with integration of transcriptomics data with other "omics" techniques are performed to gather deeper understanding of diverse cellular states.

4.6 Limitations of scRNA-seq

ScRNA-seq is a cutting-edge technology, slowly progressing towards becoming an irreplaceable pillar in the field of transcriptomics by providing researchers with cellular developmental trajectories, aiding in understanding complex tissues, and uncovering rare cell populations. The utility of this ground-breaking methodology is still limited by a number of technical hurdles. Procurement of intact single cells from complex tissues having preserved cellular transcriptome is one such challenge which affects data quality and consistency. Also, general methods of capturing single cells suffer with poor efficiency, due to which samples comprising of fewer cells yield suboptimal quality data. Use of microfluidics to segregate single cells is a common practice in scRNA-seq, however, it demands a homogenous cell size and a higher cost (Kolodziejczyk et al. 2015). Another major limitation of scRNA-seq is the use of poly (T) primers to synthesize cDNAs, which omits the quantification of nonpolyadenylated RNA species. Furthermore, dropout of poorly-expressing mRNAs is a serious issue leading to poor coverage as well as false-positivity of low-abundance transcripts in scRNA-seq experiments (Deng et al. 2014). During the process of cDNA amplification, multiple rounds of PCR add noise to the pool, rendering it difficult to accurately quantitate the transcripts. Remarkably, the amount of RNA spike-ins, which are used to differentiate between true biological divergence from noise (Brennecke et al. 2013), needs careful optimization as these are not compatible with every protocol. In addition to these limitations, researchers face several obstructions during bioinformatic analysis of the data, which affect precise quantification. In particular, storage and processing of the large and complex datasets are the primary concerns, which demand high-end systems with significant data storage and processing capacity to perform the analysis. Since, scRNA-seq sequences thousands of cells with each cell contributing to the dimension of the data, it becomes a humungous task to reduce dimensionality and process highly complex data which not only consumes time but also requires advanced skillsets. Every step of in silico analysis whether data preprocessing, normalization, cell clustering or DEGs brings about certain level of variation, thus affecting the reproducibility.

Overall, the running cost of scRNA-seq per cell per sample is exorbitantly high due to the above-mentioned issues as well as the requirement of expensive molecular-grade reagents, high-end sequencing platforms and advanced data analysis tools. As a consequence, scRNA-seq is restricted to advanced laboratories and is not affordable in resource-limited settings, thus limiting its potential applications.

4.7 Conclusions and Future Perspectives

scRNA-seq has revolutionized our understanding of biological systems, expanding our knowledge of cellular identity from a static concept to a highly dynamic continuum. By elucidating transcriptional programs at the resolution of individual cells, the scRNA-seq has unveiled concealed layers of cellular heterogeneity, delineated developmental trajectories, and illuminated the molecular underpinnings of health and disease with an unprecedented depth. From identification of novel subsets of cells to unveiling of the intricate heterogeneity prevalent within cellular microenvironments, scRNA-seq has made significant strides since its inception. This groundbreaking technology has not only advanced our understanding of biological sciences but has also paved the way for advancements in precision medicine. This field allows for the development of tailored therapeutic strategies that can be specifically designed for functionally critical and rare subpopulations. Akin to every other revolutionary technology, scRNA-seq has its share of challenges. Data interpretation is complicated by technical limitations such as batch effects, dropout events, and biases in cell capture (Svensson et al. 2018). Moreover, scRNA sequencing captures transcriptional states in a snapshot, leaving spatial and temporal dynamics of cellular behavior only partially resolved.

The next frontier in cellular research lies in integrative approaches that seamlessly combine complementary modalities such as epigenomics, proteomics, metabolomics, and spatial transcriptomics with single-cell transcriptomics. These approaches would facilitate a comprehensive reconstruction of cellular identity in situ *(*Stuart and Satija 2019*)*. Forthcoming advancements in machine learning, artificial intelligence, and computational tools are anticipated to enhance data integration and refine trajectory inference, thereby propelling us closer to predictive models of cell fate. Translation of single-cell insights into clinical practice is poised to revolutionize prognostics, diagnostics, and therapeutics. The early detection of disease through circulating single-cell transcriptomes, enabling the segregation of patients into groups based on diverse tumor subclones and the development of tailored immunotherapies targeting specific immune cell subpopulations is one such possibility which underscores the substantial potential of this field (Kinker et al. 2020). As technologies are becoming more integrative, cost-effective, and scalable, the aspiration to construct a comprehensive human cell atlas which encompasses health, developmental stages, and disease is increasingly within reach (Regev et al. 2017). In essence, scRNA-seq has ushered a potent molecular lens, capable of perceiving biological systems not as a monochromatic canvas but as a vibrant and intricate portrait of diverse cellular states and interactions. The upcoming journey of scRNA-seq anticipates enhanced resolution, profound integration, and a significant translational impact in biological and health sectors.

Acknowledgement We acknowledge core funding support from the Translational Health Science and Technology Institute, Department of Biotechnology, Govt. of India. Department of Science and Technology, Govt. of India is acknowledged for providing research fellowship to J. A.

References

Alwine JC, Kemp DJ, Stark GR (1977) Method for detection of specific RNAs in agarose gels by transfer to diazobenzyloxymethyl-paper and hybridization with DNA probes. Proc Natl Acad Sci USA 74:5350–5354

Arya A, Tripathi P, Dubey N, Aier I, Kumar Varadwaj P (2025) Navigating single-cell RNA-sequencing: protocols, tools, databases, and applications. Genomics Inform 23:13

Avraham R, Haseley N, Brown D, Penaranda C, Jijon HB, Trombetta JJ, Satija R, Shalek AK, Xavier RJ, Regev A, Hung DT (2015) Pathogen cell-to-cell variability drives heterogeneity in host immune responses. Cell 162:1309–1321

Becht E, McInnes L, Healy J, Dutertre CA, Kwok IWH, Ng LG, Ginhoux F, Newell EW (2018) Dimensionality reduction for visualizing single-cell data using UMAP. Nat Biotechnol 37:38–44

Brennecke P, Anders S, Kim JK, Kolodziejczyk AA, Zhang X, Proserpio V, Baying B, Benes V, Teichmann SA, Marioni JC, Heisler MG (2013) Accounting for technical noise in single-cell RNA-seq experiments. Nat Methods 10:1093–1095

Bustin SA, Mueller R (2005) Real-time reverse transcription PCR (qRT-PCR) and its potential use in clinical diagnosis. Clin Sci (Lond) 109:365–379

Carter B, Zhao K (2021) The epigenetic basis of cellular heterogeneity. Nat Rev Genet 22:235–250

Deng Q, Ramskold D, Reinius B, Sandberg R (2014) Single-cell RNA-seq reveals dynamic, random monoallelic gene expression in mammalian cells. Science 343:193–196

Drmanac R, Sparks AB, Callow MJ, Halpern AL, Burns NL, Kermani BG, Carnevali P, Nazarenko I, Nilsen GB, Yeung G, Dahl F, Fernandez A, Staker B, Pant KP, Baccash J, Borcherding AP, Brownley A, Cedeno R, Chen L, Chernikoff D, Cheung A, Chirita R, Curson B, Ebert JC, Hacker CR, Hartlage R, Hauser B, Huang S, Jiang Y, Karpinchyk V, Koenig M, Kong C, Landers T, Le C, Liu J, McBride CE, Morenzoni M, Morey RE, Mutch K, Perazich H, Perry K, Peters BA, Peterson J, Pethiyagoda CL, Pothuraju K, Richter C, Rosenbaum AM, Roy S, Shafto J, Sharanhovich U, Shannon KW, Sheppy CG, Sun M, Thakuria JV, Tran A, Vu D, Zaranek AW, Wu X, Drmanac S, Oliphant AR, Banyai WC, Martin B, Ballinger DG, Church GM, Reid CA (2010) Human genome sequencing using unchained base reads on self-assembling DNA nanoarrays. Science 327:78–81

Efremova M, Vento-Tormo M, Teichmann SA, Vento-Tormo R (2020) CellPhoneDB: inferring cell-cell communication from combined expression of multi-subunit ligand-receptor complexes. Nat Protoc 15:1484–1506

Evers TMJ, Hochane M, Tans SJ, Heeren RMA, Semrau S, Nemes P, Mashaghi A (2019) Deciphering metabolic heterogeneity by single-cell analysis. Anal Chem 91:13314–13323

External RNA Controls Consortium (2005) Proposed methods for testing and selecting the ERCC external RNA controls. BMC Genomics 6:150

Frumkin D, Wasserstrom A, Itzkovitz S, Harmelin A, Rechavi G, Shapiro E (2008) Amplification of multiple genomic loci from single cells isolated by laser micro-dissection of tissues. BMC Biotechnol 8:17

Guillaumet-Adkins A, Rodriguez-Esteban G, Mereu E, Mendez-Lago M, Jaitin DA, Villanueva A, Vidal A, Martinez-Marti A, Felip E, Vivancos A, Keren-Shaul H, Heath S, Gut M, Amit I, Gut I, Heyn H (2017) Single-cell transcriptome conservation in cryopreserved cells and tissues. Genome Biol 18:45

Hashimshony T, Wagner F, Sher N, Yanai I (2012) CEL-Seq: single-cell RNA-Seq by multiplexed linear amplification. Cell Rep 2:666–673

Hayashi T, Shibata N, Okumura R, Kudome T, Nishimura O, Tarui H, Agata K (2010) Single-cell gene profiling of planarian stem cells using fluorescent activated cell sorting and its "index sorting" function for stem cell research. Develop Growth Differ 52:131–144

He J, Lin L, Chen J (2022) Practical bioinformatics pipelines for single-cell RNA-seq data analysis. Biophys Rep 8:158–169

Higuchi R, Fockler C, Dollinger G, Watson R (1993) Kinetic PCR analysis: real-time monitoring of DNA amplification reactions. Biotechnology (NY) 11:1026–1030

Hippen AA, Falco MM, Weber LM, Erkan EP, Zhang K, Doherty JA, Vaharautio A, Greene CS, Hicks SC (2021) miQC: an adaptive probabilistic framework for quality control of single-cell RNA-sequencing data. PLoS Comput Biol 17:e1009290

Ilicic T, Kim JK, Kolodziejczyk AA, Bagger FO, McCarthy DJ, Marioni JC, Teichmann SA (2016) Classification of low quality cells from single-cell RNA-seq data. Genome Biol 17:29

Jaitin DA, Kenigsberg E, Keren-Shaul H, Elefant N, Paul F, Zaretsky I, Mildner A, Cohen N, Jung S, Tanay A, Amit I (2014) Massively parallel single-cell RNA-seq for marker-free decomposition of tissues into cell types. Science 343:776–779

Jiang L, Chen H, Pinello L, Yuan GC (2016) GiniClust: detecting rare cell types from single-cell gene expression data with Gini index. Genome Biol 17:144

Jiang R, Sun T, Song D, Li JJ (2022) Statistics or biology: the zero-inflation controversy about scRNA-seq data. Genome Biol 23:31

Jovic D, Liang X, Zeng H, Lin L, Xu F, Luo Y (2022) Single-cell RNA sequencing technologies and applications: a brief overview. Clin Transl Med 12:e694

Kinker GS, Greenwald AC, Tal R, Orlova Z, Cuoco MS, McFarland JM, Warren A, Rodman C, Roth JA, Bender SA, Kumar B, Rocco JW, Fernandes P, Mader CC, Keren-Shaul H, Plotnikov A, Barr H, Tsherniak A, Rozenblatt-Rosen O, Krizhanovsky V, Puram SV, Regev A, Tirosh I (2020) Pan-cancer single-cell RNA-seq identifies recurring programs of cellular heterogeneity. Nat Genet 52:1208–1218

Kivioja T, Vaharautio A, Karlsson K, Bonke M, Enge M, Linnarsson S, Taipale J (2011) Counting absolute numbers of molecules using unique molecular identifiers. Nat Methods 9:72–74

Klein AM, Treutlein B (2019) Single cell analyses of development in the modern era. Development 146

Kolodziejczyk AA, Kim JK, Svensson V, Marioni JC, Teichmann SA (2015) The technology and biology of single-cell RNA sequencing. Mol Cell 58:610–620

Krumlauf R (1994) Analysis of gene expression by northern blot. Mol Biotechnol 2:227–242

Kuijpers L, Hornung B, van den Hout-van Vroonhoven M, van IJcken WF, Grosveld F, Mulugeta E (2024) Split Pool Ligation-based Single-cell Transcriptome sequencing (SPLiT-seq) data processing pipeline comparison. BMC Genomics 25:361

Lacar B, Linker SB, Jaeger BN, Krishnaswami SR, Barron JJ, Kelder MJE, Parylak SL, Paquola ACM, Venepally P, Novotny M, O'Connor C, Fitzpatrick C, Erwin JA, Hsu JY, Husband D, McConnell MJ, Lasken R, Gage FH (2016) Nuclear RNA-seq of single neurons reveals molecular signatures of activation. Nat Commun 7:11022

Li X, Wang CY (2021) From bulk, single-cell to spatial RNA sequencing. Int J Oral Sci 13:36

Lowe R, Shirley N, Bleackley M, Dolan S, Shafee T (2017) Transcriptomics technologies. PLoS Comput Biol 13:e1005457

Lu J, Sheng Y, Qian W, Pan M, Zhao X, Ge Q (2023) scRNA-seq data analysis method to improve analysis performance. IET Nanobiotechnol 17:246–256

Macosko EZ, Basu A, Satija R, Nemesh J, Shekhar K, Goldman M, Tirosh I, Bialas AR, Kamitaki N, Martersteck EM, Trombetta JJ, Weitz DA, Sanes JR, Shalek AK, Regev A, McCarroll SA (2015) Highly parallel genome-wide expression profiling of individual cells using nanoliter droplets. Cell 161:1202–1214

Mazutis L, Gilbert J, Ung WL, Weitz DA, Griffiths AD, Heyman JA (2013) Single-cell analysis and sorting using droplet-based microfluidics. Nat Protoc 8:870–891

McDavid A, Finak G, Chattopadyay PK, Dominguez M, Lamoreaux L, Ma SS, Roederer M, Gottardo R (2013) Data exploration, quality control and testing in single-cell qPCR-based gene expression experiments. Bioinformatics 29:461–467

Meacham CE, Morrison SJ (2013) Tumour heterogeneity and cancer cell plasticity. Nature 501:328–337

Mortazavi A, Williams BA, McCue K, Schaeffer L, Wold B (2008) Mapping and quantifying mammalian transcriptomes by RNA-Seq. Nat Methods 5:621–628

Nagalakshmi U, Wang Z, Waern K, Shou C, Raha D, Gerstein M, Snyder M (2008) The transcriptional landscape of the yeast genome defined by RNA sequencing. Science 320:1344–1349

Natarajan KN (2019) Single-cell tagged reverse transcription (STRT-Seq). Methods Mol Biol 1979:133–153

Patel AP, Tirosh I, Trombetta JJ, Shalek AK, Gillespie SM, Wakimoto H, Cahill DP, Nahed BV, Curry WT, Martuza RL, Louis DN, Rozenblatt-Rosen O, Suva ML, Regev A, Bernstein BE (2014) Single-cell RNA-seq highlights intratumoral heterogeneity in primary glioblastoma. Science 344:1396–1401

Ramskold D, Luo S, Wang YC, Li R, Deng Q, Faridani OR, Daniels GA, Khrebtukova I, Loring JF, Laurent LC, Schroth GP, Sandberg R (2012) Full-length mRNA-Seq from single-cell levels of RNA and individual circulating tumor cells. Nat Biotechnol 30:777–782

Regev A, Teichmann SA, Lander ES, Amit I, Benoist C, Birney E, Bodenmiller B, Campbell P, Carninci P, Clatworthy M, Clevers H, Deplancke B, Dunham I, Eberwine J, Eils R, Enard W, Farmer A, Fugger L, Gottgens B, Hacohen N, Haniffa M, Hemberg M, Kim S, Klenerman P, Kriegstein A, Lein E, Linnarsson S, Lundberg E, Lundeberg J, Majumder P, Marioni JC, Merad M, Mhlanga M, Nawijn M, Netea M, Nolan G, Pe'er D, Phillipakis A, Ponting CP, Quake S, Reik W, Rozenblatt-Rosen O, Sanes J, Satija R, Schumacher TN, Shalek A, Shapiro E, Sharma P, Shin JW, Stegle O, Stratton M, Stubbington MJT, Theis FJ, Uhlen M, van Oudenaarden A, Wagner A, Watt F, Weissman J, Wold B, Xavier R, Yosef N. P. Human Cell Atlas Meeting(2017) The Human Cell Atlas. Elife 6

Rostom R, Svensson V, Teichmann SA, Kar G (2017) Computational approaches for interpreting scRNA-seq data. FEBS Lett 591:2213–2225

Salmen F, Stahl PL, Mollbrink A, Navarro JF, Vickovic S, Frisen J, Lundeberg J (2018) Barcoded solid-phase RNA capture for spatial Transcriptomics profiling in mammalian tissue sections. Nat Protoc 13:2501–2534

Sarma U, Hexemer L, Anyaegbunam UA, Legewie S (2020) Modelling cellular signalling variability based on single-cell data: the TGFb/SMAD signaling pathway. arxiv 2007.09093

Schena M, Shalon D, Davis RW, Brown PO (1995) Quantitative monitoring of gene expression patterns with a complementary DNA microarray. Science 270:467–470

Stark R, Grzelak M, Hadfield J (2019) RNA sequencing: the teenage years. Nat Rev Genet 20:631–656

Stuart T, Satija R (2019) Integrative single-cell analysis. Nat Rev Genet 20:257–272

Svensson V, Vento-Tormo R, Teichmann SA (2018) Exponential scaling of single-cell RNA-seq in the past decade. Nat Protoc 13:599–604

Tang F, Barbacioru C, Wang Y, Nordman E, Lee C, Xu N, Wang X, Bodeau J, Tuch BB, Siddiqui A, Lao K, Surani MA (2009) mRNA-Seq whole-transcriptome analysis of a single cell. Nat Methods 6:377–382

Thomsen ER, Mich JK, Yao Z, Hodge RD, Doyle AM, Jang S, Shehata SI, Nelson AM, Shapovalova NV, Levi BP, Ramanathan S (2016) Fixed single-cell transcriptomic characterization of human radial glial diversity. Nat Methods 13:87–93

Tirosh I, Izar B, Prakadan SM, Wadsworth MH 2nd, Treacy D, Trombetta JJ, Rotem A, Rodman C, Lian C, Murphy G, Fallahi-Sichani M, Dutton-Regester K, Lin JR, Cohen O, Shah P, Lu D, Genshaft AS, Hughes TK, Ziegler CG, Kazer SW, Gaillard A, Kolb KE, Villani AC, Johannessen CM, Andreev AY, Van Allen EM, Bertagnolli M, Sorger PK, Sullivan RJ, Flaherty KT, Frederick DT, Jane-Valbuena J, Yoon CH, Rozenblatt-Rosen O, Shalek AK, Regev A, Garraway LA (2016) Dissecting the multicellular ecosystem of metastatic melanoma by single-cell RNA-seq. Science 352:189–196

Trapnell C (2015) Defining cell types and states with single-cell genomics. Genome Res 25:1491–1498

Trapnell C, Cacchiarelli D, Grimsby J, Pokharel P, Li S, Morse M, Lennon NJ, Livak KJ, Mikkelsen TS, Rinn JL (2014) The dynamics and regulators of cell fate decisions are revealed by pseudotemporal ordering of single cells. Nat Biotechnol 32:381–386

Velculescu VE, Zhang L, Vogelstein B, Kinzler KW (1995) Serial analysis of gene expression. Science 270:484–487

Villani AC, Satija R, Reynolds G, Sarkizova S, Shekhar K, Fletcher J, Griesbeck M, Butler A, Zheng S, Lazo S, Jardine L, Dixon D, Stephenson E, Nilsson E, Grundberg I, McDonald D, Filby A, Li W, De Jager PL, Rozenblatt-Rosen O, Lane AA, Haniffa M, Regev A, Hacohen N (2017) Single-cell RNA-seq reveals new types of human blood dendritic cells, monocytes, and progenitors. Science 356

Wang Z, Gerstein M, Snyder M (2009) RNA-Seq: a revolutionary tool for transcriptomics. Nat Rev Genet 10:57–63

Yang S, Corbett SE, Koga Y, Wang Z, Johnson WE, Yajima M, Campbell JD (2020) Decontamination of ambient RNA in single-cell RNA-seq with DecontX. Genome Biol 21:57

Ye P, Ye W, Ye C, Li S, Ye L, Ji G, Wu X (2020) scHinter: imputing dropout events for single-cell RNA-seq data with limited sample size. Bioinformatics 36:789–797

Young MD, Behjati S (2020) SoupX removes ambient RNA contamination from droplet-based single-cell RNA sequencing data. Gigascience 9

Zeng W, Jiang S, Kong X, El-Ali N, Ball AR Jr, Ma CI, Hashimoto N, Yokomori K, Mortazavi A (2016) Single-nucleus RNA-seq of differentiating human myoblasts reveals the extent of fate heterogeneity. Nucleic Acids Res 44:e158

Recent Developments in Single-Cell Transcriptomics

5

Arpita Poddar, Suresh Ramakrishna, Kruthi Ashok, Raunak Kumar Das, Prashanth Prithviraj, Viswanathaiah Matam, Haiyan Li, George Kannourakis, and Aparna Jayachandran

Abstract

Single-cell transcriptomics encompasses transformative approaches that enable gene expression profiling at single-cell resolution, revealing cellular heterogeneity, rare subpopulations, lineage relationships, and context-specific programs. This chapter focuses on scRNA-seq, recognized as Science's Breakthrough of the Year in 2018, and highlights recent technological devel-

A. Poddar
Fiona Elsey Cancer Research Institute, Ballarat, VIC, Australia

Federation University, Ballarat, VIC, Australia

RMIT University, Melbourne, VIC, Australia

The University of Melbourne, Parkville, VIC, Australia

S. Ramakrishna
Graduate School of Biomedical Science and Engineering, Hanyang University, Seoul, South Korea

College of Medicine, Hanyang University, Seoul, South Korea

K. Ashok · R. K. Das
Vellore Institute of Technology, Vellore, India

P. Prithviraj · G. Kannourakis · A. Jayachandran (✉)
Fiona Elsey Cancer Research Institute, Ballarat, VIC, Australia

Federation University, Ballarat, VIC, Australia
e-mail: aparna@fecri.org.au

V. Matam
Department of Biotechnology, Alliance School of Applied Engineering, Alliance University, Bengaluru, India

H. Li
RMIT University, Melbourne, VIC, Australia

V. Singh (ed.), *Advances in Single Cell Analysis*,
https://doi.org/10.1007/978-981-95-9789-5_5

opments, major platforms, and applications in cancer, particularly aggressive and heterogeneous contexts such as triple-negative breast cancer (TNBC). Integrating scRNA-seq with spatial transcriptomics allows precise mapping of gene expression within native tissue environments, providing insights into dynamic processes such as epithelial–mesenchymal transition (EMT), a developmental program often reactivated in disease, including cancer, where it drives tumor progression, metastasis, and therapy resistance. Combining single-cell data with multiomics including chromatin accessibility, surface protein expression, and epigenetic modifications illuminates regulatory networks, cellular interactions, and context-specific signaling pathways. Furthermore, advanced computational pipelines facilitate quality control, clustering, trajectory inference, RNA velocity, and spatial niche reconstruction, revealing rare populations, signaling gradients, and dynamic processes such as EMT. Together, these approaches provide a multidimensional framework for understanding tissue complexity, guiding biomarker discovery, and informing precision therapeutic strategies, underscoring the transformative impact of single-cell technologies in cancer research and beyond.

Keywords

Single-cell transcriptomics · scRNA-seq · Spatial transcriptomics · TNBC · EMT · Biomarker · Multiomics

5.1 Introduction

The study of gene expression through RNA transcripts is fundamental to understanding cellular function, identity, and regulatory mechanisms. The transcriptome encompasses the complete set of RNA molecules within a cell, reflecting their abundance across developmental, physiological, and pathological conditions. By analyzing the transcriptome, researchers can interpret genome activity, characterize the molecular composition of cells and tissues, and uncover dynamic regulatory processes that govern cellular behavior and responses in both health and disease (Huang et al. 2024).

While bulk transcriptomic approaches provide valuable insights, they average signals across populations of cells, masking rare or functionally distinct subpopulations and obscuring transient or context-specific gene expression programs. To overcome these limitations, single-cell transcriptomics method such as single-cell RNA sequencing (scRNA-seq) has emerged as a transformative method that enables the profiling of gene expression at the resolution of individual cells (Sabti et al. 2025). By capturing cellular heterogeneity, scRNA-seq allows researchers to identify rare subpopulations, delineate lineage relationships, and track dynamic transcriptional states over time (Kotsifaki et al. 2024). This technological advance supports in-depth interrogation of molecular mechanisms that govern cellular function and adaptation. Recent applications of scRNA-seq have revealed subtle yet critical biological processes such as cellular differentiation, stress responses, and transitions between functional states that are often hidden in bulk analyses. For example, studies in complex tissues have demonstrated how scRNA-seq can uncover cells undergoing epithelial–mesenchymal transition (EMT), a process associated with changes in cell plasticity and migratory behavior, as well as mechanisms of immune modulation and responses to environmental perturbations (McDermott et al. 2025). These capabilities highlight the power of single-cell approaches to reveal previously unappreciated layers of biological complexity and to inform the identification of biomarkers and the design of targeted interventions.

The impact of scRNA-seq has been widely recognized, with the technology named Science's Breakthrough of the Year in 2018, reflecting its profound influence on biological research (Huang et al. 2024). Subsequent methodological innovations building upon scRNA-seq including single-cell multiomic profiling and spatial transcriptomics were later acknowledged as Nature Methods Technologies of the Year, underscoring the rapid evolution and expanding scope of single-cell approaches (Nature Methods 2020, 2021). Together, these developments position single-cell technologies as central tools in contemporary biology, with growing importance for dissecting complex systems and advancing precision medicine.

In this chapter, we will begin in Sect. 5.1 by introducing the concept of single-cell transcriptomics-based profiling as a transformative tool for dissecting cellular heterogeneity. We will compare scRNA-seq with traditional transcriptional profiling methods, including bulk RNA sequencing and microarrays, highlighting their limitations in resolving rare or functionally distinct cell populations and dynamic transcriptional states. In this section, we will also provide an overview of major scRNA-seq platforms, including SMART-seq, 10X Genomics, Drop-seq, and SPLiT-seq, discussing their principles, strengths, limitations, and applications in capturing cellular diversity at single-cell resolution. In Sect. 5.2, we will explore advances in spatially resolved transcriptomics including both next-generation sequencing (NGS)-based approaches and imaging-based techniques, which integrate single-cell resolution with tissue context to map gene expression within intact tissue architecture. This section will cover integration of emerging multiomic technologies and computational frameworks with scRNA-seq, illustrating how these approaches enable the study of cell–cell interactions, microenvironmental influences, and spatial heterogeneity. Finally, the chapter will highlight future directions in single-cell and spatial transcriptomics focus on enhancing throughput, sensitivity, and spatial resolution, while overcoming technical, computational, and clinical barriers. Integrative multiomics approaches, combined with advanced computational tools, promise to unravel complex regulatory networks, cellular interactions, and context-specific processes. Together, these innovations are expected to accelerate biomarker discovery, improve understanding of tissue biology, and guide precision therapeutic strategies across diverse biological contexts.

5.1.1 Beyond the Bulk: Mapping Cells with Single-Cell Transcriptomics

Historically, transcriptomic profiling relied on hybridization-based DNA microarrays, which measured predefined sets of transcripts across bulk tissue samples. Since the mid-1990s, microarrays have been widely used for large-scale gene expression studies; however, their utility is constrained by limitations in probe selection, dynamic range, and sensitivity (Wang et al. 2009; Jayachandran et al. 2014). RNA expression analysis platforms have historically developed more slowly than other molecular assay platforms, partly due to the need for precise experimental protocols and partly because quantifying RNA requires detecting subtle differences between samples, unlike assays that mainly assess the presence or absence of specific events, such as transcription factor binding. The development of bulk RNA sequencing (bulk RNA-seq) addressed many of these challenges, enabling unbiased, genome-wide quantification of transcripts across heterogeneous tissue samples and providing a powerful platform for comprehensive gene expression profiling (Han et al. 2025).

In contrast to microarray-based platforms that rely on predetermined probe sets, RNA-seq enables unbiased, de novo detection of transcripts, encompassing novel genes, alternative splice variants, gene fusions, and various noncoding RNA species (Wang et al. 2009). Moreover, RNA-seq provides a much broader dynamic range and greater sensitivity, allowing for more precise quantification of both low- and high-abundance transcripts. It is also less prone to background noise and cross-hybridization, resulting in enhanced specificity and reproducibility across experiments and platforms (Marioni et al. 2008). Importantly, RNA-seq generates absolute or near-absolute measurements of transcript abundance, facilitating more accurate comparisons between samples and studies, whereas microarray approaches are generally limited to relative expression estimates (Wang et al. 2009).

However, bulk RNA-seq is inherently limited in its ability to resolve cellular heterogeneity within complex tissues. Bulk RNA-seq approach suffers from mixing and adding up data from multiple cells. By averaging gene expression across all cells in a sample, bulk RNA-seq masks the contributions of rare or functionally distinct cell populations and obscures intrasample variability, spatial differences, and dynamic cellular states. As a result, transient or context-specific biological processes, such as subtle differentiation events, stress responses, or rare transcriptional programs, often remain undetected in bulk analyses (Wang et al. 2009). Single-cell RNA sequencing (scRNA-seq) overcomes this limitation by profiling gene expression at single-cell resolution, enabling the identification of rare subpopulations, lineage hierarchies, and dynamic transcriptional states that are otherwise obscured in bulk analyses (Zhang et al. 2022). Since its introduction in 2009, scRNA-seq has steadily advanced, enabling the characterization of cellular heterogeneity within tissues and the detection of gene expression changes at the resolution of individual cells. The key differences between microarray-based profiling, bulk RNA sequencing, and single-cell transcriptomic approaches are summarized in Table 5.1.

Table 5.1 Comparison of various transcriptomic methods

Parameter	Microarray	Bulk RNA-seq	ScRNA-seq
Resolution	Bulk	Bulk	Single-cell
Sensitivity	Medium	High	High
Detection of rare cell types	No	No	Yes
Dynamic range	Low	High	High
Ability to detect novel transcripts	No	Yes	Yes
Spatial information	No	No	No

The contrast between microarray-based platforms and RNA sequencing is particularly evident in the study of triple-negative breast cancer (TNBC), a highly aggressive and clinically heterogeneous subtype of breast cancer that lacks expression of estrogen receptor (ER), progesterone receptor (PR), and HER2 amplification. TNBC accounts for approximately 10–20% of all breast cancers and is associated with a higher risk of early recurrence, rapid progression to metastasis, and poorer overall prognosis compared with other breast cancer subtypes (Poddar et al. 2022, 2024; Prithviraj et al. 2020). Given this aggressive clinical behavior, precise molecular characterization of TNBC is critical for identifying actionable biomarkers, understanding tumor heterogeneity, and guiding therapeutic strategies.

Early efforts to categorize TNBC at the molecular level relied on bulk microarray-based gene expression profiling to define four major transcriptional subtypes of TNBC, including basal-like 1 (BL1), basal-like 2 (BL2), mesenchymal (M), and luminal androgen receptor (LAR) types. These subtypes provided foundational insight into TNBC heterogeneity and highlighted associations between transcriptional programs, tumor microenvironment features, and response to chemotherapy (Lehmann et al. 2016). Building on these findings, subsequent studies employed bulk RNA sequencing to achieve a more comprehensive and integrative view of TNBC biology. The Fudan University (FUSCC) classification utilized RNA-seq–based transcriptomics in combination with genomic data, including somatic mutations and copy number alterations (CNAs), to define molecularly actionable subtypes with clear links to potential therapeutic strategies. By leveraging RNA-seq, this study enabled seamless integration with other omics technologies and uncovered subtypes shaped not only by expression profiles but also by underlying genomic alterations, providing a more complete picture of tumor heterogeneity and highlighting previously unrecognized therapeutic possibilities (Weng et al. 2024). Overall, while microarrays laid the groundwork for transcriptional categorization in TNBC, RNA sequencing offers enhanced sensitivity, dynamic range, and integrative capability, allowing for more precise molecular classification, identification of actionable targets, and potential application in precision oncology.

scRNA-seq has significantly advanced the understanding of TNBC heterogeneity beyond the capabilities of bulk RNA-seq or microarray-based approaches. In particular, it has enabled the precise characterization of EMT, a biological process in which epithelial cells acquire mesenchymal properties, including enhanced migratory capacity, invasiveness, and resistance to apoptosis. EMT is a dynamic cell switching reversible program originally identified in development and later co-opted in pathological contexts such as cancer progression, metastasis and organ injury, and fibrosis (Jayachandran et al. 2009, 2016; Thompson et al. 2025). EMT is a key driver of tumor progression and metastasis in many cancers, including TNBC (Poddar et al. 2022; Cao et al. 2021, 2022; Shrestha et al. 2021). In TNBC, EMT programs contribute to the aggressive phenotype, early metastasis, and poor clinical outcomes (Poddar et al. 2022). Karaayvaz et al. highlighted extensive intra-tumoral heterogeneity in TNBC, identified rare aggressive cell populations exhibiting EMT-like transcriptional states, and characterized the tumor microenvironment, including immune and stromal components, providing insights into cell–cell interactions that drive therapy resistance (Karaayvaz et al. 2018). Complementing these findings, Bao et al. integrated scRNA-seq with bulk RNA-seq to map TNBC heterogeneity, revealing M2-like tumor-associated macrophage (TAM) infiltration associated with EMT-enriched tumor subpopulations, tumor aggressiveness, and poor prognosis (Bao et al. 2020). Their study also established gene signature biomarkers predictive of survival and immunotherapy response. Together, these studies illustrate that single-cell approaches in TNBC not only capture rare and transcriptionally diverse EMT-like cell populations often missed by bulk analyses but also provide detailed insights into tumor–microenvironment interactions, reinforcing their value for identifying prognostic biomarkers, therapeutic targets, and guiding precision oncology strategies.

5.1.2 Single-Cell RNA Sequencing Platforms

scRNA-seq has transformed the study of cellular heterogeneity by enabling gene expression analysis at the resolution of individual cells. The pioneering application of this technology was demonstrated by Tang et al., who successfully sequenced single blastomeres and oocytes, establishing a foundation for later methodological developments (Tang et al. 2009). Since that time, transcriptomic approaches including bulk RNA seq, scRNA seq, and spatial transcriptomics have advanced into high-throughput platforms capable of capturing the intricate complexity of cellular populations (Jovic et al. 2022).

The typical scRNA-seq workflow begins with the isolation of individual cells, which can be achieved via manual micropipette isolation, microscope-guided capillary pipettes, Fluorescent Activated Cell Sorting (FACS) using fluorescently labeled antibodies, or high-throughput microfluidic systems capable of handling microliter to milliliter sample volumes (Zhang et al. 2024). Isolated cells are then subjected to reverse transcription and cDNA synthesis, followed by amplification, library preparation, and sequencing. The efficiency of single-cell capture depends on generating high-quality suspensions free of doublets or clumps, effective tissue dissociation, and cell enrichment (Zhang et al. 2022).

Several scRNA-seq platforms have been developed to balance throughput, transcript coverage, sensitivity, and cost and are summarized in Table 5.2. Plate-based methods, such as SMART-Seq and its improved versions Smart-seq2 and

Table 5.2 Comparison of scRNA-seq Platforms

Platform	Method	Sensitivity	Transcript Coverage	Throughput	Cost
SMART-seq, SMART-seq2/3	Plate-based	High	Full-length	Low	High
10X Genomics	Droplet-based	Medium	3′ or 5′ end	High	Medium-high
Drop-seq, inDrop, Seq-Well	Microwell	Medium	3′ end	High	Low-medium
SPLiT-seq, sci-RNA-seq	Combinatorial indexing	Medium	3′ end	High	Low

Smart-seq3, as well as CEL-seq2, perform full-length or near–full-length transcript sequencing, enabling the detection of a greater number of genes per cell and facilitating the analysis of low-abundance transcripts and alternative splicing events. While highly informative, plate-based approaches typically process fewer cells per experiment. In contrast, droplet-based systems, such as 10X Genomics Chromium, utilize microfluidics to partition thousands to tens of thousands of cells into barcoded droplets, allowing large-scale profiling of cellular diversity and identification of rare subpopulations. The tradeoff is sequencing only short intervals at the 5′ or 3′ ends and higher dropout rates for low-expression genes (Azizi et al. 2018). Pico- or microwell-based and combinatorial indexing approaches including Seq-Well, Drop-seq, inDrop, SPLiT-seq, MARS-seq, and sci-RNA-seq further expand the single-cell transcriptomics toolkit, offering cost-effective and scalable alternatives with distinct tradeoffs between sensitivity, throughput, and experimental complexity (Drake et al. 2022).

These platforms have been instrumental in dissecting cellular heterogeneity in complex tissues and diseases. For example, in breast cancer, scRNA-seq has revealed intratumor heterogeneity, capturing coexisting cancer cell subpopulations with distinct genetic and transcriptional profiles that underlie differential therapy responses and disease relapse. High-resolution cell atlases integrating scRNA-seq with single-cell proteomics have mapped diverse cellular subtypes and functional states, while immune profiling and analyses of EMT provide insights into microenvironmental dynamics and cell plasticity (Gray et al. 2022; Ren et al. 2021). Collectively, these scRNA-seq platforms enable comprehensive mapping of cellular diversity and functional states, offering a powerful toolkit for both basic and translational research.

5.1.3 Computational Pipelines in Single-Cell Transcriptomics

The analysis of scRNA-seq data requires sophisticated computational pipelines to manage large-scale datasets and extract biologically meaningful insights. Prior to downstream analysis, rigorous quality control procedures are applied, including

Table 5.3 Computational Tools for scRNA-seq

Process	Computational tool	Description
Quality control	Seurat, Scanpy	Filtering low-quality cells, removing doublets, normalizing data
Clustering	k-means, graph-based, CIDR	Grouping cells into types/states
Dimensionality reduction	PCA, UMAP, t-SNE	Visualizing high-dimensional data
Trajectory/lineage	Wanderlust, SCORPIUS, RNA velocity	Inferring cell state transitions
Cell-cell communication	CellPhoneDB, Giotto	Inferring ligand-receptor interactions
Integration/batch correction	Harmony, Seurat v4, Scanpy	Combining datasets, removing batch effects

filtering low-quality cells, removing doublets, and assessing mitochondrial gene content, to ensure robust and reliable datasets. The computational tools for scRNA-seq are summarized in Table 5.3.

Cell clustering is a central analytical step, using unsupervised learning to group cells based on similar gene expression profiles and identify distinct cell types and states. This process is typically preceded by normalization, feature selection, and dimensionality reduction techniques such as principal component analysis (PCA), Uniform Manifold Approximation and Projection (UMAP), and t-distributed stochastic neighbor embedding (t-SNE). Common clustering algorithms include k-means, hierarchical clustering, graph-based methods, community-detection approaches, and deep learning models. Specialized tools such as CountClust, CIDR, SIMLR, and SAFE enhance cell-type recognition, while biclustering frameworks like DivBiclust and SAME integrate cluster solutions across multiple methods (Li et al. 2021). Following clustering, cell populations are annotated using canonical marker genes or reference-based approaches, and differential gene expression analysis is applied to uncover genes that distinguish cell populations or transcriptional states, providing insights into underlying regulatory programs.

Trajectory inference (TI) methods order cells along dynamic biological processes, revealing transitions between functional states and cell fate decisions. Linear trajectory methods include Wanderlust, MATCHER, and SCORPIUS, whereas cyclic trajectory tools such as reCAT and ElPiGraph capture recurrent or oscillatory processes. RNA velocity–based approaches, leveraging spliced and unspliced transcripts, enable inference of future cellular states and further refine dynamic modelling (Niu et al. 2024). These approaches have been pivotal in studying differentiation, lineage specification, and disease progression, including tumor evolution and therapy resistance (Hutton et al. 2025). Recent computational advances illustrate the power of these methods by applying trajectory inference and dynamic modeling to EMT and identified conserved intermediate cell states and transcriptional programs across multiple cancer types, uncovering continuous cell state changes that are obscured in bulk analyses and highlighting the capacity of computational pipelines to infer regulatory mechanisms underlying cellular plasticity (McDermott et al. 2025).

Integration tools, such as Seurat, Scanpy, and Harmony, enable alignment of scRNA-seq datasets from different technologies or experimental conditions while correcting batch effects. Harmony operates in principal component space to project cells into a shared embedding that promotes grouping by biological identity rather than batch (Korsunsky et al. 2019). Seurat, an R package developed by the Satija Lab, applies graph-based clustering for normalization, quality control, and heterogeneity analysis (Pang et al. 2024), while Scanpy provides analogous functionality in Python with seamless integration of machine learning libraries such as LIMIX, TensorFlow, and GPy/GPflow (Wolf et al. 2018).

Together, the combination of diverse scRNA-seq platforms and computational pipelines allows researchers to dissect cellular diversity, dynamic states, and rare subpopulations with unprecedented resolution. These tools provide a robust framework for mapping complex tissues, understanding functional heterogeneity, and integrating transcriptomic insights with complementary omics data, advancing both basic and translational research.

5.2 From Single Cells to Multiomics: Emerging Technologies

In this section, we explore recent developments in single-cell and multiomics technologies that go beyond conventional transcriptomic analyses. We focus on approaches that integrate multiple molecular layers such as chromatin accessibility, surface protein expression, and epigenetic modifications through techniques like scATAC-seq, CITE-seq, and combinatorial multiomics platforms, as well as spatial transcriptomics, which maps gene expression within tissue architecture to provide a comprehensive view of cellular states, interactions, and regulatory networks (Fig. 5.1).

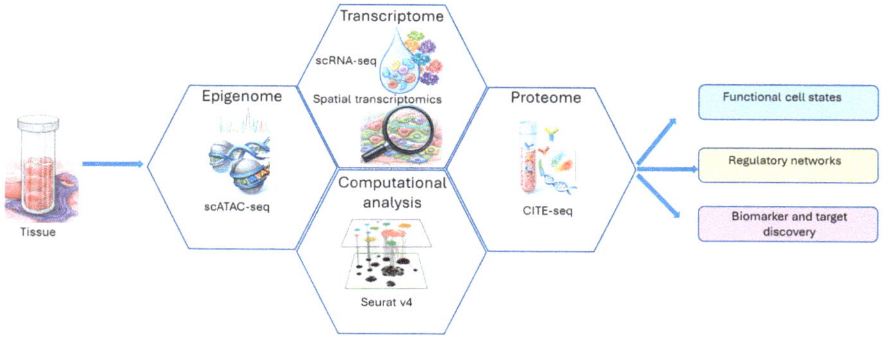

Fig. 5.1 Coupling multimodal single-cell profiling approaches. Multimodal single-cell profiling integrates transcriptomic, proteomic, and epigenomic layers using computational frameworks to define cellular identities and functional states, regulatory networks and identify novel biomarkers and drug targets

5.2.1 Integrating the Layers: Transcriptome, Proteome, and Epigenome at Single-Cell Resolution

While advanced computational pipelines including clustering, trajectory inference, and RNA velocity enable detailed mapping of cellular diversity and dynamic states, single-cell transcriptomics alone captures only one layer of molecular information. To achieve a more comprehensive understanding of cellular regulation and functional states, researchers are increasingly integrating complementary molecular layers including chromatin accessibility, surface protein expression, and epigenetic modifications through multiomics approaches. These integrative strategies, including single-cell ATAC-seq (scATAC-seq), CITE-seq, and combinatorial methods such as sci-CAR, SNARE-seq, and SHARE-seq, provide complementary perspectives that can reveal regulatory networks, identify rare cell populations, and elucidate mechanisms underlying complex biological processes and therapeutic resistance (Chen et al. 2019; Ma et al. 2020; Fang et al. 2021; Quek et al. 2024).

5.2.1.1 Single-Cell ATAC-seq (scATAC-seq)

Single-cell Assay for Transposase-Accessible Chromatin using sequencing (scATAC-seq) profiles chromatin accessibility at single-cell resolution. By mapping open chromatin regions, scATAC-seq identifies cis-regulatory elements, such as enhancers and promoters, that drive gene expression programs and oncogenic processes. It also enables study of epigenetic changes associated with drug resistance or disease progression, providing critical insights into regulatory mechanisms and cellular heterogeneity (Fang et al. 2021).

5.2.1.2 CITE-seq

Cellular Indexing of Transcriptomes and Epitopes by Sequencing (CITE-seq) combines transcriptomic and proteomic profiling in individual cells. Using oligonucleotide-tagged antibodies, it quantifies surface protein expression alongside RNA transcripts, capturing hundreds of epitopes per cell. CITE-seq has been applied to longitudinal analyses of tumors in metastatic cancer patients, uncovering heterogeneous immune states and variable responses to immunotherapy (Quek et al. 2024).

5.2.1.3 Emerging Multiomics Approaches

Other integrative platforms, such as SHARE-seq, SNARE-seq, and sci-CAR, enable simultaneous profiling of multiple molecular layers, including chromatin and transcriptome, further expanding single-cell multiomics capabilities. Computational frameworks like Seurat v4, MOFA+, and totalVI allow integration of these diverse datasets, linking transcriptional, proteomic, and epigenetic information for comprehensive cellular characterization (Chen et al. 2019; Ma et al. 2020). These approaches have been instrumental not only in cancer research but also in developmental biology, differentiation, and lineage tracing. Looking ahead, coupling multimodal single-cell profiling with spatially resolved techniques promises to advance our understanding of tissue organization and functional regulation in both health and disease.

5.2.2 Beyond the Single Cell: Spatially Resolved Transcriptomic Landscapes

scRNA-seq enables high-resolution profiling of individual cells within complex tissues, providing detailed insights into cellular heterogeneity and dynamic states. However, tissue dissociation required for scRNA-seq disrupts the native spatial organization of cells, limiting our understanding of how cells interact, their precise positioning, and how the surrounding microenvironment influences cellular behavior. This loss of spatial context obscures critical information about tissue architecture, intercellular relationships, and spatially regulated processes such as EMT, which can be studied in contexts like tumor invasion or organ development (Ahmed et al. 2022).

Spatial transcriptomics overcomes this limitation by combining high-throughput RNA sequencing with histological mapping, allowing gene expression to be quantified in situ within intact tissue sections. By preserving spatial information, this approach reveals transcriptional patterning, cell–cell interactions, and tissue-specific gene expression, offering a more comprehensive understanding of tissue architecture, morphology, and microenvironmental influences. Integrating scRNA-seq with spatial transcriptomics enables profiling of RNA expression in its native location, facilitating insights into the organization and functional states of cells, the heterogeneity of tumors and immune populations, and mechanisms underlying disease progression and therapy resistance (Williams et al. 2022; An et al. 2024).

Spatial transcriptomics methodologies can be generally divided into next-generation sequencing (NGS)-based approaches and imaging-based techniques, each characterized by unique underlying principles, advantages, and constraints.

5.2.2.1 NGS-Based Approaches

These approaches combine spatial barcoding with high-throughput sequencing. RNA molecules are captured on barcoded microarray slides, beads, or nanostructures, allowing the original tissue location of each transcript to be retained during library preparation and sequencing. Pioneering work by Stahl et al. demonstrated this principle using tissue sections from mouse olfactory bulbs (Stahl et al. 2016). Commercial platforms such as 10× Genomics' Visium have refined this approach with improved sensitivity and resolution (Park et al. 2023). Other NGS-based platforms include Slide-Seq, which uses randomly barcoded beads to achieve ~10 μm resolution, HDST (High-Definition Spatial Transcriptomics) for further enhanced spatial precision, DBiT-Seq, which employs microfluidic channels for spatial barcoding, Stereo-seq, using DNA nanoballs for nanoscale mapping, and Seq-Scope and Pixel-Seq, which provide subcellular transcript localization. These methods enable comprehensive, transcriptome-wide analysis of tissue architecture (Ahmed et al. 2022; Park et al. 2023).

5.2.2.2 Imaging-Based Approaches

These approaches leverage direct visualization of RNA molecules in situ using fluorescent probes and microscopy. These methods detect transcripts within tissue sections without dissociation, typically through sequential probe hybridization and imaging cycles. MERFISH (Multiplexed Error-Robust Fluorescence In Situ Hybridization) applies encoding probes to detect thousands of RNA species through multiple hybridization rounds, with advanced error-correction strategies improving sensitivity and reducing imaging time compared with earlier protocols (Ahmed et al. 2022). One of the most powerful imaging-based platforms is CosMx Spatial Molecular Imager (SMI), which integrates high-plex RNA and protein detection in a single instrument. CosMx uses cyclic fluorescent in situ hybridization chemistry coupled with high-resolution imaging to quantify up to thousands of RNA targets and dozens of validated proteins within formalin-fixed, paraffin-embedded (FFPE) or fresh frozen tissues at cellular and subcellular resolution. It supports high-plex spatial multiomics by enabling the simultaneous visualization of RNA and protein expression, facilitating cell typing, cell–cell interaction analysis, tissue phenotyping, and biomarker discovery within the intact tissue landscape. CosMx's high sensitivity and multi-modal capability allow detailed studies of cellular organization and functional processes such as ligand–receptor signaling in situ, making it a leading tool for spatial biology applications across oncology, immunology, neuroscience, and translational research (Park et al. 2023).

NGS-based spatial transcriptomics methods provide comprehensive, unbiased profiling of the transcriptome with spatial context, though they can be limited by spatial resolution and often require substantial sequencing depth to detect low-abundance transcripts. In contrast, imaging-based approaches, such as MERFISH and CosMx, deliver high-resolution, subcellular mapping and allow for multiomics integration (e.g., concurrent RNA and protein detection), albeit with greater technical complexity and reliance on advanced microscopy and image analysis. These two strategies are complementary: NGS-based platforms are well-suited for broad, tissue-wide discovery, while imaging-based platforms excel in high-resolution studies that interrogate cellular interactions and detailed cell states. Combining both approaches maximizes the ability to resolve complex tissue architecture, infer cell–cell communication, and link spatial gene expression patterns to functional biology and disease processes.

Recent studies have applied spatial transcriptomics to map EMT in diverse biological contexts. In cancer, EMT trajectories were reconstructed across epithelial tumors, identifying epithelial, hybrid, and mesenchymal macrostates, and demonstrating that intermediate EMT states are spatially localized in specific niches interacting with fibroblasts and immune cells (Malagoli Tagliazucchi et al. 2023). In tissue injury and repair, spatial transcriptomics has been used to map EMT-like programs in fibrotic and damaged organs, such as the lung, showing that stress-induced epithelial subsets adopt partial EMT phenotypes that localize to injury sites and interact with immune populations, thereby coordinating regeneration and inflammatory responses (Wang et al. 2025). Collectively, these examples illustrate that spatial transcriptomics captures the gradients, niches, and cell–cell interactions underpinning EMT across physiological and pathological processes; information that is lost in dissociated single-cell analyses.

Advanced computational pipelines including CellPhoneDB, Giotto, stLearn, and SpaOTsc support analysis of cell–cell communication, ligand–receptor interactions, and spatial niche organization. These integrative approaches have uncovered EMT gradients, signaling microenvironments, and cell–cell interactions that correlate with functional states and tissue organization, highlighting their broader relevance for studying development, disease, and therapeutic interventions (Dries et al. 2021).

5.3 Future Directions

Single-cell and spatial transcriptomic technologies have dramatically enhanced our ability to map cellular heterogeneity, uncover rare cell populations, and characterize dynamic transcriptional states within intact tissues. These advances have enabled unprecedented insights into complex biological processes, including tumor heterogeneity, EMT, immune regulation, and tissue regeneration. However, despite these successes, several challenges remain that must be addressed to fully realize their potential in research and clinical settings.

5.3.1 Innovations in Throughput, Sensitivity, and Spatial Mapping

Recent technological advances have significantly enhanced the throughput, sensitivity, and spatial resolution of scRNA-seq. Microfluidics and droplet-based platforms now allow tens of thousands of cells to be profiled in a single experiment, enabling detailed characterization of entire cell populations and the discovery of rare or transient cell subtypes (Du et al. 2023). Sensitivity improvements, exemplified by SMART-seq and its derivatives, capture full-length cDNA, reduce transcript loss, and improve transcriptome completeness, allowing more accurate detection of low-abundance genes.

Spatial context, lost during tissue dissociation in conventional scRNA-seq, is now being preserved through image-based spatial transcriptomics. Platforms such as MERFISH and CosMx maintain subcellular resolution within intact tissue sections, while computational integration tools like Spatial-Map combine single-cell and spatial transcriptomic data to reconstruct tissue architecture and enable comprehensive mapping of cell–cell interactions (Xiong et al. 2024; Gao et al. 2022). These integrative approaches promise to reveal dynamic processes such as EMT, immune–stromal interactions, and signaling gradients in situ, providing deeper insights into tissue biology and disease mechanisms.

5.3.2 Overcoming Barriers to Clinical Implementation

As single-cell technologies mature, their integration into therapeutic target discovery is becoming increasingly feasible, with recent studies demonstrating how scRNA-seq can directly inform immuno-oncology strategies (Birappa et al. 2026). Despite these innovations, barriers remain for routine clinical adoption. scRNA-seq data are highly sensitive to technical variations arising from tissue dissociation, RNA handling, and single-cell capture efficiency, which can introduce biases and affect reproducibility. Batch effects across platforms, lack of standardized protocols, and varying computational pipelines further complicate data interpretation. High costs, limited sample material (e.g., biopsy tissues), and infrastructure requirements restrict the scalability of these technologies in clinical workflows. Addressing these challenges through standardization, robust quality control, and cost-effective high-throughput platforms will be crucial for translating single-cell and spatial transcriptomics into precision medicine.

5.3.3 Integrative Multiomics and Spatial Analysis

Future directions include combining scRNA-seq with other molecular layers, such as chromatin accessibility (scATAC-seq), surface protein expression (CITE-seq), and epigenetic modifications, to achieve multidimensional cellular profiling. When integrated with spatial transcriptomics, multiomic datasets can illuminate regulatory networks, cellular interactions, and context-specific signaling pathways, enabling precise dissection of complex tissues and informing biomarker discovery and therapeutic strategies.

5.3.4 Computational and Analytical Advancements

The expansion of computational tools for trajectory inference, cell–cell communication, and spatial niche analysis will further enhance the resolution and interpretability of single-cell and spatial datasets. Advanced machine learning algorithms and standardized pipelines are expected to improve reproducibility, minimize batch effects, and facilitate large-scale integration of multicenter datasets.

5.4 Conclusion

In summary, ongoing innovations in throughput, sensitivity, spatial mapping, and integrative multiomics, combined with advances in computational analysis, promise to overcome current limitations and expand the impact of single-cell and spatial transcriptomics. Together, these developments will deepen our understanding of tissue complexity, enable the discovery of rare and functionally important cell populations, and accelerate the translation of spatially resolved molecular insights into clinical applications.

Acknowledgements The authors acknowledge support provided by the team at Fiona Elsey Cancer Research Institute towards this research. A. P. and A.J. acknowledge the Victorian Government's Women's Health Research Catalyst Grant. A.P. acknowledges the Irene Tellefson Research Fellowship.

References

Ahmed R, Zaman T, Chowdhury F, Mraiche F, Tariq M, Ahmad IS, Hasan A (2022) Single-cell RNA sequencing with spatial transcriptomics of cancer tissues. Int J Mol Sci 23(6):3042

An J et al (2024) Spatial transcriptomics in breast cancer: providing insight into tumor heterogeneity and promoting individualized therapy. Front Immunol 15:1499301

Azizi E, Carr AJ, Plitas G, Cornish AE, Konopacki C, Prabhakaran S, Nainys J, Wu K, Kiseliovas V, Setty M et al (2018) Single-cell map of diverse immune phenotypes in the breast tumor microenvironment. Cell 174(5):1293–1308

Bao X, Shi R, Zhao T, Wang Y, Anastasov N, Rosemann M, Fang W (2020) Integrated analysis of single-cell RNA-seq and bulk RNA-seq unravels tumour heterogeneity plus M2-like tumour-associated macrophage infiltration and aggressiveness in TNBC. Cancer Immunol Immunother 70(1):189–202

Birappa G, Perumalsamy H, Hong SH, Gowda DAA, Chandrasekaran AP, Karapurkar JK, Rajkumar S, Balusamy SR, Jayachandran A, Baek KH et al (2026) Single-cell RNA sequence analysis reveals USP32 as a therapeutic target to mitigate PD-L1-driven colorectal tumorigenesis in vitro and in vivo. Theranostics 16(2):986–1005

Cao L, Prithviraj P, Shrestha R, Sharma R, Anaka M, Bridle KR, Kannourakis G, Crawford DHG, Jayachandran A (2021) Prognostic role of immune checkpoint regulators in cholangiocarcinoma: a pilot study. J Clin Med 10(10):2191

Cao L, Bridle KR, Shrestha R, Prithviraj P, Crawford DHG, Jayachandran A (2022) CD73 and PD-L1 as potential therapeutic targets in gallbladder cancer. Int J Mol Sci 23(3):1565

Chen X, Miragaia RJ, Natarajan KN, Teichmann SA (2019) High-throughput sequencing of transcriptome and chromatin accessibility in the same cell using SNARE-seq. Nat Methods 16(1):35–39

Drake RS, Herndon JM, Zhang S (2022) Profiling transcriptional heterogeneity with seq-well S3. In: Single-cell transcriptomics: methods and protocols. Springer, New York, pp 57–104

Dries R, Zhu Q, Eng CL et al (2021) Giotto: a toolbox for integrative analysis and visualization of spatial expression data. Genome Biol 22:78

Du J, Li J, Guo X, Wang Y, Zhang Y, Liu Z (2023) Advances in spatial transcriptomics and related data analysis strategies. J Transl Med 21(1):330

Fang R, Yu M, Li G et al (2021) Comprehensive analysis of single-cell ATAC-seq data with SnapATAC. Nat Commun 12(1):1337

Gao D, Zhou Y, Huang J, Zhang Y, Wang Z, Li X (2022) SpatialMap: spatial mapping of unmeasured gene expression profiles in spatial transcriptomic data using generalized linear spatial models. Front Genet 13:893522

Gray GK, Ham S, Hachey SJ et al (2022) A human breast atlas integrating single-cell proteomics and transcriptomics. Dev Cell 57(11):1400–1420

Han X, Li X, Bai L, Zhang G (2025) Single-cell transcriptomics in metastatic breast cancer: mapping tumor evolution and therapeutic resistance. Front Genet 16:1669741

Huang K, Xu Y, Feng T, Lan H, Ling F, Xiang H, Liu Q (2024) The advancement and application of the single-cell transcriptome in biological and medical research. Biology 13(6):451

Hutton A, Meyer JG (2025) Trajectory Inference for Single Cell Omics. ArXiv, arXiv:2502.09354v1. PMID: 39990800

Jayachandran A, Königshoff M, Yu H, Rupniewska E, Hecker M, Klepetko W, Seeger W, Eickelberg O (2009) SNAI transcription factors mediate epithelial-mesenchymal transition in lung fibrosis. Thorax 64(12):1053–1061

Jayachandran A, Anaka M, Prithviraj P et al (2014) Thrombospondin-1 promotes an aggressive phenotype through epithelial-to-mesenchymal transition in human melanoma. Oncotarget 5(14):5782–5797

Jayachandran A, Dhungel B, Steel JC (2016) Epithelial-to-mesenchymal plasticity of cancer stem cells: therapeutic targets in hepatocellular carcinoma. J Hematol Oncol 9(1):74

Jovic D, Dragomir M, Wang Y (2022) Single-cell RNA sequencing technologies and applications: a brief overview. Clin Transl Med 12(3):e694

Karaayvaz M, Cristea S, Gillespie SM et al (2018) Unravelling subclonal heterogeneity and aggressive disease states in triple-negative breast cancer through single-cell RNA-seq. Nat Commun 9:3588

Korsunsky I, Millard N, Fan J et al (2019) Fast, sensitive and accurate integration of single-cell data with harmony. Nat Methods 16(12):1289–1296

Kotsifaki A, Maroulaki S, Armakolas A (2024) Exploring the immunological profile in breast cancer. Int J Mol Sci 25(9):4832

Lehmann BD, Jovanović B, Chen X et al (2016) Refinement of triple-negative breast cancer molecular subtypes. J Natl Cancer Inst 108(7):djw149

Li L et al (2021) Applications of single-cell RNA sequencing in cancer research. J Exp Clin Cancer Res 40(1):163

Ma A, McDermaid A, Xu J, Chang Y, Ma Q (2020) Simultaneous high-throughput ATAC and RNA expression with sequencing (SHARE-seq). Cell Res 30(6):512–525

Malagoli Tagliazucchi G, Wiecek AJ, Withnell E, Secrier M (2023) Genomic and microenvironmental heterogeneity shaping epithelial-to-mesenchymal trajectories in cancer. Nat Commun 14:789

Marioni JC, Mason CE, Mane SM, Stephens M, Gilad Y (2008) RNA-seq: an assessment of technical reproducibility and comparison with gene expression arrays. Genome Res 18(9):1509–1517

McDermott M, Mehta R, Roussos Torres E, MacLean AL (2025) Modeling the dynamics of EMT reveals genes associated with pan-cancer intermediate states and plasticity. NPJ Syst Biol Appl 11(1):31

Nature Methods (2020) Method of the year 2019: single-cell multimodal omics. Nat Methods 17(1):1–2

Nature Methods (2021) Method of the year 2020: spatially resolved transcriptomics. Nat Methods 18(1):1–2

Niu Y, Luo J, Zong C (2024) Single-cell total RNA profiling unveils regulatory hubs of transcription factors. Nat Commun 15:5941

Pang L et al (2024) Single-cell integrative analysis reveals consensus cancer cell states and clinical relevance in breast cancer. Sci Data 11(1):289

Park H-E, Jo SH, Lee RH et al (2023) Spatial transcriptomics: technical aspects of recent developments and applications. Adv Sci 10(7)

Poddar A, Rao SR, Prithviraj P, Kannourakis G, Jayachandran A (2022) Crosstalk between immune checkpoint modulators, metabolic reprogramming and cellular plasticity in TNBC. Curr Oncol 29(10):6847–6863

Poddar A, Ahmady F, Rao SR et al (2024) The role of pregnancy-associated plasma protein-A in TNBC. J Biomed Sci 31(1):23

Prithviraj P, Anaka M, Thompson EW et al (2020) Aberrant pregnancy-associated plasma protein-A expression in breast cancers prognosticates clinical outcomes. Sci Rep 10(1):13779

Quek C, Chua XY, Tan TZ et al (2024) Single-cell spatial multiomics reveals tumor microenvironment vulnerabilities. Cell Rep 43(7)

Ren L, Wang Y, Guo Y et al (2021) Single-cell RNA sequencing for breast cancer: present and future. Cell Death Discov 7(1):104

Sabti H, Arneth B, Pawlik TM et al (2025) Leveraging single-cell multi-omics to decode tumor microenvironment diversity. Pharmaceuticals 18(1):75

Shrestha R, Prithviraj P, Bridle KR, Crawford DHG, Jayachandran A (2021) Combined inhibition of TGF-β1-induced EMT and PD-L1 silencing. J Clin Med 10(9):1889

Stahl PL, Salmén F, Vickovic S et al (2016) Visualization and analysis of gene expression in tissue sections by spatial transcriptomics. Science 353(6294):78–82

Tang F, Barbacioru C, Wang Y et al (2009) mRNA-Seq whole-transcriptome analysis of a single cell. Nat Methods 6(5):377–382

Thompson EW, Redfern AD, Brabletz S et al (2025) EMT and cancer: what clinicians should know. Nat Rev Clin Oncol 22(10):711–733

Wang Z, Gerstein M, Snyder M (2009) RNA-Seq: a revolutionary tool for transcriptomics. Nat Rev Genet 10(1):57–63

Wang T, Hao J, Li B et al (2025) Single-cell and spatial transcriptomics reveal a stress-induced EMT-like epithelial subset. Front Immunol 16:1609616

Weng L, Zhou J, Guo S, Xu N, Ma R (2024) Molecular subtyping and precision medicine in TNBC. Cancer Cell Int 24:120

Williams CG et al (2022) An introduction to spatial transcriptomics for biomedical research. Genome Med 14(1):68

Wolf FA, Angerer P, Theis FJ (2018) SCANPY: large-scale single-cell gene expression data analysis. Genome Biol 19(1):15

Xiong X, Zhang Y, Li H, Wang J, Liu S, Chen Y (2024) Deciphering breast cancer dynamics. Biomark Res 12(1):107

Zhang Q-Y, Ho DW-H, Tsui Y-M (2022) Single-cell transcriptomics of liver cancer. Cell Mol Gastroenterol Hepatol 14(3):513–525

Zhang Z, Li H, Chen Y, Wang X (2024) Advancements in the application of scRNA-seq in breast research. Int J Mol Sci 25(24):13706

Advances and Applications of Single-Cell Proteomics

6

Ajay Kumar, Juveriya Israr, and Shabroz Alam

Abstract

Single-cell proteomics has emerged as a revolutionary domain in biomedical research, facilitating the precise analysis of cellular heterogeneity at the protein level. Recent technological advancements—such as ultra-sensitive mass spectrometry, microfluidic sample preparation, and multiplexed labeling techniques—have markedly improved the depth, throughput, and reproducibility of single-cell protein quantification. These improvements have facilitated novel understandings of cell-state dynamics, lineage trajectories, and post-translational alterations that were previously concealed in bulk analysis. This chapter offers a thorough examination of recent methodological advancements in single-cell proteomics, encompassing label-free and isobaric tagging techniques, data-independent acquisition (DIA), and novel spatial proteomics platforms. It additionally investigates integrative frameworks that amalgamate single-cell proteomics with transcriptomics and epigenomics to create multidimensional cellular atlases. Applications encompass various fields including cancer biology, immunology, neurodegeneration, and regenerative medicine, where single-cell proteome profiling is revealing new biomarkers, therapeutic targets, and mechanisms of disease development. This chapter emphasizes the significance of single-cell proteomics in enhancing precision medicine and systems biology by showcasing its present capabilities and future trajectories.

A. Kumar (✉)
Department of Biotechnology, Faculty of Engineering and Technology, Rama University, Kanpur, Uttar Pradesh, India
e-mail: ajaymtech@gmail.com

J. Israr
Institute of Biosciences and Technology, Shri Ramswaroop Memorial University, Barabanki, Uttar Pradesh, India

Department of Biotechnology, Era University, Lucknow, Uttar Pradesh, India

S. Alam
Department of Biotechnology, Era University, Lucknow, Uttar Pradesh, India

V. Singh (ed.), *Advances in Single Cell Analysis*,
https://doi.org/10.1007/978-981-95-9789-5_6

Keywords

Single-cell proteomics · Mass spectrometry · Cellular heterogeneity · Precision medicine · Systems biology

6.1 Introduction

The proteome—the complete set of proteins expressed by a cell at a given time represents the most direct molecular readout of cellular function. While genomics and transcriptomics have made significant strides in dissecting the molecular basis of biology, proteins are the primary effectors of cellular processes, directly mediating signaling pathways, enzymatic reactions, structural organization, and responses to the environment. Critically, the abundance, localization, and modification state of proteins do not always correlate with transcript levels, due to extensive post-transcriptional, translational, and post-translational regulation; direct protein measurement is thus indispensable for a complete understanding of cellular physiology (Guo et al. 2025a, 2025b). Traditional bulk proteomics methods have provided valuable insights into protein composition and abundance across diverse biological systems, typically relying on mass spectrometry (MS) to quantify proteins from large populations of cells and yielding averaged protein expression profiles. However, these ensemble measurements mask cell-to-cell heterogeneity and obscure the functional differences among subpopulations and rare cell types. This heterogeneity is critical in elucidating the outcomes of the biological processes in the majority of biological systems, such as cancer progression, immune system activation and stem cell development. This has given rise to the growing interest in methods that can potentially study the entire spectrum of proteins that is carried in individual cells (Truong and Kelly 2024). A technology that has been developed to meet this requirement is called single-cell proteomics (SCP). The use of the method simplifies the exhaustive analysis of protein expression in single cells. With these more recent advances in sample preparation, microfluidics, ultrasensitive mass spectrometry, and data analysis, the combination has been in aid of single-cell protein chromatography (SCP) that identifies and measures hundreds to thousands of proteins in a single cell. The potential provides a degree of solution such as never shown before reported in the study of cellular system diversity, adaptability and dynamics. This permits the apprehension of abnormal cell populations, acknowledgment of transient or dynamic states and stochastic variability capable of influencing phenotypic variation, even in genetically identical cells (Ahmad and Budnik 2023). The complexity of the proteome further underscores the importance of single-cell analysis. Proteins undergo a variety of post-translational modifications (PTMs) such as phosphorylation, glycosylation, ubiquitination, and acetylation which can rapidly alter their activity, interaction partners, localization, and stability. Protein functionality is intimately linked to subcellular localization, with spatial distribution determining roles in signaling and regulation. Additionally, proteomes are highly dynamic, changing in response to extrinsic cues, intrinsic regulatory programs, and disease states; bulk analyses can only provide averaged snapshots, whereas SCP offers time- and state-resolved maps of proteomic changes at the single-cell level (Guo et al. 2025a, 2025b). Given the ability of SCP to reveal molecular mechanisms underlying cellular heterogeneity, and its potential applications in precision medicine, developmental biology, immunology, and cancer research, this field is rapidly advancing. Continued methodological innovations are expected to further increase proteome coverage, quantitative accuracy, and throughput of single-cell protein measurements. As proteomic data become increasingly integrated with genomic, transcriptomic, and metabolomics information, SCP is poised to transform our systems level understanding of biology and disease (Truong and Kelly 2024). Table 6.1 represent the advances and applications of single-cell proteomics.

6.2 Historical Context and Evolution

Proteomics as a scientific field traces back to the 1970s with the advent of two-dimensional (2D) gel electrophoresis, which enabled the separation and analysis of complex protein mixtures, and was pivotal in mapping the first comprehensive protein profiles of organisms such as *Escherichia coli*. The term proteome a portmanteau of "protein" and "genome" was coined by Marc Wilkins in 1994 to describe the entire protein complement of a cell or organism. The establishment of foundational protein databases, such as Margaret Dayhoff's Atlas of Protein Sequence and Structure (1965), SWISS-PROT (1996), and UniProt (2002), further propelled the field by systematizing protein sequence and functional data. A major turning point for proteomics was the introduction of mass spectrometry (MS) techniques. Initially, protein identification and quantification were anchored in 2D gels, but by the late 1990s and early 2000s, advances in MS allowed researchers to detect and characterize proteins with greater sensitivity and throughput, without the limitations of gel-based methods. This enabled the analysis of ever more complex samples and established MS as the backbone of proteomics, powering efforts such as the Human Proteome Project, and eventually facilitating analysis at unprecedented resolution, including single-cell proteomics. Single-cell proteomics arose alongside transformative progress in single-cell genomics and transcriptomics disciplines spurred by the human genome project and the global focus on cellular heterogeneity. The maturation of single-cell RNA sequencing revealed the essential role of single-cell techniques in deciphering biological diversity, reinforcing the need for proteomic approaches with analogous resolution. Recent advances ultrasensitive MS, microfluidics, and refined sample handling have, within the last decade, rendered possible the quantification of hundreds to thousands of proteins per individual cell, revealing cell-state dynamics, rare populations, and novel regulatory mechanisms. Developments in genomics, transcriptomics, and proteomics have thus evolved in parallel, each discipline shaping and responding to the rapid technological advances of the others (Alam et al. 2024). The convergence of these single-cell approaches now enables integrated analyses across molecular layers, empowering systems biology and precision medicine with single-cell granularity.

Table 6.1 Advances and applications of single-cell proteomics

S. No.	Category	Advances/techniques	Applications in research and medicine	References
1	Analytical technologies	Mass spectrometry miniaturization (SCoPE-MS, SCoPE2)	High-throughput profiling of single-cell proteomes	Budnik et al. (2018)
		DIA-based workflows	Improved detection sensitivity	
		Ion mobility spectrometry	Reduced sample loss	
		Microfluidics and nanodroplet platforms (nanoPOTS)		
2	Sample preparation	Label-free single-cell workflows	Robust proteome coverage from scarce samples	Specht et al. (2021)
		Tandem mass tags (TMT) multiplexing	Parallel analysis of hundreds of single cells	
		Droplet-based lysis and digestion systems		
3	Quantitative improvements	Isobaric labeling with carrier proteome	Accurate quantification of low-abundance proteins	Cheung et al. (2021) and Kelly (2020)
		Enhanced low-input quantitation	Detection of signaling pathway dynamics	
		Targeted proteomics (PRM/SRM for single cells)		
4	Computational tools	Machine learning for noise reduction	Systems-level understanding of cell heterogeneity	Vanderaa and Gatto (2023)
		Data integration with single-cell RNA-seq	Identification of proteogenomic biomarkers	
		Multiomics integration pipelines (scProteomics + scTranscriptomics + scEpigenomics)		
5	Applications in biology	Mapping cellular heterogeneity	Developmental biology	Kelly (2020)
		Deciphering stem cell differentiation	Cancer research	
		Studying rare cell populations (e.g., circulating tumor cells)	Neurobiology	
6	Clinical & Translational Applications	Single-cell tumor microenvironment proteomics	Personalized medicine	Specht et al. (2021)
		Drug response profiling	Predicting therapeutic response	
		Precision oncology & biomarker discovery	Early disease detection	

6.3 Technologies Enabling Single-Cell Proteomics

Figure 6.1 explains the process for single-cell proteomics works in a loop, showing how each step connects to the next. It starts with isolating cells, then preparing the sample, labeling the peptides, separating them, analyzing with mass spectrometry, and finally interpreting the data computationally. This cycle keeps repeating as researchers improve and expand their understanding of proteins in individual cells.

The single-cell proteomics workflow commences with the isolation of individual cells via methods such as FACS, microfluidics, or laser capture microdissection. Each cell is subsequently lysed, and its proteins are reduced, alkylated, and digested into peptides in nanoliter amounts. These peptides are designated with isobaric tags (e.g., TMT, iTRAQ) or assessed without labels for multiplexed quantification. Peptide separation is accomplished by nanoLC or chip-based systems prior to identification and quantification by high-resolution mass spectrometry (Orbitrap, TOF, TIMS) utilizing DDA or DIA modes. Ultimately, computational pipelines facilitate data processing for protein identification, normalization, clustering, visualization, and biological interpretation (Fig. 6.1).

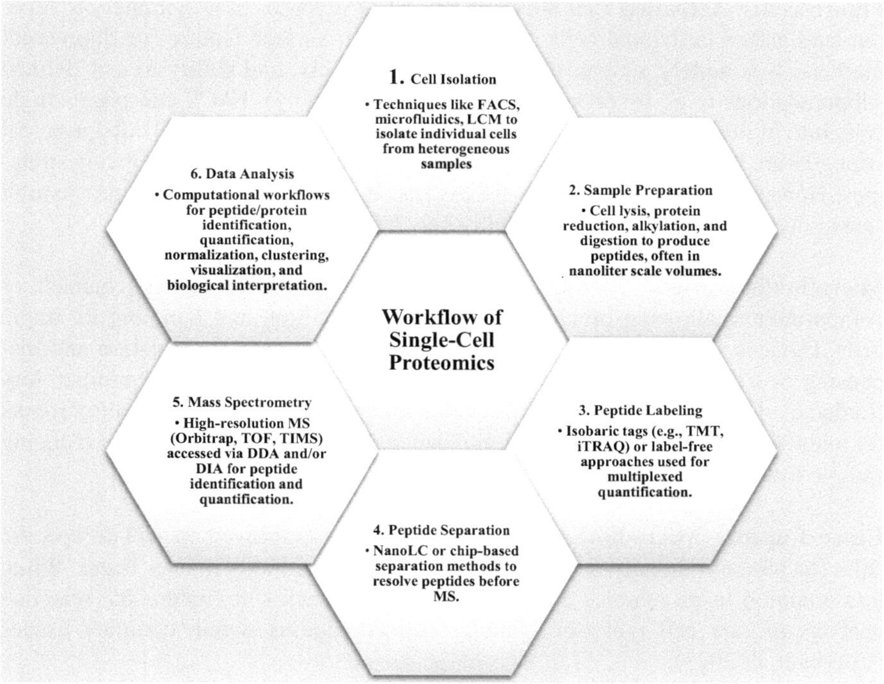

Fig. 6.1 Workflow of single-cell proteomics

6.3.1 Sample Preparation Techniques

Specialized technologies allow scientists to examine the proteins in individual cells, including their structure, function, and changes, which is the foundation of single-cell proteomics. The success of single-cell proteomic studies depends on careful and effective sample preparation steps, such as isolating cells, breaking them open, extracting proteins, breaking proteins into smaller parts, and adding tags for later analysis using mass spectrometry. This section explains current methods used in sample preparation, focusing on how cells are separated (like using FACS, microfluidics, or LCM), how proteins are extracted and broken down at the single-cell level, and the labeling techniques (such as TMT, SILAC, and iTRAQ) used to mark proteins for analysis.

6.3.1.1 Cell Isolation Methods

Single-cell proteomics requires isolating individual cells that are pure, not damaged, and ready for further analysis. The main techniques that are currently used are as follows:

Fluorescence-Activated Cell Sorting (FACS) FACS uses flow cytometry to separate and gather individual cells depending on their surface features or fluorescent markers. It is widely adopted for its speed, scalability, and ability to sort defined subpopulations (e.g., by phenotypic markers or viability). FACS can place single cells into multiwell plates for parallel processing (Eshghi et al. 2023). Its principal strengths are high throughput and the ability to integrate with complex gating strategies. However, FACS instruments are expensive, and some cell types may exhibit sensitivity to sorting pressures (Eshghi et al. 2023).

Microfluidic Devices Microfluidics manipulates cells in microscale channels or compartments, allowing precise isolation, encapsulation, and handling of single cells. Devices such as the CellenONE instrument automate the isolation and dispensing of single cells into wells or droplets with high accuracy and minimal loss (Leduc et al. 2024). These platforms enable integration with liquid handling robots for massively parallel sample preparation key to increasing throughput and reducing sample loss (Leduc et al. 2024).

Laser Capture Microdissection (LCM) LCM is employed to isolate specific cells (or regions) directly from tissue sections using a focused laser beam. While less common in proteomics compared to transcriptomics, it enables the selective analysis of rare cell types or spatially defined regions within complex tissues (Eshghi et al. 2023).

6.3.1.2 Protein Extraction and Digestion at Single-Cell Levels

Protein extraction from individual cells poses substantial challenges due to minuscule protein quantities, susceptibility to surface adsorption, and the need to preserve sample integrity. Key developments include:

Minimal Proteomic Sample Preparation (mPOP) mPOP uses freeze–heat lysis cycle to efficiently lyse single cells with minimal reagents, and without clean-up steps, therefore, reducing the loss of proteins. FACS method involves the application of flow cytometry in the isolation and collection of individual cells depending on their surface features or fluorescence. This is compatible with multiwell plates and can be executed to facilitate high-throughput and automated processing (Petalski et al. 2022). The mPOP eliminates the requirement to utilize strong chemicals and prolonged transfer guidelines and decreases the effectiveness and reliability of protein extraction per cell.

Water Droplet-in-Oil (WinO) Digestion The WinO technique works by trapping single cells in water droplets that float in oil. This method helps proteins stay in the liquid and not stick to surfaces, and it also stops the cells from being diluted. Digestion and chemical labeling occur within the confined droplet environment, with improved protein yields and compatibility for downstream MS analysis.

Nanoproteomic Sample Preparation (nPOP) nPOP enables massively parallel preparation of thousands of single cells in nanoliter-scale droplets, greatly enhancing scalability and reproducibility. Protein extraction, digestion, and labeling are all conducted within these droplets on specialized glass slides, reducing handling and sample loss. Liquid-handling robots facilitate the automated dispensing and processing of single cells using routines compatible with multiplexed MS strategies (Leduc et al. 2024).

One-Pot Sample Preparation One-pot protocols combine cell lysis, protein digestion, and labeling within the same well or vessel, minimizing transfers, pipetting, and sample handling. These approaches are prized for reducing losses associated with surface adsorption and evaporation (Eshghi et al. 2023).

6.3.1.3 Labeling Techniques: TMT, SILAC, iTRAQ

Protein and peptide labeling strategies underpin multiplexed quantification in single-cell proteomics, increasing analytical throughput, sensitivity, and flexibility.

Tandem Mass Tag (TMT) Labeling TMT reagents are isobaric chemical tags used to label peptides from different samples (including single cells), which are then pooled for simultaneous MS analysis. Each tag liberates a unique reporter ion in the MS/MS stage, enabling relative quantification across samples (Petalski et al. 2022). The use of isobaric carriers abundant tagged cells pooled with single-cell samples amplifies detectable signals, improving identification and quantitative accuracy. SCoPE2, for instance, leverages TMTpro 18-plex reagents for large-scale multiplexing with high sensitivity (Petalski et al. 2022).

iTRAQ (Isobaric Tags for Relative and Absolute Quantitation) Similar to TMT, iTRAQ labels peptides from single or grouped cells for multiplexed analysis. Although used for bulk samples extensively, iTRAQ's adoption in single-cell proteomics is limited by lower multiplexing capacity compared to next-generation TMT and its relative susceptibility to ratio distortion (Petalski et al. 2022).

SILAC (Stable Isotope Labeling by Amino Acids in Cell Culture) SILAC introduces nonradioactive, heavy isotope-labeled amino acids into the culture medium, resulting in metabolic incorporation into newly synthesized proteins. While widely used in bulk and cell culture systems, SILAC's direct application to single-cell proteomics remains challenging due to the need to grow and label each cell individually, making it relatively rare as compared to TMT/iTRAQ for true single-cell workflows (Petalski et al. 2022). Some recent adaptations opt for spike-ins or carrier-based approaches incorporating SILAC-labeled standards to calibrate quantification rather than labeling each single cell (Eshghi et al. 2023).

6.3.1.4 Integration of Sample Preparation Workflows

Most contemporary platforms combine multistep workflows to optimize sensitivity, throughput, and data quality:

- Cell isolation (via FACS, microfluidics, or LCM) is coordinated with direct transfer into low-binding nanowell plates or droplets.
- Cell lysis (such as freeze–heat cycles or WinO) is immediately followed by enzymatic digestion using trypsin in nanoliter volumes.
- Barcoding via TMT (or similar reagents) precedes pooling and MS analysis, often accompanied by the addition of isobaric carrier channels containing lysates from hundreds of cells to enhance the signal (Leduc et al. 2024).

Massively parallel, automated systems are increasingly adopted, supported by custom software for quality control and data interpretation (e.g., QuantQC for nPOP-based sample preparation) (Leduc et al. 2024).

6.3.1.5 Technical Challenges and Future Directions

Despite impressive advancements, single-cell proteomics sample preparation faces persistent challenges:

- Protein losses arising from adsorption to vessel surfaces or low concentrations.
- Inconsistent digestion efficiency owing to minute reagent volumes and variable enzyme access.
- Instrumental sensitivity limits, requiring ongoing improvements in MS hardware.
- Sample throughput—automation and multiplexing (TMTpro, plexDIA) are key avenues for future gains (Karagach et al. 2025).
- The field continues to push toward higher proteome coverage (targeting detection of thousands of proteins per cell), lower sample volumes, improved automation, and further integration of sample prep with emerging analytical technologies.

6.3.2 Mass Spectrometry (MS)

Mass spectrometry (MS) has seen significant advancements that are central to the progress of single-cell proteomics. Modern MS instruments provide high-resolution, high-sensitivity detection essential for analyzing the tiny amounts of proteins in individual cells.

6.3.2.1 Advancements in High-Resolution MS

The latest changes in the mass spectrometry technology are the high ion transfer efficiencies, reduced background noise and ultrasensitive detectors enhances the depth, protein identification and quantification and accuracy of results. Orbitrap and time-of-flight (TOF) instruments can now provide high mass accuracy and resolution to ensure the ability to sequence and measure peptides in complex single-cell samples with unquestionable certainty. Techniques, in their turn, have greatly multiplied the size of the observable proteome in the single-cell regime, now hundreds to thousands of proteins can be observed in each cell (Guo et al. 2025a, 2025b).

6.3.2.2 Data-Dependent Acquisition (DDA) vs Data-Independent Acquisition (DIA)

The old methods of DDA on-the-fly choose high-intensity precursor ions and fragment them to be analyzed using MS/MS. Whereas this method of analysis is rather well established, the DDA may not be able to quantify the less abundant peptides due to the stochastic sampling protocol. Single-cell proteomics has a limitation of repeatability and coverage of proteins, which is rather a drawback. Various methods of MS/MS spectrometry, wherein all precursors within a given mass range are fragmented, yield the DIA methods that generate wide MS/MS spectra. This enhances the consistency in detection and coverage of single-cell samples by proteomics. DIA also benefits with better software approaches and spectral libraries as they permit the separation of complex spectra. (Sanchez-Avila et al. 2025) also provide additional information by stating that DIA is an emerging popular discipline in the field of single-cell proteomics owing to its greater depth and reproducibility.

6.3.2.3 Miniaturized and Ultrasensitive LC-MS Platforms

Liquid chromatography (LC) in combination with mass spectrometry has been scaled down to allow the low sample volumes of single cells. Sensitivity can be notably increased by minimizing dead volumes in nanoflow LC columns and minimizing sample dilution in the column. Integrated microfluidic liquid chromatography devices and advancements such as low binding surfaces can be used to reduce sample loss. Recent single-cell research technologies combine ultrasensitive liquid chromatography (LC) systems with more advanced mass spectrometry (MS). These developments are typically used in automated systems which apply multiplexed isobaric labeling in order to increase detection limits (Karagach et al. 2025). The technical advances in multiple sclerosis (MS) form the basis of the wider potential of single-cell proteomics to offer robust, comprehensive, and reproducible profiles of the proteome using small amounts of biological samples. This is a move towards making the field more relevant in the area of biology and medicine.

6.3.3 Microfluidics and Nanotechnologies

Microfluidics and nanotechnologies are essential tools for single-cell proteomics (SCP), enabling the collection, processing, and analysis of individual cells with exceptional precision, sensitivity, and throughput.

6.3.3.1 Lab-on-a-Chip Technologies

consolidate many sample preparation processes including cell isolation, lysis, protein digestion, and peptide purification onto a singular microfluidic chip, therefore reducing sample loss and contamination. These devices (e.g., iProChip and SciProChip) facilitate automated, multiplexed processes enabling the capture, counting, imaging, and processing of cells directly on-chip prior to subsequent mass spectrometry (MS) analysis. The microfluidic environment reduces surface adsorption, volumes, and handling steps, improving reproducibility and sensitivity. For example, integrated microfluidic chips coupled with data-independent acquisition (DIA) MS have enabled profiling of approximately 1500 protein groups from single mammalian cells with low missing data and high quantitative accuracy (Gebreyesus et al. 2022).

6.3.3.2 Droplet-Based Proteomics

utilizes water-in-oil droplets or nanoliter-scale reaction compartments to encapsulate single cells for proteomic processing. Technologies like nano-proteomic sample preparation (nPOP) use acoustic dispensing to isolate individual cells into nanodroplets on specialized surfaces, where lysis, enzymatic digestion, and labeling occur in confined volumes. This approach minimizes sample dilution and protein losses, facilitating massively parallel sample processing and enhanced throughput. Droplet microfluidics combined with advanced liquid handling can integrate with multiplexed isobaric labeling strategies such as tandem mass tags (TMT) to boost sensitivity via isobaric carrier approaches (Gebreyesus et al. 2023).

6.3.3.3 Integration with MS Pipelines

Involves seamless coupling of these microfluidic sample preparation platforms with ultra-sensitive liquid chromatography-MS workflows. The automated on-chip processes generate peptides ready for direct MS analysis, reducing manual handling and improving consistency. By combining these technologies with advanced MS acquisition methods (e.g., DIA), SCP workflows achieve deeper proteome coverage and improved quantitative reproducibility. Moreover, chip-based systems equipped with integrated imaging and cell counting yield orthogonal phenotypic data that enhance proteomic profiles, facilitating multidimensional single-cell characterization (Gebreyesus et al. 2022). Microfluidics and nanotechnologies combine to create the technical basis of next-generation single-cell proteomics to improve sensitivity, throughput and analytical robustness. The present advances are aimed at increasing the depth of proteomics, multiplexing, and multiomics integration, therefore, improving biological and clinical applications.

6.3.4 Computational Tools and Data Analysis

Computational methods and data processing are necessary to draw valuable biological conclusions out of single-cell proteomics (SCP) datasets that are characterized by a limited input of proteins, sparse measurements, and high dimensionality. Major elements include use of machine learning to identify patterns, approaches to deconvolution of low-input data and integration of proteomics with other omics data layers.

6.3.4.1 Algorithms for Deconvoluting Low-Input Proteomics Data

The SCP data demands intensive preprocessing methods to mitigate the impact of missing measurements, batch effects, and noise, which is common to measurements of low-abundance proteins. In this method, quality checks of the data will be conducted, normalization of the data will be used to correct technical differences and the imputation techniques such as k-nearest neighbors (KNN) will be used to infer the missing protein intensity levels. Various technologies e.g. ComBat and other batch correction technologies are used to make sure that information collected in multiple trials can be compared and find similarities. SCPline and similar specialized software has been developed with built-in infrastructure to handle single-cell proteomics data acquired using mass spectrometry, antibodies, and multiomics. The application of these technologies simplifies the normalization, reduction of dimensionality of data, grouping, and visualization of the data. Peptide intensity normalization is done before aggregation causes quantifications of proteins on a protein level to ensure reliable subsequent analysis. The computational methods improve the accuracy of protein measurements of single cells and ensure uniformity of the results (Guldberg et al. 2023).

6.3.4.2 Machine Learning for Pattern Recognition

The use of machine learning approaches is essential in translating SCP data since it determines cellular subpopulations, states and paths. Dimensionality reduction methods like principal component analysis and unsupervised multidimensional analysis aim at condensing the patterns of high-dimensional protein expression into representations that are easier to handle. Clustering techniques group cells according to the levels of alike protein profiles, and therefore extract the broad diversity of cell types and states. Differentiation and activation are some examples of dynamic biological processes that can be reconstructed by methods of trajectory inference. The weighted closest neighbor (WNN) and graph-based clustering methods can be used to improve cell grouping by combining multiple types or modalities of data (Guo et al. 2025a, 2025b).

6.3.4.3 Data Integration with Other Omics

By creating the relationship between the protein expression and gene expression and regulatory processes, the combination of single-cell proteomics and genomics and transcriptomics of single cells enhances the analysis of biological data. Combined analyses of transcriptomic and proteomic data (e.g. CITE-seq) can now be performed using multiomics tools like Seurat by using weighted neighbor graphs and joint clustering. Combining the data makes it easier to correct any discrepancies between RNA and protein levels to gain a better picture of post-transcriptional regulation and cellular behavior. The standardized single-cell proteomics data help to compare the studies and build technologies to combine various biological data through the aggregation of the data through databases such as SPDB (Wang et al. 2024). Combined efforts of these tools allow researchers to analyze complex data in a more effective way, observe biological trends, and gain a deeper insight into how the cell behaves. The method can help researchers to make new strides in the fields of immunology, developmental biology, personalized medicine, and oncology.

6.4 Key Applications of Single-Cell Proteomics

6.4.1 Cancer Biology

Cancer researchers are starting to apply single-cell proteomics to better understand the variations in complexity in tumors, explain mechanisms of treatment resistance in some cancer cells, and discover small cell populations (such as circulating tumor cells) important for patient prognosis.

6.4.1.1 Tumor Heterogeneity

Tumors contain heterogeneous cellular populations, each with unique protein characteristics that can affect development, metastasis, and response to therapy. Single-cell proteomics (SCP) is a technology that enable a broad examination of differences by measuring proteins and changes occurring in tumor cells. This allows for the identification of different cellular populations that may not be evident in standard assays, such as those with differences in metastatic potential or therapeutic response. This allows for identification of cells that may respond to therapy. The use of SCP has facilitated our understanding of tumor heterogeneity to develop more personalized therapy plans (Nalla et al. 2025). These studies have been executed in highly aggressive malignancies such as triple-negative breast cancer (TNBC).

6.4.1.2 Drug Resistance Mechanisms

Because of the adaptability or longevity of specific types of cancer cells in therapy, these cells are frequently endowed with drug-resistant properties, which might lead to the re-emergence of the disease. At the cellular level, SCP, or scanning cellular proteomics, provides a method for scientists to observe changes in proteins that allow cellular acquisition of resistance to therapy. These changes may be in terms of the transmission of the signal, alterations to the protein, and organization of the protein network. This information helps identify potential markers and targets for new treatments that can overcome resistance. Keeping track of how proteins are made and broken down in each cell shows how cancer cells change their protein makeup in response to drugs.

6.4.1.3 Identifying Rare Cell Populations (e.g., Circulating Tumor Cells)

Rare cells like CTCs or cancer stem-like cells are key drivers of metastasis and disease progression but are difficult to analyze with bulk methods due to their scarcity. SCP technologies, often integrated with advanced cell isolation methods, successfully profile these rare populations, identifying unique proteomic signatures important for diagnostics and targeted therapy development. The ability to profile CTCs at the protein level yields valuable information on metastatic competence and therapeutic vulnerabilities. Together, these applications underscore the power of single-cell proteomics to reveal the functional and phenotypic diversity within tumors, offering granular insight into cancer biology that can improve diagnostics, prognostics, and precision oncology.

6.4.2 Immunology

Single-cell proteomics (SCP) has become a powerful tool in immunology for detailed profiling of immune cells such as T cells, B cells, and macrophages, advancing understanding of their heterogeneity, function, and responses at unprecedented resolution.

6.4.2.1 Immune Cell Profiling

SCP allows for high-resolution characterization of immune cell subtypes by quantifying thousands of proteins per single cell, uncovering functional diversity within populations of T cells, B cells, and macrophages. Advanced microfluidic techniques integrated with mass spectrometry provide the precise separation and analysis of proteins in individual immune cells, uncovering distinct patterns associated with their activation, development, or various activities. The application of antibodies and stable isotopes greatly enhances complex and detailed protein investigations directed towards examining differences among immune cells in healthy and diseased contexts (Xu et al. 2025). Both methods are applied to look at differences in immune cells under distinct conditions.

6.4.2.2 Cytokine Production at the Single-Cell Level

Single-cell proteomics is the method used to find cytokines and other signaling molecules made by immune cells. As a result, scientists can measure how much of these compounds different cells make. This information is very important for understanding how the immune system works, how inflammation starts, and how cells talk to each other. The combination of single-cell proteomics and spatial proteomics can be used to make maps of cytokine release in tissues. This comprehension clarifies the operation of the immune system and the origins of diseases.

6.4.2.3 Vaccine Response and Autoimmunity

SCP looks at how molecules in immune cells change before and after vaccination to learn more about how the body builds and keeps a strong immune memory. It also shows how protein levels change over time, which is related to how well vaccines work, and it helps find rare immune cell groups that respond to vaccines. SCP can also find strange things happening in immune cells, changes in communication pathways, and uncontrolled cytokine release. All of these things can lead to harmful immune responses in autoimmune diseases. The results of this study could help create more effective targeted therapies and personalized care plans. Through single-cell proteomics, scientists can learn a lot about the different types of immune cells and what they do. This is an important step in making vaccines, studying autoimmune diseases, and improving immunotherapy.

6.4.3 Neuroscience

Single-cell proteomics is a potent tool in neuroscience that enables researchers to meticulously analyze various neuron types, clarify protein patterns within brain circuits, monitor brain development, and identify mechanisms underlying neurodegenerative diseases with remarkable precision.

6.4.3.1 Neuronal Subtypes and Circuit-Level Proteomics

There are many different types of cells in the brain, such as glial cells that help and many types of neurons. These cells interact with each other in many complicated ways. SCP can help scientists look at single neurons and find unique protein patterns that show what kind of neuron it is and how it works. Proteomics provides direct information about the amount of each protein, changes that happen after synthesis (called PTMs), and how proteins interact with each other at the individual level. This elucidates the mechanisms of signal transmission between neurons and the formation of connections, distinguishing it from the examination of gene activity in isolation. New technology that uses microfluidics, isobaric labeling, and highly sensitive mass spectrometry now allows scientists to find thousands of proteins from a single neuron. This helps reveal how synapses work, how neurotransmitters function, and the unique molecular features of different cell types that help shape the brain's networks (Wu et al. 2025).

6.4.3.2 Brain Development

SCP is instrumental in mapping proteomic changes during brain development by resolving the dynamic transitions of neural progenitors into mature neuronal and glial types. This enables elucidation of molecular programs underlying neurogenesis, cell fate decisions, and circuit assembly that transcriptomic data alone cannot fully capture due to the importance of protein modifications and localization in development. Studies applying SCP to developing human and mouse brain tissues have begun to uncover temporal proteome landscapes, highlighting pathways regulating differentiation, migration, and synaptic maturation (Ahmad and Budnik 2023).

6.4.3.3 Neurodegenerative Diseases

In disease contexts like Alzheimer's, Parkinson's, and ALS, SCP can dissect disease-associated proteomic alterations at the single-cell level. This encompasses the detection of rare susceptible neuronal populations, modified protein networks, post-translational modification alterations (e.g., phosphorylation, ubiquitination), and misfolded protein species that contribute to pathology. Bulk tissue analyses do not reveal cellular heterogeneity in neurodegeneration, nor do they elucidate the molecular mechanisms driving disease progression and therapeutic responses. SCP can uncover these concealed facets of neurodegeneration. Integrative SCP and spatial proteomics can also be used to map changes in the proteome in specific brain regions and microenvironments. This can help us learn more about how diseases work in different types of cells and find biomarkers (Goto-Silva and Junqueira 2021). In general, single-cell proteomics is a game-changing method for neurobiology because it gives a functional molecular readout with cellular resolution. This readout works with transcriptomics and genomics to give a full picture of how complex the brain is in both healthy and sick people.

6.4.4 Stem Cell and Developmental Biology

Single-cell proteomics has transformed the domains of stem cell research and evolutionary biology. It enables scientists to directly quantify the abundance of each protein and its regulation within individual cells during critical processes such as determining a cell's future function and the initial phases of embryonic development.

6.4.4.1 Cell Fate Decisions and Lineage Tracing

Analyzing proteins at the single-cell level improves our understanding of how changes in certain proteins, especially those that control cell growth and signaling, change the properties of cells. SCP can directly measure how much of each protein is there and how much it changes after the protein is made. This gives us important information about how the cell is working and how it is changing as it becomes more specialized. This does not compare to the standard method of analyzing RNA. Targeted data-independent acquisition (DIA), a type of mass spectrometry, has recently improved, making it easier to measure small amounts of proteins in individual blood stem and progenitor cells. This has improved our understanding of how certain proteins control the process of making cells. When a cell starts on a certain path of development, these protein patterns act as signs that show the start of that path. The data they furnish is more accurate than that provided by genetic or RNA-based markers (Palii et al. 2019). They help us keep a better eye on how cells change and become different types.

6.4.4.2 Early Embryonic Development Studies

Also, scientists have used SCP to look at proteins in growing tissues and early embryos. This is because the cells in these tissues are different from each other and their protein levels change over time. Mass spectrometry is a technique that allows scientists to look at single cells in embryos to find the proteins that cause the first cell divisions, the differentiation into different lineages, and the changes in shape that happen during development. Researchers can improve their understanding of the temporal and spatial variability of proteins during embryonic development, allowing them to identify critical factors that influence this process. These parameters are sometimes overlooked when analyzing all proteins simultaneously. The study of proteins during the early stages of human brain development has uncovered protein networks that promote stem cell specialization and safeguard cells from apoptosis, both essential for the formation of new brain cells (Wu et al. 2025). Recent developments in SCP have made it possible for scientists to follow the fate of cells and the processes of development at the proteomic level, even within single cells. This gives a full understanding of the molecular processes that control cell specialization, plasticity, and the formation of tissues. This also opens up new possibilities for better medical treatments, research on how organisms grow, and understanding birth defects.

6.4.5 Infectious Diseases

Single-cell proteomics is a useful way to learn about how hosts and pathogens interact at the cellular level. It gives a lot of information about how the molecular connections between infectious pathogens, like bacteria and viruses, and individual host cells change during an infection.

6.4.5.1 Host–Pathogen Interactions at the Cellular Level

For hazardous germs to grow and spread after entering a cell, they must interact with cellular components in a very specific way. These interactions may involve germ proteins binding to or inhibiting cellular proteins. This gives the pathogen the ability to get into the cell or get around the body's defenses. The infected cell is controlling protein levels, where they are located, and any changes that might happen to find and kill the pathogen during this process. Single-Cell Proteomics (SCP) allows for the examination of temporal and spatial variations in the protein composition of infected cells. This shows that different cells act in different ways, which is something that more general analytical tools often miss. This is important because cells that are genetically identical can respond differently to infection, which can affect how the disease progresses (Jean Beltran et al. 2017; Biron et al. 2011).

6.4.5.2 Cellular Responses to Viral and Bacterial Infections

Through SCP, it is possible to characterize host immune defense mechanisms, such as signaling networks activated by pathogen recognition, cytokine production at the single-cell level, and the remodeling of cellular pathways during infection. Proteomics also captures pathogen-derived proteins expressed inside host cells, allowing simultaneous profiling of both host and pathogen proteomes ("dual proteomics"). This facilitates discovery of virulence factors, metabolic dependencies, and host factors essential for pathogen replication. Additionally, studying post-translational modifications (e.g., phosphorylation, ubiquitination) informs on signaling dynamics and regulation during infection (Fels et al. 2017).

6.4.5.3 Advances and Integration with Multiomics

Recent methodological advances combine SCP with complementary omics (genomics, transcriptomics, metabolomics) to provide systemic views of infectious processes (Israr et al. 2024). Chemical proteomics and spatially resolved techniques further enhance analysis of molecular interactions and pathogen localization within tissues. Computational tools enable deconvolution of complex datasets and identification of molecular signatures correlating with infection stages, immune evasion, or therapeutic targets (Muñoz et al. 2024). In summary, single-cell proteomics offers a transformative window into the molecular warfare between host and pathogen at single-cell resolution, deepening our understanding of infection biology and informing development of diagnostics, vaccines, and therapeutics.

6.5 Integration with Other Omics Approaches

Modern biological research increasingly recognizes that no single molecular layer fully captures the complexity of cellular states and processes. Thus, multiomics strategies that combine single-cell proteomics (SCP) with complementary datasets, such as single-cell RNA sequencing (scRNA-seq), epigenomics, and metabolomics, are pivotal for advancing systems-level understanding of biology and disease (Liu et al. 2024).

6.5.1 Multiomics Platforms: Combining Single-Cell RNA-seq with Proteomics

Single-cell RNA sequencing provides extensive transcriptome-wide gene expression profiles at cellular resolution, but mRNA levels only partially reflect protein abundance and functional states due to post-transcriptional and post-translational regulation. Integrating scRNA-seq with SCP bridges this gap, enabling joint analysis of the transcriptome and proteome from the same or closely matched cells. Several technological frameworks facilitate such integration:

6.5.1.1 Parallel Profiling

Techniques allow extraction and processing of nucleic acids and proteins from the same cell or matched populations to generate paired transcriptomic and proteomic data.

6.5.1.2 CITE-seq (Cellular Indexing of Transcriptomes and Epitopes by Sequencing)

Uses oligonucleotide-labeled antibodies to simultaneously profile protein markers and transcriptomes via sequencing, albeit limited to a targeted protein panel.

6.5.1.3 Mass Spectrometry-Based SCP with scRNA-seq

Emerging workflows couple microfluidics and isobaric labeling to profile unbiased protein sets alongside scRNA-seq from parallel samples, enabling cross-modal data integration and deeper biological insight.

Computational pipelines and platforms, such as Seurat and SCPline, implement joint analyses that align cell clusters, perform integrative dimensionality reduction, and identify regulatory mechanisms linking mRNA and protein expression patterns. This integrated perspective elucidates cellular heterogeneity and dynamic state transitions more comprehensively than either method alone (Ma et al. 2020).

6.5.2 Spatial Proteomics and Transcriptomics

Traditional single-cell omics approaches typically require tissue dissociation, losing spatial context critical for understanding cellular organization, microenvironment interactions, and tissue architecture. Recent advances in spatial omics combine transcriptomic and proteomic profiling with position information, preserving the native tissue landscape.

6.5.2.1 Spatial Transcriptomics

Technologies map mRNA expression with spatial coordinates using spatially barcoded probes on tissue sections (e.g., 10X Genomics Visium, Slide-seq).

6.5.2.2 Spatial Proteomics

Employ immunohistochemistry/immunofluorescence with multiplexed antibody panels, imaging mass cytometry, or mass spectrometry imaging to localize proteins within tissues at high resolution.

6.5.2.3 Integrated Spatial Multiomics

Platforms like SM-Omics leverage DNA-barcoded antibodies and transcript capture on the same tissue section, enabling simultaneous high-throughput profiling of transcripts and proteins with spatial fidelity. Methods such as Multiomics in situ pairwise sequencing (MiP-seq) further enhance multiplexing and subcellular resolution for DNA, RNA, and protein detection (Karimi et al. 2024).

Integrating spatial proteomics with spatial transcriptomics enables rich mapping of cellular phenotypes, functional states, and intercellular signaling within intact tissues, essential for understanding processes like tumor microenvironment heterogeneity, developmental patterning, and immune responses. It also facilitates the discovery of spatially restricted molecular interactions and cell niches (Khan et al. 2025; Lim et al. 2024).

6.5.3 Systems Biology Approaches and Data Integration

The combined analysis of multiomics data at single-cell and spatial resolution requires sophisticated systems biology frameworks. These methods look at how different molecular parts interact with and control each other. As a result, they help make maps of signal and gene regulatory networks and predict how cells will respond. Methodologies like network analysis, machine learning, and trajectory inference can help us find important things that affect how cells behave and new things that biological systems do. These strategies combine information from gene expression and protein levels. The amalgamation of various data types facilitates the exhibition of disparities in gene expression and protein concentrations. These discrepancies may be attributed to the mechanisms of protein synthesis or their stability, providing substantial insights into the regulatory processes subsequent to gene activation (Baião et al. 2025). This strategy makes it easier to find more accurate markers, choose important therapy targets, and improve our understanding of diseases by giving us a complete picture of how cells work and how their environment affects them.

6.5.4 Data Harmonization Challenges

Although combining single-cell and spatial multiomics data holds great promise, there are several obstacles that make it difficult to bring all the data together smoothly.

6.5.4.1 Changes in Technology

Proteomic and transcriptomic technologies operate differently regarding measurement techniques, data acquisition capabilities, and sensitivity to minor fluctuations. It is hard to compare data from different scales because many batches do not have consistent results.

6.5.4.2 There Are Differences in Resolution

The sequencing of RNA from a single cell generally exhibits increased capacity and examines a more extensive portion of the genome. It is harder to directly compare the two methods because SCP may find fewer proteins per cell and may leave out data from time to time.

6.5.4.3 Complexity

The complexity of spatial data requires sophisticated computational techniques to synchronize transcriptomic and proteomic geographic information at varying depths and resolutions. These methods include aligning images, breaking down geographical information, and combining data from different types of measurements.

6.5.4.4 Data Volume and Complexity

Integrative multiomics datasets are characterized by high dimensionality and sparsity, necessitating scalable techniques for normalization, imputation, feature selection, and visualization.

The lack of globally accepted standards and reference datasets for SCP integration hinders reproducibility and comparative analysis. Recent advancements in techniques for rectifying batch discrepancies, establishing unified data representations, aggregating data from diverse sources (such as employing weighted closest neighbor graphs), and aligning spatial information significantly enhance the precision of integrating various data kinds. Initiatives to establish standardized databases, such as SPDB for SCP, and to maintain consistency in metadata reports are enhancing the comprehensiveness and reliability of multiomics investigations (Wang et al. 2025). In summary, the integration of single-cell proteomics with transcriptomics and spatial omics provides a novel approach to comprehending cellular functions, tissue organization, and disease progression. The utilization of multiomics technologies, preservation of spatial information, and implementation of systems biology models are facilitating significant discoveries and innovative methodologies for the interpretation and integration of complicated data.

6.6 Challenges and Limitations

Even though technology has come a long way, single-cell proteomics still has some basic problems that make it not very sensitive, not very consistent, not very easy to use, and not very helpful for understanding biological concepts.

6.6.1 Sensitivity and Throughput Issues

The fact that a single cell carries a relatively small amount of protein typically only a few picograms is one of the most significant problems. In contrast to nucleic acids, proteins cannot be amplified, thus limiting detection sensitivity. The immense dynamic range of protein abundance spanning several orders of magnitude within a single cell means low-abundance proteins, many of them biologically important regulators, often remain undetected, contributing to the so-called "dark proteome" problem. Efficient sample preparation methods that minimize losses, together with ultrasensitive mass spectrometry and optimized peptide labeling (e.g., isobaric carriers), partially mitigate this but do not fully overcome it. Throughput remains limited by complex workflows and long LC-MS run times, restricting feasibility for very large-scale studies or clinical applications (Redit et al. 2023).

6.6.2 Reproducibility and Standardization

Methodological variability in sample isolation, preparation, and data acquisition can lead to batch effects and inconsistent data quality, impeding reproducibility across laboratories. The field currently lacks universally accepted standard protocols and benchmark datasets, which delays comparative analyses and adoption. While efforts are underway to establish community standards, reporting guidelines, and quality control metrics for SCP, the technological heterogeneity and rapid innovation pace pose coordination challenges. Moreover, data analysis tools are still evolving to handle sparse and noisy datasets robustly (Ahmad and Budnik 2023).

6.6.3 Cost and Accessibility

SCP requires access to cutting-edge instrumentation including high-resolution mass spectrometers coupled with nano-flow liquid chromatography, specialized microfluidic devices, and reagents for multiplexed labeling. The capital investment and operational costs, alongside the need for technical expertise, limit widespread accessibility outside specialized centers. Automated sample preparation platforms remain scarce and costly, further restricting scalability. Broader adoption will depend on industrial collaborations to develop affordable, user-friendly, and robust SCP workflows (Ahmad and Budnik 2023).

6.6.4 Data Interpretation and Biological Noise

Proteomic data at the single-cell level are intrinsically noisy due to technical variability and biological heterogeneity, complicating interpretation. Missing data points (dropouts), stochastic variation in protein expression, and post-translational modifications add layers of complexity. To tell the difference between real biological signals and measurement noise, you need advanced computer methods for normalization, imputation, and batch correction, which are still being developed. Combining SCP data with other omics layers can help clear up any confusion, but it also makes analysis harder (Guo et al. 2025a, 2025b). In short, single-cell proteomics technology is moving quickly and giving us new biological information, but there are still big problems with sensitivity, throughput, reproducibility, cost, and data interpretation. To make the most of SCP in many areas of biology and medicine, we need to keep developing new approaches, agreeing on common ideas, and moving forward with computing technology.

6.7 Future Perspectives and Innovations

In the coming years, single-cell proteomics will advance a lot, bringing major changes to the field. These advances will affect clinical applications, lead to important developments in computing, introduce new tools and techniques, and have a big impact on personalized medicine and diagnostic methods.

6.7.1 Toward Clinical Translation

Recent advancements in sensitivity, speed, and dependability have rendered SCP suitable for application in real-world medical environments. Advanced mass spectrometry instruments enable researchers to examine hundreds or even thousands of proteins from individual cells, including material from patients such as tumor biopsies, blood, and circulating tumor cells. Researchers can produce accurate maps of cellular variations and pathological conditions directly from clinical tissues by employing methods such as multiplexed isobaric labeling and spatial proteomics. SCP technologies are already facilitating the identification of biomarkers in human fluids, potentially enabling medical personnel to diagnose and monitor diseases without requiring tissue samples in the future. The current difficulty is transitioning SCP from laboratory research to practical application in hospitals and clinics. Standardized sample preparation, automation, reproducible methods, and validated computational pipelines are essential to ensure that clinicians can trust and comprehend the results (Guo et al. 2025a, 2025b).

6.7.2 AI and Computational Proteomics

Artificial intelligence (AI) and machine learning (ML) are transforming computational proteomics, particularly for sparse and noisy datasets associated with SCPs. Advanced methods optimize data preparation, normalization, dimensionality reduction, and batch correction mitigating variability and facilitating robust single-cell analysis. Platforms like SCPline automate data quality assurance, clustering, and multiomics integration, accelerating discoveries in diagnostics, developmental biology, and therapeutic response. ML-driven pattern recognition facilitates identification of rare cell populations, cell-state transitions, and disease signatures, transforming SCP into a powerful tool for precision medicine. AI approaches further enable harmonization across heterogeneous datasets, improving the rigor and scalability of SCP for research and clinical application (Guo et al. 2025a, 2025b).

6.7.2.1 Emerging Technologies

Future SCP workflows may be defined by several emerging technologies-*Nanopore-based proteomics*—Early-stage nanopore platforms are being developed for direct, single-molecule protein sequencing. Potential advantages include label-free, amplification-free analysis, high throughput, and portable devices. These platforms could eventually enable rapid clinical diagnostics or point-of-care applications by offering protein-level readouts comparable to next-generation sequencing for genomics.

6.7.2.2 In Situ Single-Cell Proteomics

Technologies combining spatial and single-cell analysis such as imaging mass cytometry, multiplexed immunofluorescence, and spatial mass spectrometry are rapidly maturing. These allow direct measurement of protein abundance, post-translational modifications, and cellular interactions within intact tissue contexts, crucial for mapping tumor microenvironments, neural circuit connectivity, or immune responses. Integrating in situ SCP with spatial transcriptomics and other modalities promises multidimensional maps of tissue function, aiding both research and diagnostics (Guo et al. 2025a, 2025b).

6.7.2.3 Automated Microfluidics and Miniaturized Systems

Integrated lab-on-a-chip devices and droplet microfluidics are streamlining SCP workflows, reducing sample losses and operational costs, and boosting throughput for population-scale clinical studies.

6.7.3 Personalized Medicine and Diagnostics

SCP is poised to drive the next wave of personalized medicine. By enabling profiling of individual cell proteomes from patient samples, SCP can uncover disease heterogeneity, rare pathological cell types, and dynamic responses to therapies. Biomarker discovery leveraging both unique protein signatures and spatial cellular context will inform patient stratification, early diagnosis, treatment monitoring, and tailored interventions. Multiomics integration (proteomics, transcriptomics, metabolomics) through AI-powered platforms will offer holistic disease understanding and precision therapeutic targeting, setting the stage for SCP-driven diagnostics and bespoke medicine (Guo et al. 2025a, 2025b; Kumar et al. 2024).

6.7.4 Outlook

As single-cell proteomics converges with emerging high-throughput and spatial technologies, fortified by AI-enabled analytical platforms, a new era of cellular and molecular medicine is materializing. While hurdles remain cost, scalability, standardization, and clinical validation—the next decade will likely see SCP become central to research and clinical practice, revolutionizing how we understand and treat disease at cellular resolution.

6.8 Conclusion

Single-cell proteomics (SCP) has emerged as a transformative technology, enabling the direct measurement of protein abundance, localization, and modifications at the resolution of individual cells. Recent advances in ultrasensitive mass spectrometry, microfluidics, nanotechnologies, and robust computational pipelines have addressed many critical challenges of sensitivity, throughput, and data interpretation. Multiplexed labeling strategies, spatially resolved workflows, and integration with other omics platforms now allow researchers to simultaneously capture molecular heterogeneity and contextual tissue organization. These innovations have opened unprecedented opportunities to study complex biological systems revealing tumor heterogeneity, immune cell diversity, neuronal subtypes, developmental trajectories, and host–pathogen interactions invisible to bulk analyses. SCP provides a unique functional readout that complements genomics and transcriptomics, uncovering post-transcriptional regulation and dynamic cellular states crucial for understanding physiology and pathology. Looking forward, SCP promises to reshape precision medicine by identifying actionable biomarkers, guiding patient stratification, and monitoring therapeutic responses at cellular resolution. As workflows become standardized, automated, and clinicians gain access to integrated spatial and multiomics platforms, SCP may evolve from a research tool into a powerful diagnostic and prognostic asset. Its capacity to link molecular detail with complex phenotypes positions it as a central driver in decoding biology and advancing targeted, personalized treatments for diverse human diseases.

References

Ahmad R, Budnik B (2023) A review of the current state of single-cell proteomics and future perspective. Anal Bioanal Chem 415(28):6889–6899. https://doi.org/10.1007/s00216-023-04759-8. Epub 2023 Jun 7. PMID: 37285026; PMCID: PMC10632274

Alam S, Israr J, Kumar A (2024) Microfluidics in genomics. In: Advances in genomics: methods and applications. Springer Nature Singapore, Singapore, pp 357–390

Baião AR, Cai Z, Poulos RC, Robinson PJ, Reddel RR, Zhong Q, Vinga S, Gonçalves E (2025) A technical review of multi-omics data integration methods: from classical statistical to deep generative approaches. Brief Bioinform 26(4):bbaf355

Biron DG, Nedelkov D, Missé D, Holzmuller P (2011) Proteomics and host—pathogen interactions: a bright future? In: Genetics and evolution of infectious disease, pp 263–303. https://doi.org/10.1016/B978-0-12-384890-1.00011-X. Epub 2010 Dec 24. PMCID: PMC7150365

Budnik B, Levy E, Harmange G et al (2018) SCoPE-MS: mass spectrometry of single mammalian cells quantifies proteome heterogeneity during cell differentiation. Genome Biol 19:161. https://doi.org/10.1186/s13059-018-1547-5

Cheung TK, Lee CY, Bayer FP, McCoy A, Kuster B, Rose CM (2021) Defining the carrier proteome limit for single-cell proteomics. Nat Methods 18(1):76–83. https://doi.org/10.1038/s41592-020-01002-5. Epub 2020 Dec 7

Eshghi A, Xie X, Hardie D, Chen MX, Izaguirre F, Newman R, Zhu Y, Kelly RT, Goodlett DR (2023) Sample preparation methods for targeted single-cell proteomics. J Proteome Res 22(6):1589–1602. https://doi.org/10.1021/acs.jproteome.2c00429. Epub 2023 Apr 24. PMID: 37093777; PMCID: PMC10243106

Fels U, Gevaert K, Van Damme P (2017) Proteogenomics in aid of host-pathogen interaction studies: A bacterial perspective. Proteomes 5(4):26. https://doi.org/10.3390/proteomes5040026. PMID: 29019919; PMCID: PMC5748561

Gebreyesus ST, Siyal AA, Kitata RB et al (2022) Streamlined single-cell proteomics by an integrated microfluidic chip and data-independent acquisition mass spectrometry. Nat Commun 13:37. https://doi.org/10.1038/s41467-021-27778-4

Gebreyesus ST, Muneer G, Huang CC, Siyal AA, Anand M, Chen YJ, Tu HL (2023) Recent advances in microfluidics for single-cell functional proteomics. Lab Chip 23(7):1726–1751

Goto-Silva L, Junqueira M (2021) Single-cell proteomics: a treasure trove in neurobiology. Biochim Biophy Acta Proteins Proteomics 1869(7):140658

Guldberg SM, Okholm TLH, McCarthy EE, Spitzer MH (2023) Computational methods for single-cell proteomics. Annu Rev Biomed Data Sci 6:47–71. https://doi.org/10.1146/annurev-biodatasci-020422-050255. Epub 2023 Apr 11. PMID: 37040735; PMCID: PMC10621466

Guo S, Zhou S, Wang G, Wang F (2025a) SCPline: an interactive framework for the single-cell proteomics data preprocessing. Brief Bioinform 26(3)

Guo T, Steen JA, Mann M (2025b) Mass-spectrometry-based proteomics: from single cells to clinical applications. Nature 638(8052):901–911. https://doi.org/10.1038/s41586-025-08584-0. Epub 2025 Feb 26. PMID: 40011722

Israr J, Alam S, Siddiqui S, Misra S, Gupta D, Kumar A (2024) Cell-free systems for genomics. In: Advances in genomics: methods and applications. Springer Nature Singapore, Singapore, pp 391–414

Jean Beltran PM, Federspiel JD, Sheng X, Cristea IM (2017) Proteomics and integrative omic approaches for understanding host–pathogen interactions and infectious diseases. Mol Syst Biol 13(3):922

Karagach S, Smollich J, Atrakchi O, Mohan V, Geiger T (2025) High throughput single-cell proteomics of in vivo cells. Mol Cell Proteomics:101018

Karimi EL, Simo N, Milet N, Te W, Alsh A, Qu N, Ail L, Abs R, Alind A, Goodma NM, Ca N (2024) Method of the year 2024: spatial proteomics. Nat Methods 21:2195–2196

Kelly RT (2020) Single-cell proteomics: progress and prospects. Mol Cell Proteomics 19(11):1739–1748

Khan M, Arslanturk S, Draghici S (2025) A comprehensive review of spatial transcriptomics data alignment and integration. Nucleic Acids Res 53(12):gkaf536. https://doi.org/10.1093/nar/gkaf536

Kumar A, Israr J, Mishra HP, Singh I (2024) Cell-free systems in metabolomics. In: Advances in metabolomics. Springer Nature Singapore, Singapore, pp 257–266

Leduc A, Khoury L, Cantlon J et al (2024) Massively parallel sample preparation for multiplexed single-cell proteomics using nPOP. Nat Protoc 19:3750–3776. https://doi.org/10.1038/s41596-024-01033-8

Lim J, Park C, Kim M et al (2024) Advances in single-cell omics and multiomics for high-resolution molecular profiling. Exp Mol Med 56:515–526. https://doi.org/10.1038/s12276-024-01186-2

Liu X, Peng T, Xu M et al (2024) Spatial multi-omics: deciphering technological landscape of integration of multi-omics and its applications. J Hematol Oncol 17:72. https://doi.org/10.1186/s13045-024-01596-9

Ma A, McDermaid A, Xu J, Chang Y, Ma Q (2020) Integrative methods and practical challenges for single-cell multi-omics. Trends Biotechnol 38(9):1007–1022. https://doi.org/10.1016/j.tibtech.2020.02.013. Epub 2020 Mar 26. PMID: 32818441; PMCID: PMC7442857

Muñoz AW, Zhao W, Sieber SA (2024) Monitoring host–pathogen interactions using chemical proteomics. RSC Chem Biol 5(2):73–89

Nalla LV, Kanukolanu A, Yeduvaka M, Gajula SNR (2025) Advancements in single-cell proteomics and mass spectrometry-based techniques for unmasking cellular diversity in triple negative breast cancer. Proteomics Clin Appl 19(1):e202400101. https://doi.org/10.1002/prca.202400101. Epub 2024 Nov 21. PMID: 39568435; PMCID: PMC11726282

Palii CG, Cheng Q, Gillespie MA, Shannon P, Mazurczyk M, Napolitani G, Price ND, Ranish JA, Morrissey E, Higgs DR, Brand M (2019) Single-cell proteomics reveal that quantitative changes in co-expressed lineage-specific transcription factors determine cell fate. Cell Stem Cell 24(5):812–820

Petalski AA, Slavov N, Specht H (2022) Single-cell proteomics preparation for mass spectrometry analysis using freeze-heat lysis and an isobaric carrier. J Vis Exp (190). https://doi.org/10.3791/63802. https://doi.org/10.3791/63802. PMID: 36571403; PMCID: PMC10027359

Redit C, Cha S, Ai N (2023) Single-cell proteomics: challenges and prospects. Nat Methods 20:317–318

Sanchez-Avila X, de Oliveira RM, Huang S, Wang C, Kelly RT (2025) Trends in mass spectrometry-based single-cell proteomics. Anal Chem 97(11):5893–5907

Specht H, Emmott E, Petelski AA et al (2021) Single-cell proteomic and transcriptomic analysis of macrophage heterogeneity using SCoPE2. Genome Biol 22:50. https://doi.org/10.1186/s13059-021-02267-5

Truong T, Kelly RT (2024) What's new in single-cell proteomics. Curr Opin Biotechnol 86:103077. https://doi.org/10.1016/j.copbio.2024.103077. Epub 2024 Feb 14. PMID: 38359605; PMCID: PMC11068367

Vanderaa C, Gatto L (2023) The current state of single-cell proteomics data analysis. Curr Protocols 3(1):e658

Wang F, Liu C, Li J, Yang F, Song J, Zang T, Yao J, Wang G (2024) SPDB: a comprehensive resource and knowledgebase for proteomic data at the single-cell resolution. Nucleic Acids Res 52(D1):D562–D571. https://doi.org/10.1093/nar/gkad1018. PMID: 37953313; PMCID: PMC10767837

Wang G, Zhao J, Lin Y et al (2025) scMODAL: a general deep learning framework for comprehensive single-cell multi-omics data alignment with feature links. Nat Commun 16:4994. https://doi.org/10.1038/s41467-025-60333-z

Wu T, Jiang L, Mukhtar T, Wang L, Jian R, Wang C, Trinh T, Kriegstein AR, Snyder M, Li J (2025) Single cell proteomics in the developing human brain. bioRxiv:2025–2006

Xu QQ, Jiang YR, Chen JB, Wu J, Chen YX, Fan QX, Wang HF, Yang Y, Pan JZ, Fang Q (2025) Single cell-pair proteomics for decoding immune-cancer cell interactions. Adv Sci (Weinh) 12(11):e2414769. https://doi.org/10.1002/advs.202414769. Epub 2025 Jan 22. PMID: 39840604; PMCID: PMC11923901

Current Progress in Single-Cell Metabolomics

7

Juveriya Israr, Shabroz Alam, and Ajay Kumar

Abstract

Single-cell metabolomics has been developed as a revolutionary method for clarifying cellular heterogeneity in metabolic activity and biochemical pathways at the individual cell level. In contrast to bulk metabolomics, which averages signals from cell populations, single-cell approaches provide high-resolution insights into individual cell metabolic states, disease processes, and microenvironmental responses. Recent improvements in analytical platforms—specifically high-sensitivity mass spectrometry, microfluidic integration, and computational data deconvolution—have greatly enhanced the scope and accuracy of single-cell metabolomics analysis. This chapter offers a thorough examination of contemporary techniques, clinical applications, and translational prospects of single-cell metabolomics, while also confronting ongoing problems like sample throughput, data standardization, and integration with multiomics frameworks.

Keywords

Single-cell metabolomics · Mass spectrometry · Microfluidics · Cellular metabolism · Multiomics integration

J. Israr
Institute of Biosciences and Technology, Shri Ramswaroop Memorial University, Barabanki, Uttar Pradesh, India

Department of Biotechnology, Era University, Lucknow, Uttar Pradesh, India

S. Alam
Department of Biotechnology, Era University, Lucknow, Uttar Pradesh, India

A. Kumar (✉)
Department of Biotechnology, Faculty of Engineering and Technology, Rama University, Kanpur, Uttar Pradesh, India
e-mail: ajaymtech@gmail.com

V. Singh (ed.), *Advances in Single Cell Analysis*,
https://doi.org/10.1007/978-981-95-9789-5_7

7.1 Introduction

The analysis of all the small molecules in a living system is known as metabolomics (Kumar et al. 2024a, 2024b). It assists the scientists in knowing how the cells and biochemical processes operate. Historically, metabolomics primarily studied groups of cells or tissues, which provided a general view of metabolism. Nevertheless, this method tended to overlook the distinctions between a single cell and some of the rare cell populations (Hajnajafi and Iqbal 2025). Since biological systems are not that simple and evolve with time, the metabolism of single cells can significantly fluctuate. This diversity is as a result of genes, environment, and disease (Owens et al. 2004). Due to this, there is increasing interest in the study of metabolism on a single-cell scale. The single-cell metabolomics is a substantial leap on the way to understand the functioning of cells (Kumar et al. 2024a, b). It provides a description of the metabolism and behavior of every cell (Rappez et al. 2021). This technique is particularly helpful in immunology, developmental biology, cancer medicine, and stem cell research, where differences in cells are a significant factor in cell functionality and cell response to therapeutic intervention (Ibáñez et al. 2013). Metabolomics, unlike other procedures, such as gene or RNA analysis, reveals the ultimate outcome of biological processes in real-time (Hajnajafi and Iqbal 2025). Despite the potential of single-cell metabolomics, it has its problems. The concentrations of metabolites are extremely low within a single cell and thus the instruments employed must therefore be highly sensitive and considerate towards the loss of sample during processing (Hu et al. 2023). Mass spectrometry, particularly secondary ion mass spectrometry (SIMS), matrix-assisted laser desorption/ionization (MALDI), and nanospray ionization (Li et al. 2025) have been improved using recent advances in mass spectrometry. The use of new technologies such as microfluidics and improved sampling methods also contribute to the sorting and processing of single cells with higher efficiency and reduced contamination and speedier results (Zhou et al. 2021). The chapter provides an in-depth introduction of the recent advances of single-cell metabolomics research. It discusses new techniques, the tools at our disposal and their application within various sciences. It further talks about the outstanding issues and the further steps that must be undertaken to fully utilize single-cell metabolomics in biological and medical studies. Recent developments in single-cell metabolomics have been categorized across different research domains to provide a clearer understanding of the field. Table 7.1 encapsulates the present advancements, emphasizing significant breakthroughs, primary obstacles, and prospective applications. The chart delineates how innovative analytical platforms and enhanced detection methodologies are advancing metabolite profiling at the single-cell level, while also highlighting ongoing limitations such as sensitivity, throughput, and data interpretation. The applications enumerated in Table 7.1 give the increasing translational significance of single-cell metabolomics in the realms of disease biomarker identification, medication response analysis, and precision medicine.

Table 7.1 Current Progress in Single-Cell Metabolomics

S. No.	Domain	Key advances	Challenges	Applications
1	Technical Foundations	Single-cell isolation via micromanipulation, LCM, microfluidics	Low metabolite abundance	Capturing rare cells (e.g., circulating tumor cells)
		Minimal-volume extraction and improved lysis protocols	Sample loss and contamination	High-fidelity metabolic signatures
		Strategies to preserve metabolite integrity	Trade-offs in temporal vs. spatial resolution	
2	Analytical Platforms	MS-based: MALDI-MS, SIMS, LC-MS, single-probe MS, nano-DESI	Sensitivity limits	High-resolution metabolic profiling
		Non-MS: NMR, Raman, fluorescence-based methods	Ion suppression	Subcellular and spatial metabolomics
			Calibration and quantification accuracy	
3	Computational Tools & Data Analysis	Data preprocessing & normalization pipelines	False positives in annotation	Systems-level cell heterogeneity
		Dimensionality reduction (PCA, t-SNE, UMAP)	Data sparsity & batch effects	Proteogenomic biomarker discovery
		Metabolite annotation using databases (HMDB, METLIN) & in silico tools (CFM-ID, MetFrag)	Multi-omics alignment across scales	Network-based pathway analysis
		Multi-omics integration (CCA, MOFA, VAEs)		
4	Biological & Clinical Applications	Identifying metabolic subpopulations	Limited throughput vs. scRNA-seq	Cancer diagnostics & precision oncology
		Drug resistance profiling	Reproducibility issues	Personalized medicine
		Immunometabolism (T-cell activation, exhaustion)	Cost and scalability constraints	Environmental & microbial studies
		Stem cell differentiation signatures		
		Host–microbe metabolic interactions		

(continued)

Table 7.1 (continued)

S. No.	Domain	Key advances	Challenges	Applications
5	Future directions	Integration with spatial omics (spatial transcriptomics, proteomics)	Standardization of workflows	Next-generation personalized medicine
		Real-time/live-cell metabolomics (nano-DESI, single-probe MS)	Regulatory validation for clinical use	Large-scale patient stratification
		AI/ML for annotation, prediction, and integration		Mechanistic insights into cellular metabolism
		Clinical translation into diagnostic pipelines		

7.2 Technical Foundations of Single-Cell Metabolomics

7.2.1 Unique Challenges at the Single-Cell Level

Single-cell metabolomics is different from bulk analysis because it has its own set of problems. These include the chance of losing samples, the very small amount of cellular material, and the fact that cellular metabolism changes all the time. To make analytical strategies that work, researcher need to understand these problems (Taylor et al. 2021).

7.2.1.1 Low Metabolite Abundance

Metabolites in the femtomole to attomole range are found in individual cells, frequently below the limit of detection of standard instruments. Due to this low abundance, very sensitive detection, technologies and procedures that limit signal loss and dilution are required (Taylor et al. 2021).

7.2.1.2 Sample Loss and Contamination

Even slight contamination can jeopardize the integrity of the data because of the tiny sample size. Furthermore, if sample processing procedures such cell sorting, lysis, and transfer are not well monitored, they may cause a considerable loss of metabolites (Hu et al. 2023).

7.2.1.3 Temporal Dynamics

The duration of metabolic processes ranges from seconds to minutes. The significance of quick quenching and real-time or nearly real-time analysis is thus highlighted by the fact that delays in sample processing might change the metabolite profile (Moing et al. 2011).

7.2.2 Sampling and Isolation Techniques

Precise and less invasive cell isolation is the first step towards accurate single-cell metabolomics. For this goal, several physical and microfluidic methods have been changed or developed (Zhou et al. 2021).

7.2.2.1 Micromanipulation

Individual cells can be picked and moved one by one under a microscope by hand using very small pipettes. This technique is very precise, even though it is not very fast, and is often used for cells that are sticky or hard to work with, like certain types of cells (Zhou et al. 2021).

7.2.2.2 Laser Capture Microdissection (LCM)

LCM allows researcher to isolate specific cells or sections of tissue slices based on their histological or molecular features. While this technique might need fixation steps that can affect the preservation of metabolites, it is especially useful in spatial metabolomics and when analyzing clinical samples (Espina et al. 2007).

7.2.2.3 Microfluidics and Droplet-Based Systems

Microfluidic technologies allow for high-throughput, automated isolation, and handling of single cells in very small volumes, like nanoliters. Droplet microfluidics also helps reduce sample loss and contamination by enclosing individual cells in droplets, which makes it possible to study them all at the same time (Feng et al. 2019).

7.2.3 Preparation and Lysis of the Cell

To extract the maximum number of metabolites without a drastic alteration in the natural metabolic condition of the cell, it is necessary to prepare the sample adequately (Dettmer et al. 2011).

7.2.3.1 Minimal-Volume Extraction

To avoid dilution, single-cell extraction methods must be effective on very small volumes, between picoliters and nanoliters. Other methods like solid-phase extraction methods and pipetting on the nanoliter level are typically used to reach precise concentrations of the analytes (Yu et al. 2018).

7.2.3.2 In-Situ and Ex-Situ Approaches

Ex-situ techniques entail removal of the cell followed by analysis on an external apparatus, whereas in-situ techniques provide the opportunity to study metabolites in their natural environment in the form of tissue samples or culture plates. Ex-situ methods may provide more flexibility and control, whereas in-situ methods are able to maintain the initial context and place data of the sample (Rubakhin et al. 2013).

7.2.3.3 Measures to Retain Metabolite Integrity

To conserve the natural metabolite profile, individuals tend to resort to enzymatic inhibitors, brief handling periods, and rapid quenching of metabolites by freezing or adding cold solvents. The sensitivity of a particular set of metabolites in most cases is taken into account when preservation techniques are developed (Carling et al. 2020).

7.3 Detection and Analytics

To detect metabolites with high accuracy at the single-cell level, analytical tools with high sensitivity, resolution, and efficiency should be utilized. The most popular techniques are based on mass spectrometry (MS), although non-MS techniques are becoming an alternative or complement in some cases. The present section describes the major technologies applied in single-cell metabolomics, their pros and cons, and the recommendations that are to be adhered to (Hu et al. 2023).

7.3.1 Mass Spectrometry-Based Techniques

Mass spectrometry remains the primary technique in single-cell metabolomics due to its capacity to detect low concentrations of metabolites, its capacity to analyze a large number of metabolites, and its application to a variety of samples. Numerous variations of the mass spectrometry techniques have been or are being improved to apply to the small cell analysis (Hajnajafi and Iqbal 2025).

7.3.1.1 Matrix-Assisted Laser Desorption/Ionization (MALDI-MS)

MALDI-MS allows for a direct, label-free way to study metabolites from cells or tissue samples in a way that shows where they are located. First, the sample is mixed with a matrix that can take in light. Then, a laser that gives off short bursts of light is used to take apart and release the chemicals for analysis. MALDI-MS is very useful in spatial metabolomics because it can image metabolite distributions at single-cell resolution without needing a lot of sample processing. Still, ion suppression, matrix selection, and the limited coverage of low-abundance metabolites are still big problems (Kaufmann 1995).

7.3.1.2 Secondary Ion Mass Spectrometry (SIMS)

Secondary ions are spewed from the sample surface using a concentrated ion beam in SIMS. It is ideal for imaging subcellular metabolites and has submicron spatial resolution. When compared to other MS techniques, SIMS frequently has lesser sensitivity and molecular specificity despite having a high spatial precision. Additionally, the detection of high-mass or fragile metabolites is limited by the hard ionization procedure (McPhail 2006).

7.3.1.3 Liquid Chromatography

Capillary Electrophoresis-MS, including LC-MS and CE-MS, along with MS, are good methods for both focused and general metabolomics studies. CE-MS is particularly well-suited for polar and charged metabolites since it separates metabolites according to size and charge, whereas LC-MS separates metabolites according to hydrophobicity or polarity. Miniaturized for single-cell processes, these methods can achieve high metabolite resolution and sensitivity. To prevent contamination and dilution, integration with single-cell sample handling requires careful optimization (Fanali et al. 2017).

7.3.1.4 Single-Probe MS and Nanospray Desorption Electrospray Ionization (Nano-DESI)

New technologies that do not require much sample preparation, like nano-DESI and Single-Probe MS, provide ionization in situ and in the air. These probes gather analyses, ionize them right away so that MS can find them, and put microdroplets of solvent on the cell surface. These methods are helpful for live-cell or real-time analysis (Yin et al. 2019), but their throughput and standardization are still being worked out.

7.3.2 Non-MS Approaches

Despite the dominance of mass spectrometry, non-MS techniques are becoming more and more relevant because of their distinct benefits, especially in label-based detection, real-time monitoring, and structural elucidation (Juryńczyk et al. 2022).

7.3.2.1 Nuclear Magnetic Resonance (NMR) Spectroscopy

Without requiring ionization or derivatization, NMR offers structural details and precise metabolite quantification. While NMR's sensitivity has historically restricted its use to bulk samples, recent developments in microcoil-based NMR, cryogenically chilled probes, and hyperpolarization have increased its suitability for single-cell or few-cell systems. It is nevertheless less sensitive than platforms that are built on Microsoft (Hatzakis 2019).

7.3.2.2 Fluorescence-Based Methods

Using chemical probes or genetically encoded sensors, fluorescence approaches enable the detection of particular metabolite types. Dynamic, real-time monitoring of cellular metabolism is made possible by these techniques, particularly for reactive species like ROS, ATP, and NADH. Nevertheless, fluorescence methods usually only detect a limited number of metabolites and necessitate prior understanding of the metabolic targets (Ciruela 2008).

7.3.2.3 Raman and Other Spectroscopic Techniques

Label-free metabolite detection using vibrational fingerprints is made possible by Raman spectroscopy, which includes surface-enhanced Raman scattering (SERS). It works well with live-cell analysis and is not overly intrusive. Despite having a high chemical specificity, its sensitivity is frequently lower than that of MS-based methods, requiring augmentation techniques such resonance Raman or nanoparticle conjugation (Shipp et al. 2017).

7.3.3 Sensitivity and Quantification

Ion suppression, technological uncertainty, and very low analyte concentrations make reliable measurement in single-cell metabolomics challenging. Robust calibration techniques and careful experimental design are necessary to meet these problems (Arriola and Hyman 2009).

7.3.3.1 Limits of Detection

The detection limit for many metabolites in single cells lies in the attomole to zeptomole range. Instrument sensitivity, ionization efficiency, and sampling techniques all influence the minimum detectable concentrations. Enhancements such as signal amplification, miniaturized detectors, and optimized ion optics have significantly improved detection limits in recent years (Oikawa and Saito 2012).

7.3.3.2 Use of Internal Standards

Internal standards are critical for normalizing variability in ionization and extraction efficiency. Isotopically labeled standards are often spiked into samples during lysis or extraction. However, in single-cell workflows, the introduction of standards must be carefully balanced to avoid dilution or perturbation of cellular content (Dittrich and Ibáñez 2015).

7.3.3.3 Ion Suppression and Calibration

When coeluting substances obstruct the ionization of target analytes, especially in intricate biological matrices, ion suppression can place. Chromatographic separation, ion source parameter optimization, and matrix-free ionization in methods such as DESI are methods to lessen ion suppression. Although they must be modified for low-volume, high-variability single-cell systems, calibration curves produced from established standards are still necessary for absolute quantification (Furey et al. 2013).

7.4 Computational Tools and Data Analysis

The effectiveness of single-cell metabolomics depends on reliable data processing and analysis pipelines in addition to sensitive detection techniques. Computational methods are essential for deriving significant biological insights from single-cell data due to its high dimensionality, intrinsic low signal-to-noise ratios, and considerable unpredictability. This section describes the major steps of the data analysis procedure, such as synthesizing the various categories of biological data, distinguishing particular metabolites, generating graphic representation of the data, and preparing the data to undergo additional examination, which is discussed by (Hu et al. 2023).

7.4.1 Data Preprocessing and Normalization

The raw data of individual cell metabolomics research is mass spectra or chromatograms. To be accurate and be compared with other samples, these data have to undergo multiple preprocessing operations. Some typical steps include baseline correction, peak detection, alignment, and noise filtering. Normalization is particularly challenging in single-cell analysis because there is significant variation between cells in terms of metabolite levels, total ion counts, and cell size. Common approaches used for normalization include:

- Total ion current (TIC) normalization.
- Probabilistic quotient normalization (PQN).
- Use of internal standards.
- Quantile normalization.

To account for physiological variations among cells, advanced techniques incorporate cell-specific metadata, including volume and protein content. Correct normalization is very important because it helps find physiologically important patterns and cuts down on technical noise (Misra 2020).

7.4.2 Dimensionality Reduction and Visualization

Single-cell metabolomics creates high-dimensional datasets that often have hundreds or thousands of traits for each cell. Dimensionality reduction techniques are used to show and group cells based on their metabolic profiles, which makes it easier to understand (Ivosev et al. 2008).

7.4.2.1 Principal Component Analysis (PCA)

A linear method that reduces dimensionality by projecting data onto directions of greatest variance. PCA is often utilized for exploratory research and quality assurance (Nemes et al. 2012).

7.4.2.2 t-distributed Stochastic Neighbor Embedding (t-SNE)

A nonlinear technique that preserves local structure and aids in demonstrating cellular metabolic variability and subpopulations (Cieslak et al. 2020).

7.4.2.3 Uniform Manifold Approximation and Projection (UMAP)

UMAP is an alternative to t-SNE that is becoming more and more popular for visualizing biological data (Healy and McInnes 2024) because it can be scaled up or down and understood. It keeps both local and global structures. These methods help us see how things change over time, find metabolic subtypes, and link metabolism to phenotypic traits.

7.4.3 Metabolite Annotation and Identification

Accurately identifying observed characteristics is a major problem in single-cell metabolomics. Single-cell annotation requires sophisticated methodologies and high-quality reference libraries due to the limited availability of spectrum data and the presence of isobaric compounds. Some important methods are:

7.4.3.1 Mass-Based Matching

Contrasting measured m/z values with those found in reference databases like METLIN or HMDB.

7.4.3.2 MS/MS Fragmentation Matching

By contrasting experimental spectra with theoretical or library spectra, specificity is increased.

7.4.3.3 In Silico Prediction Tools

To help in presumptive identification, tools such as CFM-ID and MetFrag can forecast fragmentation trends.

False positives are still a problem, particularly when low-intensity signals are used. Identification reliability is frequently reported using annotation confidence levels, such as the MSI (Metabolomics Standards Initiative) scale (Heinemann and Zenobi 2011).

7.4.4 Integration with Transcriptomics and Proteomics (Multiomics)

Metabolomics data are increasingly being combined with transcriptome and proteomic data to obtain a comprehensive picture of cellular function. Researchers can connect metabolic output, protein abundance, and gene expression at the single-cell level using multiomics techniques (Kumar et al. 2016; Kaur et al. 2024). Key integration strategies include:

7.4.4.1 Correlation-Based Methods

Direct pairwise correlation between metabolite levels and gene/protein expression (Michalak and Kwasnicka 2010).

7.4.4.2 Network-Based Approaches

Construction of metabolic or regulatory networks to understand pathway interactions (Barabási et al. 2011).

7.4.4.3 Machine Learning Models

Algorithms such as canonical correlation analysis (CCA), multiomics factor analysis (MOFA), and variational autoencoders are used to uncover latent variables linking omics layers (Varoquaux and Colliot 2023).

Challenges in multiomics integration include data sparsity, differences in scale and dynamic range, and aligning measurements from different platforms or time points.

7.4.5 Databases and Software Tools

A burgeoning ecosystem of databases and computational tools has been established or modified to support single-cell metabolomics research, addressing the distinct problems of low-abundance metabolite identification, high-dimensional data integration, and cellular heterogeneity. These instruments facilitate multiple phases of the analytical workflow, encompassing data preprocessing, metabolite identification, statistical analysis, and multiomics integration.

Fig. 7.1 Key computational tools and databases relevant to single-cell metabolomics

The widespread implementation of single-cell metabolomics in biomedical and therapeutic applications will rely on the creation of scalable, interoperable, and user-friendly computational systems (Misra 2019). Figure 7.1 summarizes the essential resources presently utilized in the field.

A schematic representation of principal software platforms and databases classified by function, from raw data processing to multiomics integration.

The broader use of single-cell metabolomics in research and therapeutic contexts may depend on the creation of scalable and user-friendly computational tools (Misra 2019).

7.5 Applications in Biology and Medicine

By allowing the direct measurement of metabolic characteristics at the level of individual cells, single-cell metabolomics has created new opportunities in biomedical research. This skill is revolutionizing our knowledge of microbial ecology, immunological response, cellular differentiation, and disease causes. Below, we go over some important uses for single-cell metabolomics that have started to have a significant influence in immunology, developmental biology, cancer research, and microbiology (Katam et al. 2022).

7.5.1 Cancer Metabolism and Tumor Heterogeneity

7.5.1.1 Identifying Metabolic Subpopulations

The physiologically varied cell populations that make up tumors aid in their growth, invasion, and resistance to treatment. Different metabolic subpopulations within malignancies that would normally be obscured in bulk analysis can be found because of single-cell metabolomics. For example, spatially defined metabolic zones that correlate to hypoxic, proliferative, and invasive niches within tumors have been identified using MALDI-MS imaging. It is possible to categorize tumor aggressiveness and forecast clinical outcomes using these metabolic "fingerprints" (DeGennaro et al. 2016).

7.5.1.2 Drug Resistance Profiling

One characteristic of cancer cells that are resistant to drugs is metabolic reprogramming. Researchers can find uncommon subpopulations with changed metabolic pathways—like elevated glycolysis, lipid metabolism, or antioxidant activity—that confer resistance to chemotherapy or targeted medicines by using single-cell metabolomics. Researchers can monitor metabolic changes in response to therapy and spot possible metabolic weaknesses for combination treatment approaches by comparing the metabolic profiles of cancer cells before and after treatment (Chen et al. 2022).

7.5.2 Stem Cell and Developmental Biology

7.5.2.1 Metabolic Signatures of Differentiation

The destiny of stem cells is intimately related to metabolic status. Differentiated cells move toward oxidative phosphorylation, whereas pluripotent stem cells usually rely on glycolysis. The resolution required to record these metabolic shifts during lineage specification is offered by single-cell metabolomics. Researchers can identify the metabolic checkpoints that control differentiation by comparing metabolite profiles with transcriptional states. This helps them create ideal culture conditions for regenerative medicine (Intlekofer and Finley 2019).

7.5.2.2 Embryonic Development Studies

Rapid cell division, migration, and differentiation occur throughout the early stages of embryonic development, and these processes are accompanied by dynamic changes in metabolism. This temporal and spatial heterogeneity is not captured by conventional approaches. Metabolite gradients in early embryos have been characterized by recent single-cell MS imaging investigations, providing insights into how localized metabolic environments affect decisions about cell fate. Additionally, this method is being applied to 3D developmental models and organoids (Pedersen 1994).

7.5.3 Immunometabolism

Immunometabolism denotes the dynamic interaction between immune cell activity and cellular metabolism. It examines the influence of metabolic pathways on immune cell behavior, destiny, and functionality, especially in relation to health, illness, and therapeutic scenarios. This section emphasizes two important elements:

7.5.3.1 Single-Cell Metabolic Phenotype in Immune Cells

The linkage between the way in which cells remodel their metabolism and immune cell behavior is unmistakable. As an example, active macrophages and T cells consume energy differently, depending on the type. Single-cell metabolomics assists in placing the immune cells into categories according to their metabolic characteristics, without referring to standard markers located on the cell surface. This is particularly effective to locate various immune cell types in complex settings, including within tumors or in long-term inflammatory regions, as demonstrated in one study (Hartmann et al. 2021).

7.5.3.2 Metabolism during T Cell Activation and Exhaustion

T-cell activation is linked to a switch in energy generation between oxidative phosphorylation to aerobic glycolysis. Conversely, lipid metabolic alterations and impaired mitochondrial processes are related to T cell exhaustion which is a typical feature in cancer and chronic infections. The analysis of these changes in the individual cells allows scientists to comprehend better the metabolic background of the immune system failure and devise ways to restore immune function, e.g. metabolic checkpoint inhibitors or personalized nutritional sustenance (Franco et al. 2020).

7.5.4 Microbial and Environmental Applications

7.5.4.1 Metabolism in Uncultivable Microbes

The vast majority of micro-organisms in the natural environment cannot be grown using standard laboratory methods. Single-cell metabolomics is a unique chance to examine the metabolism of these elusive microbes in their natural ecosystems. Instruments such as nanoSIMS and live-cell MS can be used by scientists to investigate microbial communities and environmental chemical processes. Such techniques assist in the realization of the way microbes access nutrients and decompose organic matter (Ng 2019).

7.5.4.2 Host–Microbe Interactions

To comprehend the effects of the microbiome on health and disease, it is essential to realize how single microbes respond to the tissues of the body. Single-cell metabolomics assists in demonstrating the impact of microbial chemicals on the activity of host cells and the effect of host cells on the microbial chemicals. Investigations based on single-cell technologies have revealed that gut cells that are subjected to chemicals of the bacteria are able to alter their own chemical activity, which is pertinent to the work of the immune system and the efficiency of the barrier of the gut. This constant interplay between microbes and the host contributes significantly to such diseases as cancer, metabolic syndrome, inflammatory bowel disease (Soares and Weiss 2015).

Concisely, single-cell metabolomics represents a novel method in examining cellular energy and chemical utilization within the complex organisms. This method can be of great assistance to various disciplines such as environmental microbiology, study of immune system, and tailor-made cancer treatments by learning how various cells manage their metabolism. With the improvement of the instruments used to measure these changes, and with the growth of the power of computers, the utility and impact of these techniques may multiply rapidly.

7.6 Existing Challenges and Limitations

In spite of certain advances, a lot of conceptual and technical issues remain to be overcome in single-cell metabolomics. These restrictions still make it impossible to use it in clinical and research settings on a regular basis. In order to popularize this technology and better understand biologically, it is desirable to detect and address these issues (Ali et al. 2019).

7.6.1 Standardization and Reproducibility

The largest problem with reproducibility in single-cell metabolomics is that there are no common methods. Various labs follow different methods in data processing, instrumentation, cell lysis and cell separation, contributing to the incomparability of their results. Single-cell metabolomics does not have common practices and benchmarks as transcriptomics, which has had well-known platforms such as 10× Genomics (Israr et al. 2024). There is not so much consensus on the most excellent techniques to employ and rarely do laboratories compare their findings directly.

7.6.2 Metabolite Identification Bottlenecks

One of the biggest challenges is still accurately identifying metabolites. Metabolites lack a regular structure, and it might be challenging to resolve isobaric or isomeric molecules, in contrast to nucleic acid sequences. Because of the intricacy of fragmentation spectra at low concentrations and the restricted coverage of current databases, many identified characteristics are still unannotated. To get around this obstacle, advancements in spectral libraries, high-resolution MS/MS, and in silico prediction methods are crucial.

7.6.3 Spatial and Temporal Resolution Tradeoffs

Chemical, temporal, and geographical resolution are frequently compromised in single-cell metabolomics. Lower metabolite coverage or quantification accuracy may be provided by methods like MALDI or SIMS that maintain spatial context. On the other hand, cell lysis is usually necessary for extremely sensitive techniques like LC-MS or CE-MS, which remove spatial information. Additionally, the majority of methods only take a static picture of metabolism, failing to capture the dynamic changes that take place in real time.

7.6.4 Cost, Scalability, and Throughput

The scalability of single-cell metabolomics is limited by its expense and intricacy. The majority of metabolomics workflows are labor-intensive and low-throughput, in contrast to single-cell transcriptomics, which can handle thousands of cells each run. Expensive equipment, knowledgeable operators, and labor-intensive sample preparation all add to the final cost. Miniaturized workflows, automation, and microfluidic integration are required to increase accessibility and throughput.

7.7 Future Directions and Emerging Trends

Ongoing efforts are concentrated on technological innovation, integration with other omics platforms, and translational applications in order to fully exploit the potential of single-cell metabolomics (Ali et al. 2022).

7.7.1 Integration with Spatial Omics

Integrating single-cell metabolomics with spatial omics technologies, such as spatial transcriptomics and proteomics, is a significant frontier. The microenvironmental setting that influences cellular function or malfunction can be revealed with the use of spatially resolved metabolic data. Spatial transcriptomics is being used in conjunction with methods like ambient ionization MS and MALDI imaging to map molecular and metabolic characteristics within tissues.

7.7.2 Real-Time and Live-Cell Metabolomics

The goal of real-time metabolomics is to track dynamic metabolic processes in living cells as they take place. Methods such as live-cell nano-DESI and single-probe MS are being developed to allow for minimally perturbed time-resolved analysis. These developments may provide insight into rapidly occurring processes like metabolic oscillations, cell division, and signaling responses.

7.7.3 AI and Machine Learning for Metabolomics

High-dimensional, sparse, and noisy single-cell metabolomics data are increasingly being analyzed using artificial intelligence (AI) and machine learning (ML). Among the applications are:

- Feature selection and dimensionality reduction
- Metabolite annotation from complex spectra
- Prediction of cell states and phenotypes
- Multiomics data integration

By improving the metabolomics research' interpretability and prediction ability, these tools can help identify new biomarkers and pathways.

7.7.4 Clinical Translation and Personalized Medicine

Single-cell metabolomics could be very helpful for precision medicine. In oncology and immunology specifically, the metabolic profile of patient-derived cells can assist in tailoring individualized treatments, monitoring therapeutic responses, and aiding in diagnosis. For clinical use, though, diagnostic platforms need to be approved by regulators, tested on large groups of people, and have procedures that are reliable and repeatable.

7.8 Conclusion

Single-cell metabolomics is a groundbreaking advancement in systems biology that enables scientists to examine cellular metabolism with unprecedented precision. Despite the numerous technological challenges that remain, particularly regarding standardization, sensitivity, and data interpretation, recent progress has established a robust foundation for forthcoming innovations. The next step in development is expected to come from combining spatial and multiomics platforms, as well as advances in computational modeling and real-time analysis. To fully realize the potential of single-cell metabolomics, interdisciplinary cooperation spanning analytical chemistry, cell biology, data science, and clinical research will be essential as the area develops. In addition to advancing basic science, this convergence has the potential to change personalized medicine's methods of diagnosis and treatment.

References

Ali A, Abouleila Y, Shimizu Y, Hiyama E, Emara S, Mashaghi A, Hankemeier T (2019) Single-cell metabolomics by mass spectrometry: advances, challenges, and future applications. TrAC Trends Anal Chem 120:115436

Ali A, Davidson S, Fraenkel E, Gilmore I, Hankemeier T, Kirwan JA, Lane AN, Lanekoff I, Larion M, McCall LI, Murphy M (2022) Single cell metabolism: current and future trends. Metabolomics 18(10):77

Arriola L, Hyman JM (2009) Sensitivity analysis for uncertainty quantification in mathematical models. In: Mathematical and statistical estimation approaches in epidemiology. Springer Netherlands, Dordrecht, pp 195–247

Barabási AL, Gulbahce N, Loscalzo J (2011) Network medicine: a network-based approach to human disease. Nat Rev Genet 12(1):56–68

Carling RS, McDonald BA, Austin D, Burden D, Correia J, Leung J, Mayers B, John C (2020) Challenging the status quo: a comparison of ion exchange chromatography with liquid chromatography-mass spectrometry and liquid chromatography-tandem mass spectrometry methods for the measurement of amino acids in human plasma. Ann Clin Biochem 57(4):277–290. https://doi.org/10.1177/0004563220933303. PMID: 32438818

Chen X, Sun M, Yang Z (2022) Single cell mass spectrometry analysis of drug-resistant cancer cells: metabolomics studies of synergetic effect of combinational treatment. Anal Chim Acta 1201:339621

Cieslak MC, Castelfranco AM, Roncalli V, Lenz PH, Hartline DK (2020) T-distributed stochastic neighbor embedding (t-SNE): a tool for eco-physiological transcriptomic analysis. Mar Genomics 51:100723

Ciruela F (2008) Fluorescence-based methods in the study of protein–protein interactions in living cells. Curr Opin Biotechnol 19(4):338–343

DeGennaro CM, Savir Y, Springer M (2016) Identifying metabolic subpopulations from population level mass spectrometry. PLoS One 11(3):e0151659

Dettmer K, Nürnberger N, Kaspar H, Gruber MA, Almstetter MF, Oefner PJ (2011) Metabolite extraction from adherently growing mammalian cells for metabolomics studies: optimization of harvesting and extraction protocols. Anal Bioanal Chem 399(3):1127–1139. https://doi.org/10.1007/s00216-010-4425-x. Epub 2010 Dec 2. PMID: 21125262

Dittrich P, Ibáñez AJ (2015) Analysis of metabolites in single cells—what is the best micro-platform? Electrophoresis 36(18):2196–2206

Espina V, Heiby M, Pierobon M, Liotta LA (2007) Laser capture microdissection technology. Expert Rev Mol Diagn 7(5):647–657

Fanali S, Haddad PR, Poole C, Riekkola ML (eds) (2017) Liquid chromatography: applications. Elsevier

Feng H, Zheng T, Li M, Wu J, Ji H, Zhang J, Zhao W, Guo J (2019) Droplet-based microfluidics systems in biomedical applications. Electrophoresis 40(11):1580–1590

Franco F, Jaccard A, Romero P, Yu YR, Ho PC (2020) Metabolic and epigenetic regulation of T-cell exhaustion. Nat Metab 2(10):1001–1012

Furey A, Moriarty M, Bane V, Kinsella B, Lehane M (2013) Ion suppression; a critical review on causes, evaluation, prevention and applications. Talanta 115:104–122

Hajnajafi K, Iqbal MA (2025) Mass-spectrometry based metabolomics: an overview of workflows, strategies, data analysis and applications. Proteome Sci 23(1):5. https://doi.org/10.1186/s12953-025-00241-8. PMID: 40420110; PMCID: PMC12105183

Hartmann FJ, Mrdjen D, McCaffrey E, Glass DR, Greenwald NF, Bharadwaj A, Khair Z, Verberk SG, Baranski A, Baskar R, Graf W (2021) Single-cell metabolic profiling of human cytotoxic T cells. Nat Biotechnol 39(2):186–197

Hatzakis E (2019) Nuclear magnetic resonance (NMR) spectroscopy in food science: a comprehensive review. Compr Rev Food Sci Food Saf 18(1):189–220

Healy J, McInnes L (2024) Uniform manifold approximation and projection. Nat Rev Methods Primers 4(1):82

Heinemann M, Zenobi R (2011) Single cell metabolomics. Curr Opin Biotechnol 22(1):26–31

Hu R, Li Y, Yang Y, Liu M (2023) Mass spectrometry-based strategies for single-cell metabolomics. Mass Spectrom Rev 42(1):67–94. https://doi.org/10.1002/mas.21704. Epub 2021 May 24. PMID: 34028064

Ibáñez AJ, Fagerer SR, Schmidt AM, Urban PL, Jefimovs K, Geiger P, Dechant R, Heinemann M, Zenobi R (2013) Mass spectrometry-based metabolomics of single yeast cells. Proc Natl Acad Sci USA 110(22):8790–8794. https://doi.org/10.1073/pnas.1209302110. Epub 2013 May 13. PMID: 23671112; PMCID: PMC3670324

Intlekofer AM, Finley LW (2019) Metabolic signatures of cancer cells and stem cells. Nat Metab 1(2):177–188

Israr J, Alam S, Siddiqui S, Misra S, Gupta D, Kumar A (2024) Cell-free systems for genomics. In: Singh V (ed) Advances in genomics. Springer, Singapore. https://doi.org/10.1007/978-981-97-3169-5_17

Ivosev G, Burton L, Bonner R (2008) Dimensionality reduction and visualization in principal component analysis. Anal Chem 80(13):4933–4944

Juryńczyk M, Jakuszyk P, Kurkowska-Jastrzębska I, Palace J (2022) Increasing role of imaging in differentiating MS from non-MS and defining indeterminate borderline cases. Neurol Neurochir Pol 56(3):210–219

Katam R, Lin C, Grant K, Katam CS, Chen S (2022) Advances in plant metabolomics and its applications in stress and single-cell biology. Int J Mol Sci 23(13):6985

Kaufmann R (1995) Matrix-assisted laser desorption ionization (MALDI) mass spectrometry: a novel analytical tool in molecular biology and biotechnology. J Biotechnol 41(2–3):155–175

Kaur R, Mandal D, Potdar V, Gosavi MS, Kumar A (2024) Multi-omics and its clinical application. In: Mani I, Singh V (eds) Multi-omics analysis of the human microbiome. Springer, Singapore. https://doi.org/10.1007/978-981-97-1844-3_14

Kumar D, Bansal G, Narang A, Basak T, Abbas T, Dash D (2016) Integrating transcriptome and proteome profiling: strategies and applications. Proteomics 16(19):2533–2544

Kumar A, Israr J, Mishra HP, Singh I (2024a) Cell-free systems in metabolomics. In: Singh V (ed) Advances in metabolomics. Springer, Singapore. https://doi.org/10.1007/978-981-97-7459-3_11

Kumar A, Israr J, Singh I (2024b) Design of biosynthetic pathway for production of metabolites. In: Singh V (ed) Advances in metabolomics. Springer, Singapore. https://doi.org/10.1007/978-981-97-7459-3_3

Li Z, Wang H, Yang R, Jin X, Han Q, She Z, Ge P (2025) Plasma-Derived Exosomal tRF-Phe-GAA-001 and tRF-Gly-GCC-037 as Novel Diagnostic Biomarkers for Cervical Cancer. Indian J Clin Biochem 40(4):683–690. https://doi.org/10.1007/s12291-024-01235-7. Epub 2024 May 25. PMID: 40937397; PMCID: PMC12420557

McPhail DS (2006) Applications of secondary ion mass spectrometry (SIMS) in materials science. J Mater Sci 41(3):873–903

Michalak K, Kwasnicka H (2010) Correlation based feature selection method. Int J Bio-Inspired Comput 2(5):319–332

Misra BB (2019) Open-source software tools, databases, and resources for single-cell and single-cell-type metabolomics. In: Single cell metabolism: methods and protocols. Springer New York, New York, pp 191–217

Misra BB (2020) Data normalization strategies in metabolomics: current challenges, approaches, and tools. Eur J Mass Spectrom 26(3):165–174

Moing A, Aharoni A, Biais B, Rogachev I, Meir S, Brodsky L, Allwood JW, Erban A, Dunn WB, Kay L, de Koning S (2011) Extensive metabolic cross-talk in melon fruit revealed by spatial and developmental combinatorial metabolomics. New Phytol 190(3):683–696

Nemes P, Knolhoff AM, Rubakhin SS, Sweedler JV (2012) Single-cell metabolomics: changes in the metabolome of freshly isolated and cultured neurons. ACS Chem Neurosci 3(10):782–792

Ng W (2019) Cultivating the uncultivable environmental microbe using insights from whole genome sequence that define a growth medium. PeerJ PrePrints

Oikawa A, Saito K (2012) Metabolite analyses of single cells. Plant J 70(1):30–38

Owens GK, Kumar MS, Wamhoff BR (2004) Molecular regulation of vascular smooth muscle cell differentiation in development and disease. Physiol Rev 84(3):767–801. https://doi.org/10.1152/physrev.00041.2003. PMID: 15269336

Pedersen RA (1994) Studies of in vitro differentiation with embryonic stem cells. Reprod Fertil Dev 6(5):543–552

Rappez L, Stadler M, Triana S, Gathungu RM, Ovchinnikova K, Phapale P, Heikenwalder M, Alexandrov T (2021) SpaceM reveals metabolic states of single cells. Nat Methods 18(7):799–805

Rubakhin SS, Lanni EJ, Sweedler JV (2013) Progress toward single cell metabolomics. Curr Opin Biotechnol 24(1):95–104

Shipp DW, Sinjab F, Notingher I (2017) Raman spectroscopy: techniques and applications in the life sciences. Adv Opt Photon 9(2):315–428

Soares MP, Weiss G (2015) The iron age of host–microbe interactions. EMBO Rep 16(11):1482–1500

Taylor MJ, Lukowski JK, Anderton CR (2021) Spatially resolved mass spectrometry at the single cell: recent innovations in proteomics and metabolomics. J Am Soc Mass Spectrom 32(4):872–894

Varoquaux G, Colliot O (2023) Evaluating machine learning models and their diagnostic value. In: Machine learning for brain disorders, pp 601–630

Yin R, Burnum-Johnson KE, Sun X, Dey SK, Laskin J (2019) High spatial resolution imaging of biological tissues using nanospray desorption electrospray ionization mass spectrometry. Nat Protoc 14(12):3445–3470

Yu D, Zhou L, Xuan Q, Wang L, Zhao X, Lu X, Xu G (2018) Strategy for comprehensive identification of Acylcarnitines based on liquid chromatography-high-resolution mass spectrometry. Anal Chem 90(9):5712–5718. https://doi.org/10.1021/acs.analchem.7b05471. Epub 2018 Apr 20

Zhou WM, Yan YY, Guo QR, Ji H, Wang H, Xu TT, Makabel B, Pilarsky C, He G, Yu XY, Zhang JY (2021) Microfluidics applications for high-throughput single cell sequencing. J Nanobiotechnology 19(1):312. https://doi.org/10.1186/s12951-021-01045-6. PMID: 34635104; PMCID: PMC8507141

Microfluidics for Generation and Analysis of Single Cell

8

Tesfaye Kassaw Bedru, Beteley Tekola Meshesha, Shegaw Ahmed Mohammed, Abayneh Getachew Demesa, and Mani Jayakumar

Abstract

Single-cell analysis has revolutionized our understanding of cellular heterogeneity, disease mechanism, and personalized medicine. Traditional bulk assays are bound to mask minute but critical differences between cells, necessitating the need for tools that single out, manipulate, and read out cells with single-cell resolution. Microfluidics is now an unparalleled platform for single-cell research, with unheralded control over the microenvironment of cells, reagent consumption, and high-throughput manipulation. This chapter provides a general overview of microfluidic methods in single-cell manufacturing, encapsulation,

T. K. Bedru
School of Chemical and Bio Engineering, College of Technology and Built Environment, Addis Ababa, Ethiopia

School of Chemical and Mechanical Engineering, Kombolcha Institute of Technology, Wollo University, Kombolcha, Ethiopia

LUT School of Engineering Sciences, LUT University, Kouvola, Finland

B. T. Meshesha
School of Chemical and Bio Engineering, College of Technology and Built Environment, Addis Ababa, Ethiopia

Africa Center of Excellence for Water Management, Addis Ababa University, Addis Ababa, Ethiopia

S. A. Mohammed
School of Chemical and Bio Engineering, College of Technology and Built Environment, Addis Ababa, Ethiopia

A. G. Demesa
LUT School of Engineering Sciences, LUT University, Kouvola, Finland

M. Jayakumar (✉)
Department of Biotechnology, Faculty of Engineering, Karpagam Academy of Higher Education, Coimbatore, Tamil Nadu, India

Centre for Natural Products and Functional Foods, Karpagam Academy of Higher Education, Coimbatore, Tamil Nadu, India

V. Singh (ed.), *Advances in Single Cell Analysis*,
https://doi.org/10.1007/978-981-95-9789-5_8

isolation, and analysis. Focus is on droplet-based, valve-based, and inertial microfluidic systems, each of which has been optimized to address specific biological and biomedical challenges. Furthermore, we explore the integration of microfluidics with downstream analytical devices such as single-cell RNA sequencing, proteomics, and real-time imaging. We also survey recent advances in on-chip cell culture, viability screening, and intracellular sensing. Throughput, cell viability, and integration with omics technologies are examined critically as the challenges. Finally, we highlight upcoming microfluidics-enabled single-cell analysis trends, including organ-on-chip platforms, artificial intelligence collaboration, and clinical applications in diagnostics and therapeutics.

Keywords

Microfluidics · Single-cell analysis · Droplet microfluidics · Cell encapsulation · Cell sorting

8.1 Introduction

Single-cell analysis is a revolution in biomedical science that offers vital insights into cellular heterogeneity, disease mechanisms, and personalized therapy. Traditional bulk assays average signals across large numbers of cells, masking rare but biologically important cell activities (such as stem-like cancer cells or immune microenvironments). This limitation spurred the development of analytical tools that can unlock biological processes at single-cell resolution, essential to defining intricate cellular diversity in tissues and diseases (Han et al. 2023; Zhou et al. 2021).

Microfluidics has emerged as an enabling platform for single-cell analysis by overcoming the limitations of bulk methods with precise control over microenvironments, minimal reagent usage, and scalability of throughput. Its emergence represents the integration of engineering, physics, and life sciences with unprecedented performance in cell manipulation, assay integration, and detection. In particular, microfluidic devices operate at microliter to picoliter scale volumes, similar to the size of the cells themselves and thus enable massively parallel, high-sensitivity assays with little sample loss (Zhou et al. 2021).

Among the microfluidic techniques, droplet-based microfluidics is the most widely used technique. In this arrangement, individual cells are encapsulated in submicrometer-diameter droplets that act as isolated microreactors, which allow for precise control, incubation, and analysis of single cells. The technology can produce thousands to millions of uniform-sized droplets per minute, making high-throughput single-cell assays possible even in low-resource labs (Li et al. 2023a; Liu et al. 2022). These droplets are not only applicable to a variety of applications but also include genetic analysis, drug screening, protein assays, and multiplexed functional testing (Liu et al. 2022).

Droplet microfluidics integrated with the downstream omics platforms. For instance, droplet-based single-cell RNA sequencing (scRNA-seq) like the

VASA-seq method have pushed transcriptomic coverage to whole-length transcripts, alternative splicing, and noncoding RNAs in thousands of cells in one run ("Profiling the Total Single-Cell Transciptome Using Droplet Microfluidics" 2022). Microfluidic modules for cell lysis, barcoding, and molecular partitioning also reduce technical noise and accelerate library preparation, enabling high-fidelity sequencing to support precision medicine targets (Han et al. 2023; Zhou et al. 2021).

In addition to droplets, valve-based and hydrodynamic microfluidics offer alternative methods. Valve-based systems offer reagent flow control and cell location dynamically, enabling on-chip combinatorial screening and time-resolved stimulation. Hydrodynamic traps can trap single cells to longitudinally observe cell state or behavior under highly controlled environmental conditions (Maurya et al. 2022).

Collectively, these microfluidic platforms address the urgent demand for devices that are capable of measuring single-cell phenotypes, genotypes, and functional states to unravel heterogeneity and dynamic responses. They reduce reagent usage, increase throughput and efficiency, and are readily compatible with optical, sequencing, and analytical downstream approaches, rendering systematic single-cell studies increasingly prevalent in research and clinical applications (Kumari et al. 2023).

Despite significant progress, challenges remain. Scaling up microfluidic platforms from research models to reliable commercial systems depends on further developments in material engineering, user-friendly design, and integration into standard laboratory practices. Maintaining cell viability, precise control over microenvironmental conditions, and reproducibility across platforms are also important for clinical application (Wu et al. 2018).

This chapter gives an overview of microfluidic technologies for single-cell analysis. The chapter begins with the building blocks of fluid dynamics at the microscale and device fabrication, followed by the description of prominent platforms such as droplet-based, valve-controlled, and inertial systems, used for precise single-cell encapsulation and sorting. On-chip cell culture methods and cell viability measurements are described, followed by integration of microfluidics with single-cell omics, optical imaging, and intracellular sensing. Immune profiling, cancer heterogeneity, and drug screening case studies show their clinical relevance. In addition concludes with immediate challenges and future directions, including biocompatibility, throughput, data integration, organ-on-chip technologies, and clinical applications.

8.2 Microfluidic Principles Relevant to Single-Cell Applications

8.2.1 Fluid Dynamics at the Microscale

In microfluidic devices, the characteristic dimensions, namely channel heights/widths, lie in the range of a few to a few hundred micrometers. At these length scales and with aqueous buffers of moderate viscosity, viscous forces overwhelm inertial forces, driving flows that are steady and predictable. This regime is quantified by the Reynolds number, $Re = \rho UL/\mu$, where density ρ, velocity U, length scale

L, and dynamic viscosity μ together determine whether a flow is laminar or turbulent. For most cell-handling devices, Re ≤ 1–100, so laminar flow dominates, allowing deterministic advection, stable interfaces between co-flowing streams, and diffusion-limited mixing, features underpinning single-cell encapsulation, on-chip reactions, and precise dosing. Understanding and exploiting low-Re behavior is at the foundation of microfluidic control over single cells, from deterministic positioning to metered reagent exchange and shear-aware culture environments. Recent methodological primers and overviews reaffirm these basics and show how designers use scaling analyses to select channel dimensions and flow rates that will maintain cell viability while achieving desired residence times and shear profiles (Ferreira et al. 2024).

At very low Reynolds numbers typical of microfluidic flows, mixing is governed by diffusion, making homogenization an intrinsically slow process. To overcome this challenge, passive mixers are widely exploited in microfluidic systems, such as the geometrical arrangement of staggered herringbone structures that increases mixing performance by inducing chaotic advection without any external power supply. Other approaches involve active mixing strategies based on electrokinetic actuation and acoustic or pressure pulsation to achieve controlled mixing dynamics where required. Such hybrid strategies merge simplicity in design with tenability for any particular assay requirement, from single-cell encapsulation to reagent delivery (Ward and Fan 2015).

By contrast, for inertial single-cell manipulation, microchannel are designed using intermediate Reynolds number regimes approximately between 10 and 200 where inertial lift forces and Dean drag forces balance to focus cells by size, shape, or deformability. In curved or spiral micro channel, this interplay establishes a set of stable streamlines for high-throughput, label-free focusing, and sorting of cells, hence making inertial microfluidics a powerful tool in single-cell preparation pipelines (Peng et al. 2024; Zhao et al. 2020).

8.2.2 Surface Tension, Capillarity, and Laminar Flow Behavior

Interfacial physics is central to microfluidics. Surface tension and wettability (via contact angle) critically control droplet formation, stability, and break-up, but they also enable capillary-driven, pump-free flow in passive microfluidic devices. Droplet microfluidics involves shearing two immiscible phases, such as water-in-oil at T-junctions or flow-focusing orifices. The balance between viscous shear forces and interfacial tension-quantified via capillary number $Ca = \mu U/\gamma$, where μ represents viscosity, U is velocity, and γ is surface tension-dictates droplet size and its monodispersity. To perform consistent single-cell encapsulation for further assays and splitting/merging with high precision in complex workflows, mastering these parameters is required. Recent comprehensive reviews systematically outline the design principles for achieving stable, high-throughput droplet generation that supports single-cell applications (Fergola et al. 2025).

In capillary-driven systems, flow arises from curvature-induced Laplace pressures in the absence of external pumps. These are useful for portable diagnostics and long-term culture modules that would greatly benefit from low shear. Current research maps how surface tension, channel geometry, and wettability, including paper and porous substrates-control start/stop valves, flow rates, and timing networks to simplify fluid handling for on-chip assays and to enable cost-effective platforms (Jang et al. 2021; Khanjani et al. 2025; Wang et al. 2021).

These interfacial controls coexist with the low-Re, laminar baseline of microflows. Designers utilize laminar co-flows to meter reagents to single cells, create stable chemical gradients for chemotaxis or drug response, and confine shear stress within narrow bounds, thereby maintaining cell viability. Up-to-date engineering research emphasize that careful selection of Re, Ca, and contact angles yields predictable droplet sizes, focusing positions, and residence times, key to robust single-cell analytics (Ferreira et al. 2024).

8.2.3 Material Selection and Fabrication Techniques

Material selection and fabrication technique play major roles in the design and performance of single-cell microfluidic devices, impacting everything from biocompatibility and optical clarity, through molecular sorption and gas permeability, to manufacturability and cost. Polydimethylsiloxane (PDMS) is often favored for rapid prototyping via soft lithography due to its excellent optical transparency and gas permeability, which are beneficial for cell oxygenation. However, it also suffers from several known drawbacks: it absorbs hydrophobic molecules and various proteins, leading to assay bias. Recent strategies such as vapor-deposited Nano adhesive layers have been shown to greatly reduce protein adsorption while preserving the ease of PDMS fabrication and sealing properties (You et al. 2020).

In contrast, the value of glass and silicon lies in their chemical inertness, low auto-fluorescence, and stable surface chemistry, which renders them particularly well-suited for applications that involve the use of strong solvents, electrophoresis, or long-term imaging. These substrates remain central when analytic rigor is required, despite often higher fabrication costs (Aralekallu et al. 2023).

More recently, thermoplastics such as Polymethyl methacrylate (PMMA), Cyclic Olefin Copolymer/Cyclic Olefin Polymer (COC/COP), Polycarbonate (PC), and Polystyrene (PS) have become popular owing to their compatibility with high-volume processing techniques, including injection molding, hot embossing, and solvent or thermal bonding. Remarkably, in this regard, several studies underscore how strict process control in terms of temperature, pressure, and cooling time allows for the proper replication and scalability essential for transitioning devices from the lab to the clinic (Shakeri et al. 2022; Zhou et al. 2024).

Fabrication techniques have also diversified with the wide range of prototype and production needs. Soft lithography remains the workhorse for the preparation of PDMS-based designs, enabling the realization of complex multilayer components, including pneumatic valves and peristaltic pumps. Hot embossing and injection molding of thermoplastics are rapidly becoming methods of choice for reproducible and robust chip fabrication at scale. Innovations such as self-sealing thermoplastic elastomers enable rapid device assembly without adhesives-yielding chemically resistant, optically clear constructs suitable for modular droplet generation and other flow chemistry applications (McMillan et al. 2021).

8.3 Microfluidic Platforms for Single-Cell Generation and Handling

8.3.1 Droplet-Based Microfluidics

Droplet-based microfluidics has emerged as a game-changing platform for the analysis of single cells, enabling the encapsulation of individual cells in picoliter-sized droplets into isolated micro reactors. This approach enables high-throughput screening, precise control over microenvironments, and integration with downstream molecular analyses. The fundamental principle for generating droplets is based on the manipulation of fluid dynamics inside micro channel, where aqueous phases containing cells are sheared into droplets by immiscible oil phases. Recent innovations have refined these processes, enhancing droplet uniformity and stability, crucial for reproducible single-cell assays. For example, Tang et al. (2024) presented a new droplet microfluidic chip that contains a double spiral focusing unit and a flow resistance-based sample enrichment module, and was able to create a near-equidistant linear arrangement of cells encapsulated within droplets, hence enhancing encapsulation efficiency (Tang et al. 2024).

Single-cell encapsulation into droplets follows Poisson statistics, this leads to a distribution where a portion of droplets will have more than one cell, some with zero, and optimally just one cell per droplet. Recent studies, including the optimization of flow rates, inertial focusing, and acoustic waves positioning were improved efficiency of encapsulation. The optimization process increases the chance of single-cell occupancy per droplet, which is critical for downstream analysis. Thus, the integration of droplet microfluidics with artificial intelligence (AI) has enabled multiple enhancements in these systems. AI algorithms analyze the images captured in real time and classify droplets in function of cellular features such as size, morphology, or fluorescence intensity. These features enable the automation of sorting and further analysis of droplets, allowing an increase in throughput and a decrease in human error. As an example, DropAI platforms use AI analysis to optimize cell-free expression systems, showing the potential of AI in droplet-based assays (Zhu et al. 2025).

Applications of droplet-based microfluidics to single-cell analysis are immense and keep on growing. In genomics, for example, Drop-seq and in Drop enabled single-cell RNA sequencing by encapsulating individual cells with barcoded beads within droplets. This allows the parallel processing of thousands of single cells, providing information on cellular heterogeneity and gene expression profiles. More recently, the development of new methods has led to the realization of multimodal sequencing techniques that combine transcriptomic and epigenomic analyses in the same droplet, hence providing a more holistic understanding of cellular states. For example, Li et al. recently presented UDA-seq (Li et al. 2025), a universal droplet microfluidics-based single-cell multimodal sequencing workflow connecting droplet microfluidics with combinatorial indexing, achieving very high throughput and data quality similar to standard droplet microfluidic procedures (Li et al. 2025).

Along with genomic applications, droplet microfluidics has been applied to high-throughput screening (HTS) in drug discovery and enzyme engineering. Individual cells or enzymes can be encapsulated in droplets and subjected to different conditions; then, using various detection methods, researchers can identify those showing desired traits, such as resistance to certain drugs or enhanced catalytic activity. This significantly speeds up the process of finding what is of interest and at much lower reagent consumption compared to traditional bulk assays. The use of microfluidic chips for directed evolution of biomolecules was reviewed by Vladisavljević (2024), focusing on different strategies for performing fluidic operations on-chip (Vladisavljević 2024).

Beyond this, droplet-based systems have also been utilized within immunology to investigate single-cell responses to stimuli. Examples include the encapsulation of immune cells within droplets, which enables the observation of cytokine release following antigen presentation;a common method through which immune cell activation and function are studied. These applications are of particular importance when elucidating complex immune responses and in the identification of targeted therapies. Li et al. (2023a) present the channel design, droplet generation, and control technology of droplet microfluidics. In addition, the paper elaborates on how droplet microfluidics is applied in single-cell culture, single-cell screening, and single-cell detection in detail (Li et al. 2023b).

8.3.2 Valve-Based and Pneumatic Systems

Valve-based and pneumatic microfluidic systems have been powerful platforms in single-cell analysis due to the possible precise manipulation of fluids that enable controlled trapping and dynamic interrogation of cells within micro channel. These systems use elastomeric micro valves, which open and close microchannel in real-time, typically actuated by pneumatic pressure. Such a mechanism allows researchers to program fluid routing, isolate individual cells, and perform multiplexed assays with high reproducibility. The principle has been pushed to high-throughput levels with remarkable integration. For instance, Briones et al. (2021) designed a microfluidic platform with close to 5000 hydrodynamic traps coupled to micro valves operating at low sealing pressures of 0.04 MPa that demonstrated for enzymatic assays on both Jurkat and Peripheral Blood Mononuclear Cell (PBMC) single cells (Briones et al. 2021). Thus, it shows how large valve arrays can facilitate automation and scalability in single-cell workflows.

Beyond throughput, these systems also provide accurate temporal control critical for investigating dynamics of biological processes. Sinha et al. (2022) presented a completely automated microfluidic chip able to stimulate and monitor single cells in real time sequentially. Their platform allowed controlled trapping of suspension and adherent cells and uncovered dynamic responses such as the activation of caspase-3 in K562 leukemia cells and STAT-1 translocation in fibroblasts, providing unparalleled insights into the dynamics of single-cell signaling (Sinha et al. 2022). In another instance, Liu et al. (2021) summarized the recent developments of valve-based nucleic acid analytical platforms, including the capability to simplify the workflow from isolation through the analysis and amplification process, and thus overcome the challenge of cellular heterogeneity in diagnostics and genomics (Liu et al. 2021).

More recent innovations extend to secretion assays and barcoded analysis. A recent 2025 review in Lab on a Chip highlighted the role of valve-based platforms for time-resolved optical detection of cellular secretions, where precise stimulation combined with temporal barcoding enables scalable and dynamic profiling of single-cell outputs (Xu et al. 2025). Simultaneously, integration with artificial intelligence is transforming system performance. Taha et al. (2025) discussed how machine learning and AI tools can be combined with valve-based microfluidics to classify, monitor, and sort cells in real time, significantly enhancing throughput, precision, and reproducibility in both basic research and clinical diagnostics (Taha et al. 2025).

8.3.3 Inertial Microfluidics

Inertial microfluidics has been a well-liked means of carrying out cell sorting without chemical labeling. It generally relies on inherent differences in cell size, shape, and density for rapidly focusing cells into separate streams in a continuous manner. The technique may also rely on physical inertial lift and Dean drag forces within specially designed channel geometries, such as spiral or serpentine shapes. These physical forces allow focusing of cells into separate streams, enabling clog-free high-speed sorting (Kalyan et al. 2021). For instance, Xu et al. (2022) present a 3D multilayer chip by integrating spiral and serpentine channel designs for (circulating tumor cell) CTC collection. At 1.3 mL/min, it separated 5 μm and 15 μm particles with about 92% efficiency and 98% purity, respectively. It also separated tumor cells from red blood cells with 80% separation efficiency and approximately 90% purity while maintaining high speed (Xu et al. 2022a).

More recently, Zhu et al. (2024) demonstrated a continuous-flow inertial device integrating slanted spiral channels and contraction-expansion geometries, achieving ternary separation of single tumor cells, tumor cell clusters, and white blood cells from malignant effusions. Remarkably, at 3.5 mL/min, the system removed 94% of White Blood Cells (WBCs), recovered more than 97% of the M. D. Anderson–Mammary/Breast 231 (MDA-MB-231) tumor cells, and preserved over 90% of the tumor cell clusters-intact for downstream cytological diagnostics (Zhu et al. 2024). These works highlight the unique strengths of inertial microfluidics: label-free, high-speed sorting based on inherent biophysical properties, with robust scalability and transparency toward clinical translation. Table 8.1 summarizes the major advantages and limitations of different microfluidic platforms for single-cell analysis, while Fig. 8.1 depicts the simplified workflow of a general microfluidic single-cell analysis process.

Table 8.1 Key advantages and limitation of microfluidic platform

Microfluidic platform	Key advantages	Limitations	Typical applications	Reference
Droplet-based	High-throughput (10^3–10^6 droplets/min), compartmentalized reactions, reduced reagent use	Poisson distribution may lead to empty/multi-cell droplets; oil–water interfaces may affect viability	scRNA-seq, drug screening, single-cell proteomics	Li et al. (2025)
Valve-based (pneumatic)	Precise temporal control, multiplexed assays, dynamic stimulation	Complex fabrication, less portable, higher cost	Sequential stimulation, secretion assays, barcoded analysis	Sinha et al. (2022)
Inertial microfluidics	Label-free sorting, continuous high-throughput operation, minimal reagents	Limited to size/deformability differences, clogging possible	CTC isolation, immune cell sorting	Xu et al. (2022b)
Digital microfluidics	Programmable droplet actuation, integration with omics workflows	Limited throughput, evaporation challenges	Single-cell proteomics, sample prep for MS	Lamanna et al. (2020)

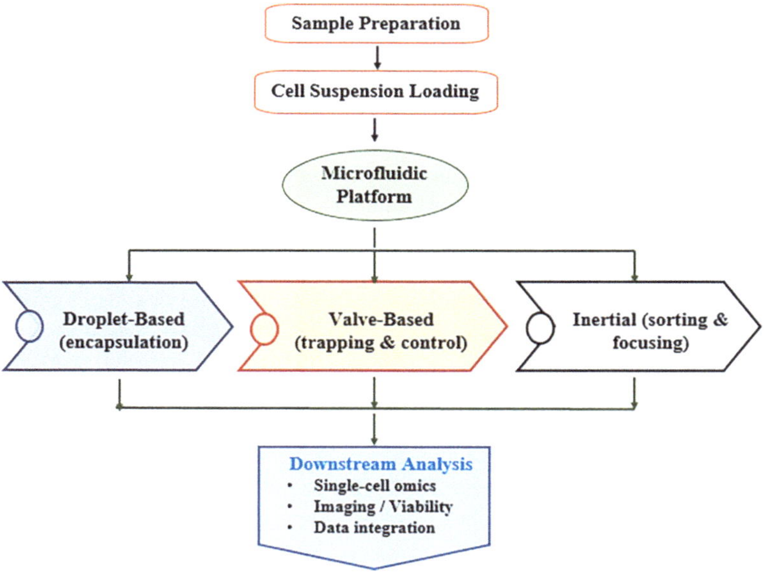

Fig. 8.1 Simplified workflow of a typical microfluidic single-cell analysis process

8.4 On-Chip Cell Culture and Viability Assessment

On-chip cell culture platforms utilize the power of microengineering in order to recapitulate key physiological parameters that are usually missed in traditional cell culture. In most systems, these include microchambers and perfusion networks that enable long-term cultures, real-time monitoring of viability, and precise control of environmental conditions, advances that greatly heighten the biological relevance and rigor of the models.

Recent innovations have overcome some of the traditional problems with the diffusion of nutrients and oxygen in 3D constructs such as spheroids. One such example is that of Takagi et al. (2023), who developed a vertical-flow perfusion system with porous microchambers fabricated using salt-leaching methods. These enable the medium to directly perfuse through the chamber walls, providing uniform nutrition and preventing the formation of necrotic cores. Their system, while culturing approximately 200 μm HepG2 spheroids at 0.5 μL/min, showed significantly improved proliferation compared with horizontal-flow controls due to enhanced vertical perfusion (Takagi et al. 2023).

Continuous perfusion allows for sustained viability and enables dynamic monitoring. In the recently developed skin-on-chip models, full-thickness skin equivalents were cultured under perfusion and their viability was monitored through LDH assays. With the two-day measurement of stable LDH levels in the perfused system, as opposed to increased cytotoxicity in conventional biopsies, the present system demonstrates its long-term tissue culture capabilities with valid viability readouts (Rhee et al. 2024). Furthermore, the possibility of integrating biosensors or fluorescence monitoring within transparent channel geometries allows for real-time observation of cell growth kinetics and metabolic activity without interfering with the culture, thereby assessing viability in an immediate and noninvasive way.

The environmental control, in particular oxygen, pH, and shear stress, is an indispensable feature of these microfluidic culture systems. The study investigated thermoplastic micro channel and found that an increase in channel height delays the oxygen depletion under static conditions, whereas increased flow rates of medium improve oxygen delivery but increase shear stress. The authors proposed a pulsatile medium flow regime intermittently supplying the medium only when the oxygen falls below threshold levels.

This approach minimizes continuous shear while maintaining physiologically adequate oxygenation. Other designs using PDMS microfluidics utilize the gas permeability of the material, delivering controlled oxygen levels through adjoining gas channels, separated by membranes, without disturbing the fluid flow over cells (Sheidaei et al. 2022).

Shear stress remains a critical design consideration, with excessive shear having the potential to compromise sensitive cell types, and proper shear stimulating mechano-transduction pathways. Devices have incorporated pulsatile or stop-and-flow regimes to reduce overall shear exposure and recapitulate patterns in physiological blood flow. This not only serves to protect cells but also encourages physiologically relevant mechanical stimuli, such as endothelial alignment or eNOS production in vascular models (Fallon et al. 2022; Salimi-Afjani et al. 2025; Sheidaei et al. 2022).

Together, these microfluidic platform innovations offer three key advantages: long-term perfused culture, dynamic viability, and growth monitoring, and fine environmental control (oxygen, pH, shear) (Farhang Doost and Srivastava 2024; Sheidaei et al. 2022; Takahashi et al. 2023; Wei et al. 2020). These advances facilitate various applications, from tissue modeling and drug screening to personalized medicine, in a well-grounded way with minimal reliance on either crude in vitro models or animal studies (Farhang Doost and Srivastava 2024).

8.5 Integration with Analytical Techniques

8.5.1 Single-Cell Omics

Microfluidic technologies have become the backbone for the development of single-cell omics, including transcriptomics, proteomics, and metabolomics, with unprecedented throughput, sensitivity, and minimal sample loss. A recent paper in micromachines underlines how microfluidic chips enables massively parallel, compartmentalized processing of single cells for RNA-seq and spatial multi-omics while significantly reducing reagent consumption and batch variability through nanoliter-scale reaction volumes and integrated barcoding systems (Sun et al. 2025).

In proteomics, digital microfluidic (DMF) platforms now perform streamlined protocols of sample preparation steps: lysis, digestion, labeling, and even interface directly to high-performance LC-MS/MS systems. In 2023, automated on-chip processing of limited samples, such as small tumor biopsies, resulted in robust protein quantification with minimal contamination and manual handling (Sun et al. 2025).

Single-cell metabolomics is still challenging due to extremely low sample sizes and huge molecular diversity, whereas mass spectrometry (MS) approaches are rapidly developing. A study highlighted recent developments including live-cell sampling and spatial metabolomics imaging but emphasized significant challenges for throughput, standardization, and integration with other omics layers (Petrova and Guler 2025; Zhang et al. 2023).

In all omics domains, sample preparation and analysis are problematic. Single cells present vanishingly small volumes, and processes such as lysis, molecule capture, and transfer are highly susceptible to losses or contamination. Microfluidic platforms can reduce these problems by allowing compartmentalization on-chip and decreasing the number of transfer steps involved, thus enabling direct coupling with detection systems. Combining multiple omics on one chip clearly increases complexity, involves sophisticated engineering, and introduces new modes of error, a tradeoff between integration and reliability (Petrova and Guler 2025; Wevers et al. 2023).

Together, these microfluidic innovations push single-cell omics toward high-throughput, sensitive, and integrated workflows. Transcriptomic, proteomic, and metabolomics analyses are increasingly achievable on compact microfluidic platforms, offering unprecedented molecular insight from individual cells. Remaining challenges, in particular with regard to workflow integration and reliable operation with minimal risk of contamination, set the next frontier in microscale omics.

8.5.2 Optical and Fluorescence Imaging

Microfluidic platforms combined with optical and fluorescence imaging have transformed single-cell analysis into the capability for real-time tracking and quantification of cellular behaviors at high spatial and temporal resolution. A study demonstrated a label-free tracking technique inside of antibody-coated microfluidic channels: through computational imaging, the authors were able to track individual T-cells with success rates up to 98% and differentiate between CD8-positive and CD8-negative cells based on transient motion delays, without any need for prior labeling or extensive sample prep (Roebroek et al. 2025). This is a good example of how high-fidelity cell phenotyping can be directly carried out in-flow using purely label-free optical methods.

Fluorescence imaging brings added molecular specificity and sensitivity to such systems. In one recent example, the introduction of a microfluidic flow cytometer with a uniform optical field; quantitative single-cell protein analysis by fluorescence imaging enabled leukemia profiling and cancer-stem-cell hierarchies (Gao et al. 2023). This illustrates the power of fluorescence-integrated microfluidics in delivering robust bimolecular insights at single-cell resolution.

Other label-free techniques are also effective. A low-cost microfluidic scanning flow cytometer utilizes angle-resolved light scattering for accurate cell size and refractive index measurements. This offers a simple and inexpensive way for high-throughput analysis without fluorescent labels (Reale et al. 2023). Similarly, the combination of scanning electrochemical microscopy (SECM) with programmable microwell arrays enabled the fast, label-free measurement of enzyme activity such as alkaline phosphatase in hundreds of single cells, showing the flexibility of optical-electrochemical methods (Shi et al. 2021).

These imaging frameworks face dual modes of detection-label-based fluorescence and label-free optical/spectral-each with various tradeoffs. Fluorescent labeling provides molecular specificity but can perturb cell physiology and entail elaborate staining protocols (Gao et al. 2023). In contrast, label-free imaging maintains native cell states and reduces preparatory overhead, possibly at the cost of lacking the molecular detail achievable with fluorophores (Reale et al. 2023).

Across both modalities, the integration of microfluidics with imaging enables real-time tracking and thus allows time-resolved studies of cell migration, deformation, and signaling. These approaches have been used to monitor cell mechanics, metabolic activity, and signal transduction in microenvironments emulating physiological conditions (Gao et al. 2023; Roebroek et al. 2025). Whether quantifying protein localization, cell motion tracking, or enzyme activity screening, these imaging techniques elevate microfluidic platforms into powerful tools for dynamic high-content single-cell phenotyping.

8.5.3 Intracellular Sensing

The integration of electrophysiological recording and intracellular delivery in microfluidic platforms continues to redefine how we can probe cell function with precision and throughput. A study in Lab on a Chip introduced an innovative microfluidic chip seamlessly integrated with CMOS-based microelectrode arrays. This system enabled real-time electrophysiological monitoring of human iPSC-derived cardiac microtissues in both hanging-drop and standing-drop culture modes, capturing clear depolarization and repolarization signals, and precisely quantifying beating rates before and after drug (epinephrine) administration, all within a dynamic, flow-compatible environment (Bounik et al. 2025). This example once again underlines how microfluidics, combined with dense electrode layouts, provides high-fidelity, spatially resolved electrophysiological readouts.

Complementing electrophysiology, advances in intracellular delivery of probes and sensors have enabled direct access to cell interiors. One recent development uses a combination of microfluidic mechanoporation with electroporation for single-cell DNA delivery by forcing cells through constriction channels of diameters less than their own, mechanically disrupting the membrane, followed by targeted electric pulses to breach the nuclear envelope. This dual approach significantly enhanced DNA expression efficiency and rapid nuclear localization, demonstrating how integrated microfluidic systems can facilitate the delivery of molecular probes to the nucleus with minimal toxicity (Kaladharan et al. 2021). Bridging both domains, platforms that combine intracellular delivery with electrophysiological recording have begun to emerge. Though still at an early stage, a Microfluidic-Microelectrode Array (MEA) hybrid device successfully delivered molecules to individual cells while simultaneously recording electrical activity, thus enabling a synchronized view of molecular perturbation and resultant electrophysiological response.

These approaches face several technical challenges. Reliable electrophysiological readouts require high electrode sensitivity, consistent cell-electrode contact, and low-noise environments-each being difficult to maintain within flowing microfluidic systems (Bounik et al. 2025). Simultaneously, intracellular delivery needs to overcome natural resilience in cell membranes without causing irreversible damage or loss of viability: a balance that the microfluidic mechanoporation, Nanostraw, and flow-through electroporation platforms are trying to achieve (Wang et al. 2024). Integration of both functionalities onto a single platform introduces additional challenges in biocompatibility, scalability, and integration of systems; simultaneously demanding high-fidelity recording of delivery, careful engineering of flow, electrode layout, and interfaces of delivery is required (Li et al. 2023b).

Nevertheless, the combination of real-time electrophysiology with focused intracellular delivery represents a critical advancement in single-cell analysis.

Table 8.2 Comparison of analytical integration techniques in microfluidic single-cell systems

Analytical technique	Integration mode (with microfluidics)	Target parameter/ readout	Typical throughput (order of magnitude)	Typical detection sensitivity	References
Single-cell transcriptomics (scRNA-seq)	Droplet microfluidics with barcoded bead encapsulation	Gene expression, splicing	10^3–10^5 cells/run	High; low transcript detection	De Jonghe et al. (2023)
Proteomics (on-chip prep)	Digital microfluidics for lysis, digestion, LC–MS coupling	Protein quantification	10^2–10^3 cells/run	Femtomole-attomole; sample loss	Steinbach et al. (2025)
Single-cell metabolomics	Open microfluidic or nano-ESI MS coupling	Small-molecule metabolites	10–10^2 cells/run	Attomole-femtomole; limited by MS rate	Petrova and Guler (2025)
Optical / fluorescence imaging	Flow-through micro channels with uniform illumination	Morphology, fluorescence markers	10^3–10^5 cells/h	Submicron; single-molecule quantification	Gao et al. (2023)
Electrophysiology	CMOS-MEA integrated microfluidic chamber	Action potentials, drug response	10–10^2 cultures/ run	µV–mV; high temporal (kHz)	Bounik et al. (2025)

Microfluidic systems that integrate such capabilities can provide high-temporal-resolution functional readouts after controlled molecular or electrical manipulations, allowing for precise inquiries in drug screening, neurotoxicity testing, and single-cell electrophysiology (Bounik et al. 2025; Wang et al. 2024). Device designs are maturing, and as integration improves further, such hybrid systems will likely become critical components in dynamic, high-content single-cell assays with simultaneous perturbation and readout capabilities in compact, high-throughput formats (Hur and Chung 2021). Table 8.2 presents the comparison of some of the analytical integration techniques used in microfluidic single-cell systems.

8.6 Emerging Applications and Case Studies

Microfluidic innovations are quickly changing the frontline applications of various important applications in the field of single-cell immunology, oncology, and pharmacology. A recent study used a pillar-lattice-array microfluidic platform to produce uniform core-tumor-stromal spheroids and allowed high-throughput tracking of T-cell infiltration and cytotoxicity across thousands of structures, thus enabling the automated screening of immunomodulatory compounds in a physiologically relevant 3D context (Ao et al. 2022). In the realm of cancer heterogeneity and circulating tumor cells (CTCs), microfluidic chips have been the mainstay for enriching very rare CTCs from blood and performing downstream single-cell molecular analyses and allowed highlighting diverse phenotypes, mutations, and genomic rearrangements across individual patient-derived CTCs (Lim et al. 2019). Finally, drug response assays at the level of single cells have been dramatically upgraded by the "SMART" microfluidic platform sporting approximately 4320 microchambers and a concentration-gradient generator. This device allowed administering multiple doses of drugs into thousands of leukemia cells and clones, unveiled heterogeneity in IC_{50} values, and revealed differential resistance profiles even among clones derived from the same patient (Wang et al. 2022).

These case studies emphasize microfluidics as an ever-emerging high-throughput and high-resolution engine for immune-oncology, CTC-based diagnostics, and precision drug testing: basic ideas and premises of single-cell resolution and real-world clinical relevance.

8.7 Challenges and Limitations

Despite considerable advances, microfluidic single-cell platforms still face critical throughput vs. resolution trade-offs. For example, serial microfluidic mass sensor arrays achieve excellent measurement fidelity at the expense of throughput, underscoring a basic balance where enhanced resolution often reduces processing speed (Cermak et al. 2016). Material biocompatibility and sustained viability also remain concerns; whereas MEMS-based microfluidic devices can control cultured cells with great precision, their diverse material composition and microscale structures contribute to challenges in ensuring consistent biocompatibility, stability over long periods of time, and low levels of cell stress (Zainal Abidin et al. 2025). Lastly, integration of rich high-dimensional data analytics, including AI-driven multiparameter profiling, into microfluidic data streams has computational bottlenecks and interpretability issues, especially when datasets from devices are large and increasingly complex (Taha et al. 2025). These limitations together call for designs carefully balanced, more biocompatible fabrication approaches, and wiser data analysis frameworks in order to realize the full potential behind the microfluidic single-cell technologies.

8.8 Future Directions

Looking ahead, microfluidic technologies continue to be very promising. Organs-on-chip with single-cell precision are now under rapid development: modern platforms create tissue-tissue interfaces and mechanical functions while enabling thousands of quantitative, single-cell-resolution assays for pharmacokinetics, disease modeling, and personalized medicine (An et al. 2025). At the same time, the integration of AI and machine learning in the microfluidic workflows is promoting a revolution in data processing: adaptive algorithms now automatically design experiments, optimize chip performance, and enable real-time cell characterization and classification from huge datasets of multimodal type (Park et al. 2024). Lastly, clinical translation and point-of-care (POC) diagnostics now become possible due to a microfluidic platform integrating compact form factors with AI-enhanced detection, thus paving the way for rapid and accurate diagnostics directly at the bedside or in resource-limited settings (Behera et al. 2023; Doganay et al. 2024). Altogether, these developments position microfluidics as the backbone of future precision medicine, where dynamic single-cell monitoring, smart analytics, and immediate clinical deployment converge.

8.9 Conclusion

Microfluidics have rapidly evolved into indispensable tools for single-cell analysis, offering precise fluid control, high-throughput manipulation, and seamless integration with advanced analytical techniques. Droplet systems, valve-based architectures, inertial sorting, and on-chip culture have pushed the boundaries of what can be measured and controlled at the level of a single cell. Their application to immune profiling, cancer heterogeneity, circulating tumor cells, and drug response assays showcases their translational importance; meanwhile, integrations with omics, imaging, and intracellular sensing further deepen biological insight. Significant challenges remain in balancing throughput with resolution, ensuring material biocompatibility, and managing the flood of high-dimensional data. Looking ahead, the intersection of organs-on-chip with single-cell precision, AI-driven analytics, and clinically deployable point-of-care diagnostics cements microfluidics as a central driver of both discovery and precision medicine toward a future in which cellular diversity is not only studied but also harnessed for therapeutic impact.

References

An L, Liu Y, Liu Y (2025) Organ-on-a-chip applications in microfluidic platforms. Micromachines 16(2):201. https://doi.org/10.3390/mi16020201

Ao Z, Cai H, Wu Z, Hu L, Nunez A, Zhou Z, Liu H, Bondesson M, Lu X, Lu X, Dao M, Guo F (2022) Microfluidics guided by deep learning for cancer immunotherapy screening. Proc Natl Acad Sci USA 119(46):e2214569119. https://doi.org/10.1073/pnas.2214569119

Aralekallu S, Boddula R, Singh V (2023) Development of glass-based microfluidic devices: a review on its fabrication and biologic applications. Mater Des 225:111517. https://doi.org/10.1016/j.matdes.2022.111517

Behera PP, Kumar N, Kumari M, Kumar S, Mondal PK, Arun RK (2023) Integrated microfluidic devices for point-of-care detection of bio-analytes and disease. Sens Diagn 2(6):1437–1459. https://doi.org/10.1039/D3SD00170A

Bounik R, Landolt AE, Lee J, Viswam V, Cardes F, Modena MM, Hierlemann A (2025) Seamless integration of CMOS microsensors into open microfluidic systems. Lab Chip 25(9):2205–2221. https://doi.org/10.1039/D4LC01000K

Briones J, Espulgar W, Koyama S, Takamatsu H, Tamiya E, Saito M (2021) A design and optimization of a high throughput valve based microfluidic device for single cell compartmentalization and analysis. Sci Rep 11(1):12995. https://doi.org/10.1038/s41598-021-92472-w

Cermak N, Olcum S, Delgado FF, Wasserman SC, Payer KR, A Murakami M, Knudsen SM, Kimmerling RJ, Stevens MM, Kikuchi Y, Sandikci A, Ogawa M, Agache V, Baléras F, Weinstock DM, Manalis SR (2016) High-throughput measurement of single-cell growth rates using serial microfluidic mass sensor arrays. Nat Biotechnol 34(10):1052–1059. https://doi.org/10.1038/nbt.3666

De Jonghe J, Kaminski TS, Morse DB, Tabaka M, Ellermann AL, Kohler TN, Amadei G, Handford CE, Findlay GM, Zernicka-Goetz M, Teichmann SA, Hollfelder F (2023) spinDrop: a droplet microfluidic platform to maximise single-cell sequencing information content. Nat Commun 14(1):4788. https://doi.org/10.1038/s41467-023-40322-w

Doganay MT, Chakraborty P, Bommakanti SM, Jammalamadaka S, Battalapalli D, Madabhushi A, Draz MS (2024) Artificial intelligence performance in testing microfluidics for point-of-care. Lab Chip 24(21):4998–5008. https://doi.org/10.1039/D4LC00671B

Fallon ME, Mathews R, Hinds MT (2022) In vitro flow chamber design for the study of endothelial cell (patho)physiology. J Biomech Eng 144(2). https://doi.org/10.1115/1.4051765

Farhang Doost N, Srivastava SK (2024) A comprehensive review of organ-on-a-chip technology and its applications. Biosensors 14(5):225. https://doi.org/10.3390/bios14050225

Fergola A, Ballesio A, Frascella F, Napione L, Cocuzza M, Marasso SL (2025) Droplet generation and manipulation in microfluidics: a comprehensive overview of passive and active strategies. Biosensors 15(6). https://doi.org/10.3390/bios15060345

Ferreira M, Carvalho V, Ribeiro J, Lima RA, Teixeira S, Pinho D (2024) Advances in microfluidic systems and numerical modeling in biomedical applications: a review. Micromachines 15(7). https://doi.org/10.3390/mi15070873

Gao C, Zhang T, Wei Y, Liu Q, Ma L, Gao M, Zhao X, Wang Y, Chen D, Sun L, Wang J, Chen J (2023) Development of a microfluidic flow cytometer with a uniform optical field (Uni-μFCM) enabling quantitative analysis of single-cell proteins and its applications in leukemia gating, tumor classification, and hierarchy of cancer stem cells. ACS Sens 8(9):3498–3509. https://doi.org/10.1021/acssensors.3c01060

Han X, Xu X, Yang C, Liu G (2023) Microfluidic design in single-cell sequencing and application to cancer precision medicine. Cell Rep Methods 3(9). https://doi.org/10.1016/j.crmeth.2023.100591

Hur J, Chung AJ (2021) Microfluidic and nanofluidic intracellular delivery. Adv Sci (Weinheim, Baden-Wurttemberg, Germany) 8(15):e2004595. https://doi.org/10.1002/advs.202004595

Jang I, Kang H, Song S, Dandy DS, Geiss BJ, Henry CS (2021) Flow control in a laminate capillary-driven microfluidic device. Analyst 146(6):1932–1939. https://doi.org/10.1039/D0AN02279A

Kaladharan K, Kumar A, Gupta P, Illath K, Santra TS, Tseng F-G (2021) Microfluidic based physical approaches towards single-cell intracellular delivery and analysis. Micromachines 12(6). https://doi.org/10.3390/mi12060631

Kalyan S, Torabi C, Khoo H, Sung HW, Choi S-E, Wang W, Treutler B, Kim D, Hur SC (2021) Inertial microfluidics enabling clinical research. Micromachines 12(3). https://doi.org/10.3390/mi12030257

Khanjani E, Fergola A, López Martínez JA, Nazarnezhad S, Casals Terre J, Marasso SL, Aghajanloo B (2025) Capillary microfluidics for diagnostic applications: fundamentals, mechanisms, and capillarics. Front Lab Chip Technol 4. https://doi.org/10.3389/frlct.2025.1502127

Kumari S, Saha U, Bose M, Murugan D, Pachauri V, Sai VVR, Madaboosi N (2023) Microfluidic platforms for single cell analysis: applications in cellular manipulation and optical biosensing. Chemosensors 11(2):107. https://doi.org/10.3390/chemosensors11020107

Lamanna J, Scott EY, Edwards HS, Chamberlain MD, Dryden MDM, Peng J, Mair B, Lee A, Chan C, Sklavounos AA, Heffernan A, Abbas F, Lam C, Olson ME, Moffat J, Wheeler AR (2020) Digital microfluidic isolation of single cells for -omics. Nat Commun 11(1):5632. https://doi.org/10.1038/s41467-020-19394-5

Li B, Ma X, Cheng J, Tian T, Guo J, Wang Y, Pang L (2023a) Droplets microfluidics platform—a tool for single cell research. Front Bioeng Biotechnol 11. https://doi.org/10.3389/fbioe.2023.1121870
Li X, Ma Y, Xue Y, Zhang X, Lv L, Quan Q, Chen Y, Yu G, Liang Z, Zhang X, Weng D, Chen L, Chen K, Han X, Wang J (2023b) High-throughput and efficient intracellular delivery method via a vibration-assisted nanoneedle/microfluidic composite system. ACS Nano 17(3):2101–2113. https://doi.org/10.1021/acsnano.2c07852
Li Y, Huang Z, Xu L, Fan Y, Ping J, Li G, Chen Y, Yu C, Wang Q, Song T, Lin T, Liu M, Xu Y, Ai N, Meng X, Qiao Q, Ji H, Qin Z, Jin S et al (2025) UDA-seq: universal droplet microfluidics-based combinatorial indexing for massive-scale multimodal single-cell sequencing. Nat Methods 22(6):1199–1212. https://doi.org/10.1038/s41592-024-02586-y
Lim SB, Lim CT, Lim W-T (2019) Single-cell analysis of circulating tumor cells: why heterogeneity matters. Cancer 11(10). https://doi.org/10.3390/cancers11101595
Liu L, Dong X, Tu Y, Miao G, Zhang Z, Zhang L, Wei Z, Yu D, Qiu X (2021) Methods and platforms for analysis of nucleic acids from single-cell based on microfluidics. Microfluid Nanofluid 25(11):87. https://doi.org/10.1007/s10404-021-02485-0
Liu D, Sun M, Zhang J, Hu R, Fu W, Xuanyuan T, Liu W (2022) Single-cell droplet microfluidics for biomedical applications. Analyst 147(11):2294–2316. https://doi.org/10.1039/D1AN02321G
Maurya R, Gohil N, Bhattacharjee G, Alzahrani KJ, Ramakrishna S, Singh V (2022) Chapter eight—Microfluidics for single cell analysis. In: Pandya A, Singh V (eds) Micro/nanofluidics and lab-on-chip based emerging technologies for biomedical and translational research applications—Part A, vol 186. Academic, pp 203–215. https://doi.org/10.1016/bs.pmbts.2021.07.013
McMillan AH, Mora-Macías J, Teyssandier J, Thür R, Roy E, Ochoa I, De Feyter S, Vankelecom IFJ, Roeffaers MBJ, Lesher-Pérez SC (2021) Self-sealing thermoplastic fluoroelastomer enables rapid fabrication of modular microreactors. Nano Select 2(7):1385–1402. https://doi.org/10.1002/nano.202000241
Park J, Kim YW, Jeon H-J (2024) Machine learning-driven innovations in microfluidics. Biosensors 14(12). https://doi.org/10.3390/bios14120613
Peng T, Qiang J, Yuan S (2024) Sheathless inertial particle focusing methods within microfluidic devices: a review. Front Bioeng Biotechnol 11. https://doi.org/10.3389/fbioe.2023.1331968
Petrova B, Guler AT (2025) Recent developments in single-cell metabolomics by mass spectrometry—a perspective. J Proteome Res 24(4):1493–1518. https://doi.org/10.1021/acs.jproteome.4c00646
Profiling the total single-cell transciptome using droplet microfluidics (2022) Nat Biotechnol 40(12):1766–1767. https://doi.org/10.1038/s41587-022-01370-7
Reale R, Peruzzi G, Ghoreishi M, Stabile H, Ruocco G, Leonetti M (2023) A low-cost, label-free microfluidic scanning flow cytometer for high-accuracy quantification of size and refractive index of particles. Lab Chip 23(8):2039–2047. https://doi.org/10.1039/d2lc01179d
Rhee S, Xia C, Chandra A, Hamon M, Lee G, Yang C, Guo Z, Sun B (2024) Full-thickness perfused skin-on-a-chip with in vivo-like drug response for drug and cosmetics testing. Bioengineering 11(11):1055. https://doi.org/10.3390/bioengineering11111055
Roebroek T, Van Roy W, Roth S, Chacon Orellana L, Luo Z, El Jerrari Y, Arnett C, Claes K, Ha S, Jans K, Labie R, Lin Z, Obst M, Origuella D, Pham V, Van Bellinghen F, Girikumar Krishna AV, Vaezzadeh E, Vanhove W et al (2025) Continuous, label-free phenotyping of single cells based on antibody interaction profiling in microfluidic channels. Anal Chem 97(16):8975–8983. https://doi.org/10.1021/acs.analchem.5c00385
Salimi-Afjani N, Rieben R, Obrist D (2025) Pulsatile-flow culture: a novel system for assessing vascular-cell dynamics. Lab Chip 25(7):1755–1766. https://doi.org/10.1039/D4LC00949E
Shakeri A, Khan S, Jarad NA, Didar TF (2022) The fabrication and bonding of thermoplastic microfluidics: a review. Materials 15(18):6478. https://doi.org/10.3390/ma15186478
Sheidaei Z, Akbarzadeh P, Nguyen N-T, Kashaninejad N (2022) A new insight into a thermoplastic microfluidic device aimed at improvement of oxygenation process and avoidance of shear stress during cell culture. Biomed Microdevices 24(2):15. https://doi.org/10.1007/s10544-022-00615-1

Shi M, Wang L, Xie Z, Zhao L, Zhang X, Zhang M (2021) High-content label-free single-cell analysis with a microfluidic device using programmable scanning electrochemical microscopy. Anal Chem 93(36):12417–12425. https://doi.org/10.1021/acs.analchem.1c02507
Sinha N, Yang H, Janse D, Hendriks L, Rand U, Hauser H, Köster M, van de Vosse FN, de Greef TFA, Tel J (2022) Microfluidic chip for precise trapping of single cells and temporal analysis of signaling dynamics. Commun Eng 1(1):18. https://doi.org/10.1038/s44172-022-00019-2
Steinbach MK, Leipert J, Matzanke T, Tholey A (2025) Digital microfluidics for sample preparation in low-input proteomics. Small Methods 9(1):2400495. https://doi.org/10.1002/smtd.202400495
Sun Y, Yu N, Zhang J, Yang B (2025) Advances in microfluidic single-cell RNA sequencing and spatial transcriptomics. Micromachines 16(4):426. https://doi.org/10.3390/mi16040426
Taha M, Pappa A-M, Saleh H, Alazzam A (2025) Enhancing cell characterization with microfluidics and AI: a comprehensive review of mechanical, electrical, and hybrid techniques. Biotechnol Rep (Amst) 47:e00905. https://doi.org/10.1016/j.btre.2025.e00905
Takagi M, Yamada M, Utoh R, Seki M (2023) A multiscale, vertical-flow perfusion system with integrated porous microchambers for upgrading multicellular spheroid culture. Lab Chip 23(9):2257–2267. https://doi.org/10.1039/D3LC00168G
Takahashi N, Yoshino D, Sugahara R, Hirose S, Sone K, Rieu J-P, Funamoto K (2023) Microfluidic platform for the reproduction of hypoxic vascular microenvironments. Sci Rep 13(1):5428. https://doi.org/10.1038/s41598-023-32334-9
Tang T, Zhao H, Shen S, Yang L, Lim CT (2024) Enhancing single-cell encapsulation in droplet microfluidics with fine-tunable on-chip sample enrichment. Microsyst Nanoeng 10(1):3. https://doi.org/10.1038/s41378-023-00631-y
Vladisavljević GT (2024) Droplet microfluidics for high-throughput screening and directed evolution of biomolecules. Micromachines 15(8). https://doi.org/10.3390/mi15080971
Wang S, Zhang X, Ma C, Yan S, Inglis D, Feng S (2021) A review of capillary pressure control valves in microfluidics. Biosensors 11(10). https://doi.org/10.3390/bios11100405
Wang B, He B-S, Ruan X-L, Zhu J, Hu R, Wang J, Li Y, Yang Y-H, Liu M-L (2022) An integrated microfluidics platform with high-throughput single-cell cloning array and concentration gradient generator for efficient cancer drug effect screening. Mil Med Res 9(1):51. https://doi.org/10.1186/s40779-022-00409-9
Wang R, Wang Z, Tong L, Wang R, Yao S, Chen D, Hu H (2024) Microfluidic Mechanoporation: current Progress and applications in stem cells. Biosensors 14(5). https://doi.org/10.3390/bios14050256
Ward K, Fan ZH (2015) Mixing in microfluidic devices and enhancement methods. J Micromech Microeng 25(9):094001. https://doi.org/10.1088/0960-1317/25/9/094001
Wei L, Li W, Entcheva E, Li Z (2020) Microfluidics-enabled 96-well perfusion system for high-throughput tissue engineering and long-term all-optical electrophysiology. Lab Chip 20(21):4031–4042. https://doi.org/10.1039/D0LC00615G
Wevers D, Ramautar R, Clark C, Hankemeier T, Ali A (2023) Opportunities and challenges for sample preparation and enrichment in mass spectrometry for single-cell metabolomics. Electrophoresis 44(24):2000–2024. https://doi.org/10.1002/elps.202300105
Wu H, Zhu J, Huang Y, Wu D, Sun J (2018) Microfluidic-based single-cell study: current status and future perspective. Molecules 23(9):2347. https://doi.org/10.3390/molecules23092347
Xu X, Huang X, Sun J, Chen J, Wu G, Yao Y, Zhou N, Wang S, Sun L (2022a) 3D-stacked multistage inertial microfluidic Chip for high-throughput enrichment of circulating tumor cells. Cyborg Bionic Syst (Washington, DC) 2022:9829287. https://doi.org/10.34133/2022/9829287
Xu X, Huang X, Sun J, Chen J, Wu G, Yao Y, Zhou N, Wang S, Sun L (2022b) 3D-stacked multistage inertial microfluidic Chip for high-throughput enrichment of circulating tumor cells. Cyborg Bionic Syst 2022:9829287. https://doi.org/10.34133/2022/9829287
Xu Y, Chan MTJ, Yang M, Meng H, Chen C-H (2025) Time-resolved single-cell secretion analysis via microfluidics. Lab Chip 25(5):1282–1295. https://doi.org/10.1039/D4LC00904E

You JB, Lee B, Choi Y, Lee C-S, Peter M, Im SG, Lee SS (2020) Nanoadhesive layer to prevent protein absorption in a poly(dimethylsiloxane) microfluidic device. BioTechniques 69(1):46–51. https://doi.org/10.2144/btn-2020-0025

Zainal Abidin AR, Zambry NS, Ibrahim F, Nor Rashid N, Jamaluddin NF, Wan Kamarul Zaman WS (2025) Current advances in biocompatibility assessment for MEMS in biomedical applications. Microfluid Nanofluid 29(6):32. https://doi.org/10.1007/s10404-025-02806-7

Zhang C, Dévédec SEL, Ali A, Hankemeier T (2023) Single-cell metabolomics by mass spectrometry: ready for primetime? Curr Opin Biotechnol 82:102963. https://doi.org/10.1016/j.copbio.2023.102963

Zhao Q, Yuan D, Zhang J, Li W (2020) A review of secondary flow in inertial microfluidics. Micromachines 11(5). https://doi.org/10.3390/mi11050461

Zhou W, Yan Y, Guo Q, Ji H, Wang H, Xu T, Makabel B, Pilarsky C, He G, Yu X, Zhang J (2021) Microfluidics applications for high-throughput single cell sequencing. J Nanobiotechnol 19(1):312. https://doi.org/10.1186/s12951-021-01045-6

Zhou M, Jiang S, Yang X, Li Q, Jiang B, Zhou L, Zhang L (2024) The design and fabrication of thermoplastic microfluidic chips with integrated micropillars for particle separation. J Mater Res Technol 33:874–883. https://doi.org/10.1016/j.jmrt.2024.09.135

Zhu Z, Ren H, Wu D, Ni Z, Xiang N (2024) High-throughput and simultaneous inertial separation of tumor cells and clusters from malignant effusions using spiral-contraction-expansion channels. Microsyst Nanoeng 10(1):36. https://doi.org/10.1038/s41378-024-00661-0

Zhu J, Meng Y, Gao W, Yang S, Zhu W, Ji X, Zhai X, Liu W-Q, Luo Y, Ling S, Li J, Liu Y (2025) AI-driven high-throughput droplet screening of cell-free gene expression. Nat Commun 16(1):2720. https://doi.org/10.1038/s41467-025-58139-0

Single-Cell Analysis for Bacterial Infections

9

Mansi Acharya and Indra Mani

Abstract

Infectious diseases are caused by bacteria which involve in the complex interactions between microbial pathogens and host cells. Historically, studies of host–pathogen interactions relied on bulk measurements (e.g., population-level transcriptomics, flow cytometry of mixed cells, or colony counts), which average signals over millions of cells. Such bulk assays obscure critical cell-to-cell differences and rare subpopulations. In fact, both microbial populations and host immune cells are intrinsically heterogeneous: genetically identical bacteria can exhibit striking phenotypic variation (e.g., in metabolism, virulence gene expression, or stress response), and clonal immune cells adopt diverse activation states during infection. Single-cell techniques overcome these limitations by profiling individual cells, revealing distinct subsets of host or pathogen cells. Recent advances in single-cell genomics, transcriptomics, proteomics, and imaging now enable the capture of the full diversity of infection processes at unprecedented resolution. This chapter reviews how single-cell methods have transformed the study of bacterial infections by uncovering hidden heterogeneity, mapping host–pathogen dynamics, and revealing spatial and temporal patterns in microbial communities and immune responses.

Keywords

Single-cell RNA sequencing · Imaging flow cytometry · Host–pathogen interactions · Cellular heterogeneity · Microbiome · Infectious disease

M. Acharya
School of Computational and Integrative Sciences, Jawaharlal Nehru University, New Delhi, India

I. Mani (✉)
Department of Microbiology, Gargi College, University of Delhi, New Delhi, India
e-mail: indra.mani@gargi.du.ac.in

V. Singh (ed.), *Advances in Single Cell Analysis*,
https://doi.org/10.1007/978-981-95-9789-5_9

9.1 Introduction

Bacterial infections trigger complex, multi-cellular immune responses and involve intricate interactions between pathogens and host cells. Historically, investigations of host–pathogen dynamics have relied on bulk measurements (e.g., population-level transcriptomics, flow cytometry of mixed cell samples, or colony counts) that average signals over millions of cells. Such bulk approaches inherently mask cellular variability and rare subpopulations (Davis and Isberg 2016; Sesen and Whyte 2020; Huang et al. 2021). In contrast, single-cell technologies capture each cell's molecular profile, revealing the diversity of states that contribute to infection. These methods can identify uncommon immune cell subsets, distinct activation states, or stochastic pathogen behaviors that would be obscured in averaged data (Papalexi and Satija 2018; Penaranda and Hung 2019).

Single-cell RNA sequencing (scRNA-seq) has emerged as a powerful tool to interrogate cellular heterogeneity with unprecedented resolution (Cheng et al. 2023). By sequencing the transcriptome of thousands of individual cells in parallel, scRNA-seq uncovers novel cell types and dynamic programs within immune populations (Papalexi and Satija 2018; Cheng et al. 2023). In bacterial infection studies, this has already yielded new insights. For example, recent scRNA-seq analyses of bacterial pneumonia and *Helicobacter*-associated gastric lesions have identified discrete immune and epithelial cell subsets correlated with disease severity (Xiao et al. 2025; Li et al. 2025). Similarly, the entire blood immune landscape in sepsis was deconvolved into distinct cell clusters (e.g., an expanded NKT population) that were not apparent by bulk assays (Lei et al. 2021). High-resolution imaging and cytometry complement these findings by tracking infection events in live cells or by linking microbial identity to cell morphology (Li et al. 2017; Moran et al. 2023).

In this chapter, we review the main single-cell approaches used to study bacterial infections. We first briefly describe the key methods including scRNA-seq and advanced imaging/flow cytometry techniques that enable single-cell profiling. We then highlight applications of these tools: how they uncover cellular heterogeneity in infected tissues, resolve the composition and behavior of microbial communities, and illuminate dynamic host–pathogen interactions. Throughout, we emphasize findings that were only possible at single-cell resolution and discuss how these high-resolution data complement traditional approaches. Finally, we consider future directions, such as multi-omics and spatial single-cell analyses, which promise to deepen our understanding of infectious disease biology (Papalexi and Satija 2018; Penaranda and Hung 2019).

9.2 Methods for Single-Cell Analysis

9.2.1 Single-Cell RNA Sequencing (scRNA-seq)

Single-cell RNA sequencing (scRNA-seq) profiles the transcriptome of individual cells, providing an unbiased snapshot of gene expression. Protocols such as

Smart-seq2, Drop-seq, inDrop, and 10x Genomics Chromium use microfluidics or cell sorting to isolate single cells, reverse transcribe mRNA, and add molecular barcodes (UMIs) to each transcript (Huang et al. 2021; Lu et al. 2023). These methods differ in throughput and sensitivity: high-throughput droplet methods (e.g., Drop-seq, 10x) can capture thousands to tens of thousands of cells per run, albeit often with shallow coverage per cell, whereas plate-based methods (e.g., Smart-seq2) profile fewer cells (hundreds) with higher gene coverage (Huang et al. 2021; Lu et al. 2023). In either case, scRNA-seq measures gene expression (typically detecting hundreds to thousands of genes per cell), enabling identification of cell types, states, and rare subpopulations that would be masked in bulk assays (Cheng et al. 2023; Huang et al. 2021).

In infectious settings, scRNA-seq is applied to host cells (e.g., immune cells, epithelial cells) and, with specialized protocols, even to pathogens. For example, dual scRNA-seq methods attempt to recover both host and bacterial transcripts from the same infected cell. This approach has revealed "myriad of host transcriptional states during infection," illustrating that each host cell responds differently to invading bacteria (Penaranda and Hung 2019). In practice, detecting bacterial mRNA in scRNA-seq is challenging because many bacteria lack polyadenylated transcripts; methods such as random-priming or targeted enrichment are under development. Nevertheless, researchers have begun profiling infected host cells and the pathogens within them (so-called dual RNA-seq), highlighting dynamic crosstalk at single-cell resolution (Penaranda and Hung 2019).

Single-cell transcriptomics has been applied to study tuberculosis, *Salmonella*, and other bacterial infections, identifying, for instance, subpopulations of host T cells with dysfunctional gene signatures in active disease. In one study of peripheral blood from tuberculosis patients, single-cell RNA-seq revealed subsets of myeloid and lymphoid cells with impaired antigen presentation and cytotoxic programs—insights that bulk analyses cannot provide. As Penaranda and Hung note, single-cell profiling "enabled a greater appreciation of the cellular diversity" in host responses to bacteria (Penaranda and Hung 2019). Likewise, scRNA-seq has uncovered transcriptional heterogeneity in bacterial populations themselves. For example, scRNA-seq or related single-molecule FISH in *Salmonella* revealed that only a fraction of bacteria express key virulence genes within a population, accounting for differences in infectivity. Thus, scRNA-seq illuminates the heterogeneity of both host and pathogen transcripts during infection, revealing rare activated cells and microbial phenotypes that drive disease outcomes (Penaranda and Hung 2019; Davis and Isberg 2016).

Computationally, scRNA-seq data require specialized pipelines. Raw sequencing reads are aligned (often with tools like STAR or kallisto) and unique molecular identifiers used to count transcripts (Lu et al. 2023). Quality control, normalization, and batch correction are crucial to mitigating technical noise and dropouts. Dimensionality reduction and clustering then identify cell subtypes, and differential expression reveals pathways activated in infection. Many bioinformatic tools now integrate multi-dataset analyses and assign cell types with reference atlases (Lu et al. 2023; Cheng et al. 2023). Despite these advances, challenges remain, including sparsity of data, annotation of infected versus bystander cells, and distinguishing signal from contaminating transcripts—all active areas of research (Cheng et al. 2023; Penaranda and Hung 2019).

9.2.2 Mass Cytometry Time-Of-Flight (CyTOF)

Mass cytometry time-of-flight (CyTOF) is a protein-based single-cell method that uses heavy-metal-tagged antibodies to quantify dozens of protein markers per cell. Essentially, CyTOF extends flow cytometry to much higher multiplexing by replacing fluorescent labels with mass tags that are resolved by time-of-flight mass spectrometry. A seminal review notes that CyTOF "provides measurement of over 40 simultaneous cellular parameters at single-cell resolution," dramatically exceeding conventional flow cytometry (Spitzer and Nolan 2016). In a typical CyTOF experiment, immune cells (often from blood or tissue) are stained with panels of 20–50 metal-tagged antibodies against cell-surface and intracellular proteins. The cells are then nebulized into an inductively coupled plasma, ionized, and the metal tags quantified. The result is a high-dimensional proteomic profile of each cell.

CyTOF's high parameterization makes it ideal for dissecting complex immune responses during infection. For example, one can simultaneously measure canonical markers of T cells, B cells, monocytes, etc., along with activation markers and cytokines, all in the same sample. Because CyTOF analyzes individual cells at high throughput (often 10^5–10^6 cells per sample), it can reveal rare cell subsets and subtle shifts in protein expression that occur in infection or inflammation (Spitzer and Nolan 2016). A useful comparison notes that CyTOF has the throughput "required to survey millions of cells from an individual sample," enabling comprehensive immune profiling (Spitzer and Nolan 2016). In practice, typical CyTOF throughput is similar to flow cytometry (up to a few thousand cells per second), so a sample of millions of cells can be acquired in an hour or two.

In CyTOF, data analysis resembles high-dimensional flow cytometry. Specialized tools (SPADE, viSNE, FlowSOM, etc.) are used to cluster millions of cells into immunophenotypes (Table 9.1).

Recent studies have used CyTOF to chart immune responses during infection (e.g., identifying expansion of activated monocytes or cytotoxic T cells), though these reports often focus on host phenotyping rather than direct microbial features. Because CyTOF profiles proteins, it complements scRNA-seq by capturing signaling states, phosphorylation, and surface markers that mRNA assays miss.

9.2.3 Single-Cell Proteomics

Mass-spectrometry-based single-cell proteomics is an emerging field that aims to identify and quantify hundreds to thousands of proteins per cell. Unlike CyTOF, it does not rely on antibodies; rather, it extracts proteins from an isolated single cell, digests them into peptides, and analyzes them by high-sensitivity liquid

Table 9.1 Comparison of key single-cell techniques

Technique	Measured analyte	Resolution/Features	Throughput (cells)	Cell types/ applicability	Notes
ScRNA-seq	Single-cell transcriptomes (mRNA)	~10^2–10^4 genes/cell; genome-wide	High (10^3–10^5 per run)	Eukaryotic cells (host); emerging for bacteria	Detects rare cell types; dropout/noise; limited to polyA (mostly)
Mass Cytometry (CyTOF)	Proteins (antibody binding)	~40–50 markers/cell simultaneously	High (10^5–10^6 per sample)	Immune cells (host); require single-cell suspension	Deep phenotyping of immune populations; requires antibody panels
Single-cell Proteomics	Proteins (mass spec)	Up to 10^3 proteins/cell (emerging)	Low (10–10^2 per day)	Any cell; requires cell lysis	Measures post-translational forms; currently low throughput; technical complexity
Imaging (Microscopy)	Spatial patterns of RNA/protein	Subcellular resolution (~200 nm); spatial context	Low-to-moderate (imaging fields)	All cell types; intact tissues	For example, FISH, MERFISH, immunofluorescence; captures location of cells/molecules
Flow Cytometry/ FACS	Proteins, fluorescent reporters	~10–18 markers; cell scatter; reporter intensity	Very high (10^4–10^5/s)	Bacteria (via fluorogenic reporters) or host cells	Fast quantification of known markers; can sort cells; limited multiplex (vs. CyTOF)

chromatography–tandem mass spectrometry (LC-MS/MS). Recent advances (e.g., SCoPE-MS, nanoPOTS, microfluidic chips) have pushed detection limits to a few hundred proteins per mammalian cell, albeit at low throughput (Nalehua and Zaia 2024).

A key advantage of single-cell proteomics is the ability to detect post-translational modifications and protein isoforms that are invisible to RNA assays. As one review notes, characterizing the proteome of single cells "allows us to gain a greater understanding than afforded by transcriptomics alone" (Nalehua and Zaia 2024). This is especially relevant in infections where bacterial toxins, host signaling, and immune responses are regulated at the protein level. For example, protein-level activity of cytokines or bacterial effectors could be measured directly if isolation methods improve.

However, single-cell proteomics currently faces major technical hurdles. Throughput is very limited, often only tens to hundreds of cells can be analyzed per day, and sensitivity is challenged by the tiny amount of material (a typical mammalian cell has ~0.5–1 ng of total protein) (Nalehua and Zaia 2024). To achieve higher throughput, strategies such as tandem mass tag (TMT)-based isobaric labeling and multiplexing (e.g., plex-DIA) are used, but these can compromise quantitation accuracy (Nalehua and Zaia 2024). There is a trade-off between throughput and depth: increasing speed often sacrifices coverage of low-abundance proteins (Nalehua and Zaia 2024). In practice, single-cell proteomics has mostly been applied to relatively large mammalian cells; extending it to small bacteria (with ~100× less protein) is an area of ongoing development.

In summary, single-cell proteomics holds great promise to complement scRNA and CyTOF by profiling the actual effectors of cellular response. It has already begun to reveal heterogeneity in signaling networks, but widespread application to infection biology awaits improvements in sensitivity and scalability (Nalehua and Zaia 2024).

9.2.4 Microscopy (Spatial and Live-Cell Imaging)

Microscopy encompasses a variety of single-cell and spatial techniques that visualize cells and molecules with high resolution. In infection studies, microscopy can be used both as an imaging readout and as a preparatory step for molecular analysis. Traditional fluorescence microscopy (confocal, widefield, super-resolution) can locate and quantify fluorescently tagged bacteria or host markers in tissue or cell culture. Spatial transcriptomics methods (e.g., MERFISH, seqFISH, Slide-seq, Visium) map gene expression within intact tissue sections, preserving the microenvironment of infection. For example, spatial methods have revealed how gut microbiota form distinct niches in the intestinal mucosa (Madhu et al. 2023). In situ hybridization techniques (RNA-FISH or RNAscope) can detect individual bacterial rRNA or mRNA molecules in host tissues, identifying infected cells and bacterial load.

Fluorescence in situ hybridization (FISH) is especially useful for identifying unculturable pathogens. In a notable example, researchers used rRNA-targeted FISH probes in combination with electron microscopy to link the phylogeny and biomineral structure of single magnetotactic bacteria cells (Li et al. 2017). Although not in an infection context, this "correlated FISH-EM" approach shows the power of single-cell microscopy: one can pinpoint an individual microbe by sequence signature and then examine its morphology by EM. In bacterial infections, a similar approach can label specific pathogen subgroups (via rRNA or gene probes) and then analyze their localization or ultrastructure within host tissues.

Live-cell time-lapse microscopy is another powerful single-cell method for infection. By labeling host cells (e.g., with dyes for organelles) and pathogens (e.g., expressing GFP in bacteria), one can record movies of infection dynamics at single-cell resolution. Recent studies have used high-content imaging to track thousands of infected host cells in parallel. For instance, Dewangan et al. tracked the interactions of *Pseudomonas aeruginosa* and *E. coli* with mammalian cells in real time. They found that *P. aeruginosa* exhibited diverse behaviors (swimming, rotation, firm adhesion) on host surfaces, unlike the more passive *E. coli* (Dewangan et al. 2024). Such experiments reveal that even isogenic bacteria display a variety of motility and adherence patterns at the single-cell level, and that host cell surface interactions are highly heterogeneous.

Another example is shown in Fig. 9.1, where primary macrophages were infected with GFP-labeled *Listeria monocytogenes* and imaged by time-lapse confocal microscopy every hour for approximately 5 hours (Moran et al. 2023). Automated image analysis segmented each macrophage and revealed three distinct outcomes at the single-cell level (Fig. 9.1). Quantification of 358 independent infection events showed that about 32% (±6% SD) of individual bacteria underwent intracellular replication, reaching up to 30 cells by 5 h. A similar fraction (32%, ±8%) of bacteria persisted as single, non-dividing intracellular cells, while the remaining 36% (±8%) of bacteria disappeared (interpreted as clearance in the phagosome) (Moran et al. 2023). After infection, the researchers stained the macrophages to understand where the bacteria were located inside the cells. They used two markers: actin, which surrounds bacteria that escape into the cytoplasm, and LAMP1, which marks the membrane of lysosomes that trap bacteria. Most replicating *Listeria* were surrounded by actin filaments, indicating that they had escaped the phagosome and were growing in the cytoplasm. In contrast, the non-replicating bacteria were often found inside compartments rich in LAMP1, meaning they remained trapped within the phagosome and were unable to divide. A small number of bacteria showed mixed features, suggesting that some could persist for a while without multiplying (Moran et al. 2023). Thus, single-cell microscopy of *Listeria monocytogenes* infection uncovers a spectrum of outcomes, i.e., replication, persistence, or clearance, that arise from heterogeneous host–pathogen interactions and the specific intracellular niche occupied by each bacterium (Moran et al. 2023).

In the context of spatial profiling, advanced methods like spatial host–microbiome sequencing have recently emerged. For example, Lötstedt et al. developed an all-sequencing-based method to capture tissue histology, host transcripts, and

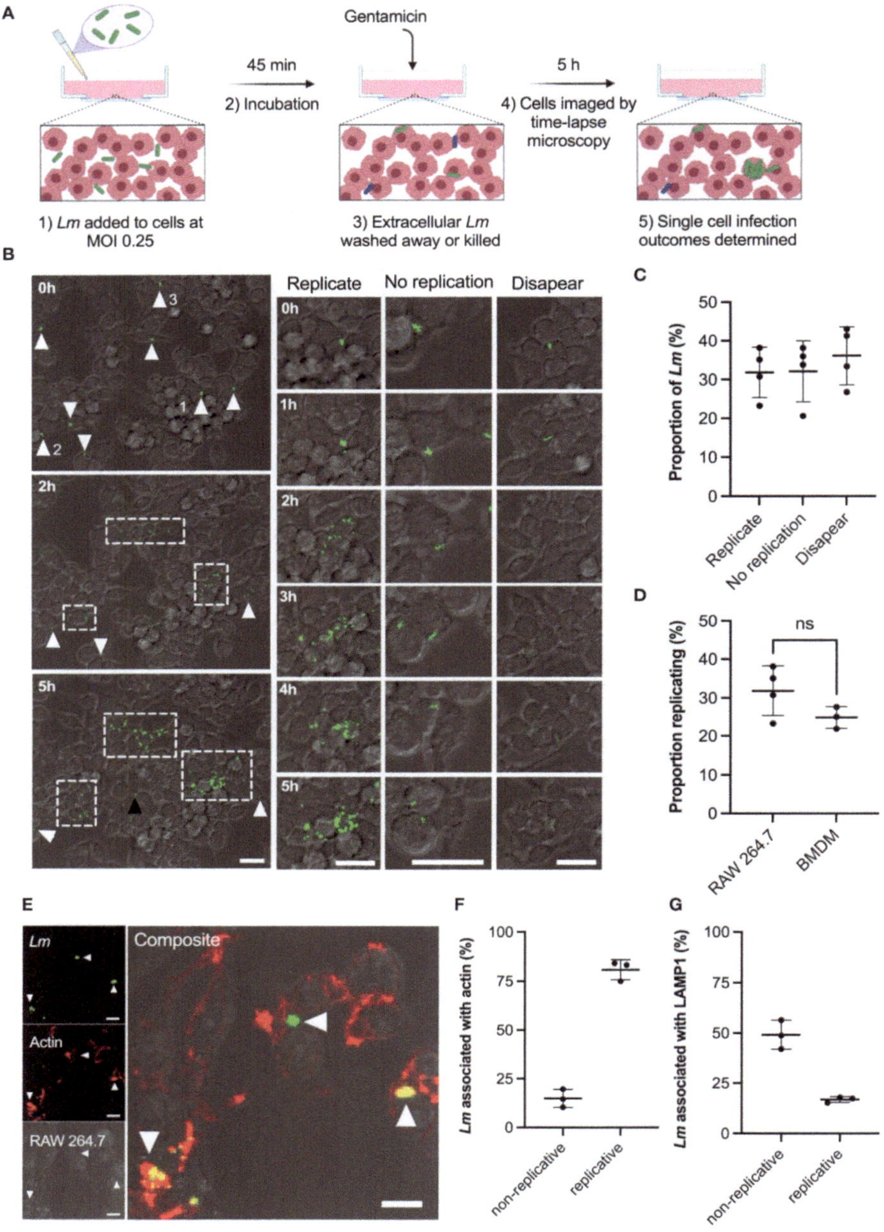

Fig. 9.1 Single-cell live imaging of *L. monocytogenes* (Lm) infection in macrophages reveals divergent fates. (**a**) Infection protocol schematic: RAW 264.7 macrophages were infected with GFP-expressing Lm at MOI 0.25, incubated 45 min, then washed and treated with gentamicin ($t = 0$) to kill extracellular bacteria. Cells were imaged by time-lapse confocal microscopy for 5 h. (**b**) Representative frames from live-cell imaging (time indicated) of infected macrophages

bacterial 16S RNA together on the same slide, enabling mapping of microbial niches in tissues (Lötstedt et al. 2024). Similarly, Madhu et al. review spatial heterogeneity of the gut microbiome, emphasizing that disrupted spatial organization is a common feature of intestinal disease and that single-cell genomics is being used to map microbiome structure and species in situ (Madhu et al. 2023). These spatial imaging approaches complement scRNA-seq by adding anatomical context to the single-cell data.

9.2.5 Flow Cytometry and Cell Sorting

Conventional flow cytometry remains a workhorse for single-cell analysis of both microbes and host cells. In flow cytometry, cells in suspension pass one by one through lasers, and scattered light plus fluorescent labels are recorded. Standard flow can measure on the order of 10–18 fluorescence channels simultaneously (depending on instrument), allowing phenotyping of thousands of cells per second. For bacterial research, flow cytometry is often used with fluorescent reporter strains or dyes. Galbusera et al. note that "fluorescence flow cytometry is increasingly being used to quantify single-cell expression distributions in bacteria in high-throughput" (Galbusera et al. 2020). For example, a bacterial population engineered to express GFP under a certain promoter can be analyzed by flow to reveal the distribution of expression levels across millions of cells. When combined with statistical models, flow data can provide quantitative measures of gene expression noise and detect subpopulations (e.g., persister cells, biofilm producers) (Galbusera et al. 2020).

Flow cytometry also plays a key role in host studies. Cell sorting (FACS) can isolate specific host cell subsets (e.g., infected vs. uninfected, or $CD4^+$ vs. $CD8^+$ T cells) for downstream single-cell analysis. For instance, during infection, one might label phagocytes with a fluorescent bacterial tracker to sort only the cells that have ingested bacteria, then perform scRNA-seq on those cells. This coupling of flow sorting with genomic methods adds an extra layer of selection. However, flow-based techniques do not provide spatial context and are limited to known marker panels. They also cannot directly measure the pathogen's genome or transcriptome (except via reporter fluorescence).

Fig. 9.1 (continued) (brightfield) with Lm-GFP (green). Examples of single bacteria that replicate (left column), remain non-replicative (middle), or disappear (right) over time are highlighted (arrowheads/boxes). (**c**) Quantification of single-bacterium infection outcomes at 5 h: percentage of events leading to bacterial replication, persistence, or disappearance (mean ± SD; $n = 4$ replicates). (**d**) Similar analysis in primary bone-marrow-derived macrophages ($n = 3$) showing the proportion of infections that became replicative (mean ± SD, not significantly different by Mann–Whitney test). (**e**) Fixed-cell fluorescence image (5 h) showing Lm-GFP (green) and F-actin (phalloidin, red). Arrowheads indicate a replicating Lm with a clear actin tail (↓), a non-replicating Lm associated with actin (↑), and a non-replicating Lm lacking actin (←). (f, g) Summary of host marker association: (**f**) Percent of replicating and non-replicating Lm-GFP that co-localize with F-actin (mean ± SD; $n = 3$). (**g**) Percent of replicating and non-replicating Lm-GFP that co-localize with LAMP1 (mean ± SD; $n = 3$). Scale bars: B, 10 μm; E, 10 μm. (Adapted from Moran et al. 2023)

Recent enhancements blur the line between flow cytometry and multi-omics. Index sorting records fluorescence intensities for each sorted cell, allowing matched correlation of phenotype and downstream RNA profiles. More recently, techniques like CITE-seq use oligo-tagged antibodies so that a single scRNA-seq experiment simultaneously measures the transcriptome and dozens of protein markers (captured as sequenced DNA tags) (Cheng et al. 2023). These multimodal assays integrate the strengths of flow cytometry (protein markers) with sequencing.

9.3 Applications of Single-Cell Techniques in Infection

Single-cell methods have begun to reshape our understanding of infectious disease in several key ways. Here we discuss how these techniques have been applied to reveal cellular heterogeneity, map host–pathogen dynamics, analyze microbiomes, and profile infections over time and space.

9.3.1 Revealing Immune and Microbial Heterogeneity

A major impact of single-cell analysis has been the discovery of previously unrecognized cellular diversity. In the immune system, scRNA-seq and CyTOF have uncovered new subtypes of cells that emerge during infection. For example, scRNA-seq of peripheral blood from tuberculosis patients (an active bacterial infection) identified subsets of T cells and monocytes with distinct functional impairments (reduced cytotoxic gene expression, altered signaling pathways) that were not apparent in bulk data (Xiao et al. 2025). Similarly, during sepsis or pneumonia, single-cell profiling has revealed a spectrum of monocyte and neutrophil states ranging from hyper-inflammatory to immunosuppressive phenotypes (Xiao et al. 2025). CyTOF studies of infected tissues (e.g., *Listeria*-infected livers) have detected rare dendritic cell or innate lymphoid cell subsets that correlate with protection or pathology (Spitzer and Nolan 2016).

Conversely, single-cell approaches have also characterized bacterial population heterogeneity during infection. Even within a clonal pathogen population, stochastic gene expression and microenvironmental differences lead to phenotypic variation. More broadly, Davis and Isberg's review emphasizes that bacterial heterogeneity, genetic or phenotypic, often provides a selective advantage during stress, and multiple single-cell approaches have begun to chart this diversity (Davis and Isberg 2016).

Microbiome studies at the single-cell level extend this concept. Instead of analyzing a community as a bulk "soup," single-cell genomics isolates individual microbial cells from a sample (e.g., gut, oral cavity) and sequences their genomes or transcriptomes. Lloréns-Rico et al. note that such single-cell techniques are rapidly being developed "to isolate, culture, and characterize the genomes and transcriptomes of individual microbes in complex communities" (Lloréns-Rico et al. 2022). This allows the discovery of low-abundance or uncultured species, and linking of function to specific cells. In infection contexts (e.g., dysbiosis-associated colitis or lung microbiota in pneumonia), single-cell microbiomics can identify which bacterial strains expand or express virulence factors, adding resolution to the bulk signal.

Single-cell analysis unmasks the heterogeneity hidden in infection. Instead of treating infections as uniform events, researchers can now map how each cell (host or microbe) uniquely participates. This has revealed, for example, that among an infected patient's macrophages, only some cells mount a strong antimicrobial response while others become "permissive niches" for bacteria (Dramé et al. 2025). It also shows that within a bacterial infection, a fraction of cells may produce toxins or enter dormancy, leading to relapse or antibiotic resistance. Understanding these rare but critical subpopulations helps explain why infections can have unpredictable outcomes.

9.3.2 Understanding Host–Pathogen Dynamics

Single-cell techniques enable dynamic views of infection by characterizing the interactions between individual host cells and pathogens. One approach is time-resolved single-cell imaging, as illustrated by the *Listeria monocytogenes* example in Fig. 9.1. By continuously tracking infected host cells, researchers observe that infection outcomes (replication vs. clearance) vary widely from cell to cell. Another such study revealed that *L. pneumophila* replicates "in only a small subset of permissive macrophages" (~15%) (Dramé et al. 2025). This finding challenges the view of infection as a deterministic process; instead, chance differences among host cells play a major role. Tracking also identified early metabolic signatures: cells destined to support bacterial growth showed distinct mitochondrial dynamics (TMRM signals) hours before bacteria began dividing (Dramé et al. 2025). These results suggest that host cell metabolism can precondition infection, insights only visible at single-cell resolution.

In addition to host imaging, single-cell tracking of bacteria on host surfaces provides new perspectives on pathogen behavior. The study by Dewangan et al. employed high-speed microscopy and image analysis to follow *P. aeruginosa* and *E. coli* interacting with mammalian cells (Dewangan et al. 2024). They found that *P. aeruginosa* displayed heterogeneous motility patterns (swimming, adhesion, rotation) and generally higher adherence than *E. coli*, whose cells were more uniformly mobile (Dewangan et al. 2024). Genetic perturbation (deleting the flagellar protein FliD) altered *P. aeruginosa*'s behavior and reduced antibiotic tolerance, indicating that single-cell motion dynamics correlate with infection phenotypes. This study highlights that what one observes on average (bulk adhesion rate, for example) may hide a spectrum of single-cell behaviors, each potentially important for pathogenesis.

Dual single-cell profiling approaches also elucidate host–pathogen crosstalk. For example, simultaneous scRNA-seq of host cells and pathogens (dual RNA-seq) can reveal how individual host immune states are coupled to specific bacterial responses.

Penaranda and Hung discuss how single-cell RNA-seq "within the context of host–pathogen interactions" can clarify these dynamics (Penaranda and Hung 2019). In bacterial infections, a dual single-cell view could similarly pinpoint which host cells preferentially engulf and kill bacteria versus those that fail to respond, and how bacterial gene expression adapts in real time within each cell.

Another dynamic perspective comes from genetic lineage tracing (a non-single-cell technique that nevertheless offers cellular resolution). For instance, barcoding bacteria with unique DNA tags and tracking their descendants in a host can quantify bottlenecks and dissemination paths (Campbell et al. 2026). While not single-cell in measurement, lineage tracing complements single-cell data by showing the clonal spread of pathogens (which single-cell profiling alone might miss). Looking forward, integrated approaches that combine lineage barcodes, scRNA-seq of hosts, and imaging could provide a holistic map of infection dynamics at the single-cell level.

9.3.3 Single-Cell Microbiome Analysis

Beyond acute infections, single-cell approaches are illuminating the microbiome's complexity—the vast communities of bacteria living on and within us that influence health and disease. Single-cell microbial genomics and transcriptomics allow characterization of individual microbiome members without culturing. In practice, microbes from a community sample are isolated (often by flow sorting or microfluidic droplets) and sequenced individually. Lloréns-Rico et al. explain that single-cell techniques are providing "unique insights into the biology and behavior of human microbiomes" by sequencing genomes and transcriptomes of individual microbes (Lloréns-Rico et al. 2022). For example, single-cell genomics has uncovered novel bacterial lineages in the gut and oral cavity whose genomes and metabolic capabilities were previously unknown.

Spatial single-cell microbiome methods add another layer. Madhu et al. review how gut bacteria are spatially organized and that "disrupted spatial organization" is linked to disease (Madhu et al. 2023). By sequencing microbes with their location (using techniques like FISH-seq or spatial metagenomics), researchers can see which bacteria co-localize with inflamed tissues or which niche-specific pathogens occupy. Spatial mapping of the microbiome thus bridges the gap between ecology and host tissue context.

Another microbiome application is tracking pathogen emergence within a community. For example, in a polymicrobial infection (like an abscess), single-cell sequencing could identify which bacteria express virulence genes or produce toxins in vivo. It could also reveal how individual commensals respond to infection or antibiotics. Such detailed profiling may lead to targeted microbiome therapies.

Finally, single-cell analysis is used in synthetic biology and phage therapy contexts. Researchers use single-cell techniques to engineer and track bacteria with novel properties (e.g., engineered probiotics) and monitor their behavior in hosts. Flow cytometry and imaging can detect single engineered cells in the microbiome and ensure they behave as intended. In summary, single-cell microbiome analysis complements infection studies by elucidating which community members are active in disease and how they spatially and functionally interact with the host (Lloréns-Rico et al. 2022; Madhu et al. 2023).

9.3.4 Temporal and Spatial Profiling During Infection

Infections unfold over time and within specific anatomical spaces. Single-cell methods are increasingly used to capture the temporal evolution and spatial architecture of infections.

Temporally, researchers have begun applying scRNA-seq and imaging at multiple timepoints of infection. For example, one might sample infected tissues (lung, spleen, gut) at early, middle, and late stages of bacterial infection and perform single-cell profiling on each. This reveals the trajectory of immune cell differentiation and pathogen gene expression over time. In some studies, time-lapse microscopy directly captures hours-long infection dynamics in single cells, but scRNA-seq can complement by providing "snapshots" at distinct phases (Fig. 9.1). By comparing time-resolved single-cell data, it is possible to infer cell state transitions (e.g., naive → activated → exhausted T cells) and pathogen adaptation events (Dramé et al. 2025).

Spatially, single-cell techniques preserve the location of cells within tissue. In spatial transcriptomics, thin tissue sections are assayed such that the coordinates of each sequenced spot or cell are known. For bacterial infections, spatial methods can map where pathogens localize (e.g., which cells harbor the bacteria) and how nearby immune or stromal cells respond. One pioneering method, spatial host–microbiome RNA-seq (SHM-seq), simultaneously captures host mRNAs and bacterial 16S rRNA on the same tissue grid (Lötstedt et al. 2024). This allows researchers to generate a "map" showing, for example, that specific pathogenic bacteria congregate near certain immune cell subsets, or that host cells adjacent to bacteria express inflammatory genes.

The integration of temporal and spatial data is particularly powerful. The overarching goal is to understand infections as dynamic processes in space and time: for example, identifying that a wave of interferon-producing cells arrives at a granuloma only after 3 days of infection, or that an invasive bacterial colony expands preferentially along a vascular niche. Single-cell methods give the resolution needed to observe these spatiotemporal patterns.

9.3.5 Integration with Multi-omics

Real biological systems involve multiple layers of regulation (genome, epigenome, transcriptome, proteome, metabolome). A major frontier is integrating single-cell measurements across these layers. For instance, techniques like CITE-seq or REAP-seq allow simultaneous measurement of single-cell RNA and dozens of proteins (using oligo-tagged antibodies) (Cheng et al. 2023). Emerging methods also profile the epigenome (scATAC-seq) or CRISPR perturbations alongside the transcriptome. Integrating such datasets can reveal how, say, chromatin accessibility changes underlie transcriptional responses in infection, or how surface receptor levels (measured by CyTOF) correlate with gene expression states in the same cell (Spitzer and Nolan 2016).

Cheng et al. highlight new computational tools that integrate multimodal single-cell data (transcriptomic, epigenomic, proteomic) and even infer cell–cell communication networks (Cheng et al. 2023). In host–pathogen contexts, such multi-omics might involve simultaneously sequencing host RNA and the metabolome of the same cell, or linking immune cell epigenetic state to bacterial burden. Similarly, single-cell proteogenomics aims to relate mRNA and protein from the same cell, clarifying post-transcriptional regulation during infection.

Another integrated approach is multi-organ or systems-level single-cell analysis. For instance, during systemic infection, one might profile blood, spleen, and brain simultaneously at single-cell resolution to see how immune responses propagate through the body. These integrative studies benefit from advanced bioinformatics that can fuse data from different assays and conditions. The result is a more holistic view of infection, connecting molecular changes to cellular behaviors.

9.4 Challenges and Limitations

While single-cell techniques are powerful, they have limitations and pitfalls that must be carefully managed.

9.4.1 Technical and Experimental Issues

Single-cell assays often require dissociating tissues into suspensions, which can perturb cells. For instance, lung or gut dissociation can activate stress pathways or cause the loss of certain cell types. Handling of pathogens poses biosafety issues and can affect RNA integrity. Dropout (missing detection of expressed genes) is inherent in scRNA-seq, especially for low-copy transcripts (like many bacterial genes) (Cheng et al. 2023). CyTOF and flow cytometry are limited to predefined marker panels; if an important protein is not included, it will be missed (Spitzer and Nolan 2016). Single-cell proteomics suffers from low sensitivity; only abundant proteins are reliably detected (Nalehua and Zaia 2024). Imaging approaches typically have lower throughput and may suffer phototoxicity in live-cell experiments.

9.4.2 Computational Challenges

Analyzing single-cell data is nontrivial. ScRNA-seq data are sparse (lots of zeros) and require normalization and batch-correction. Distinguishing true biological variation from technical noise (doublets, ambient RNA contamination) is critical. Tools are rapidly evolving, but incorrect parameter choices can lead to spurious clusters or overlooked subpopulations (Cheng et al. 2023; Lu et al. 2023). Integrating host and pathogen data adds complexity: for dual RNA-seq, one must reliably separate human and bacterial reads, which can be tricky if sequences are similar or if there is environmental contamination. Inferring cell–cell communication (e.g., ligand–receptor analysis) from scRNA data is also inherently indirect and model-dependent.

9.4.3 Statistical Power and Representativeness

Single-cell studies often sample thousands of cells, which is much smaller than bulk populations. Rare infection events (e.g., a very small number of tissue-resident macrophages that control bacteria) may still be missed by limited sampling. Conversely, observed heterogeneity can sometimes reflect random sampling of very few cells rather than true biological variability. Care must be taken to include sufficient replicates and control samples. Validation of single-cell findings with orthogonal methods (e.g., flow cytometry or imaging of larger cohorts) is often necessary.

9.4.4 Interpretation

Single-cell "snapshots" show correlation but not causation. For example, one might observe that a subpopulation of neutrophils expresses a certain gene signature during infection, but whether those cells are driving disease or just responding to it requires further experiment. Similarly, inferring cell trajectories or fate decisions from static scRNA data involves assumptions (e.g., pseudotime inference) that may not fully capture the biology. Spatial data give location but can be lower resolution (e.g., some spatial methods capture multi-cell spots rather than single cells).

9.4.5 Pathogen-Specific Limitations

Applying single-cell methods to the pathogen itself can be challenging. Bacterial cells are much smaller than mammalian cells and contain less nucleic acid, making molecular capture harder. There is also the question of viability: isolating intact single bacteria from tissue for omics is very difficult. Some solutions use microfluidic droplet lysis or whole-genome amplification, but these can introduce biases. For intracellular pathogens, distinguishing intracellular RNA from the host is another hurdle.

Despite these challenges, careful experimental design and methodological improvements are rapidly mitigating many issues. Robust data-processing pipelines and cross-validation have become standard. Importantly, even with noise and dropouts, single-cell methods consistently uncover qualitative phenomena (e.g., existence of subpopulations) that are real and reproducible. It remains essential to corroborating key findings by independent assays.

9.5 Future Directions and Conclusion

Single-cell analysis of bacterial infections is a rapidly evolving field. Technological trends suggest even deeper and faster profiling: droplet-based methods that handle tens of thousands of cells per sample, combined with more gene expression capture, and new microfluidic devices for bacteria. On the proteomic side, improvements in mass spectrometry (e.g., enhanced ionization, carrier proteomes) may soon enable routine quantification of protein networks in hundreds of single cells per day. Spatially resolved omics will continue to mature, with higher resolution and throughput, allowing nearly entire tissue sections to be mapped at single-cell detail.

On the biological front, single-cell atlases of infection are already being constructed. For example, efforts to profile immune cells from many patients across stages of sepsis or COVID-19 (a viral infection) have shown that this approach can identify biomarkers of disease severity. Similar bacterial infection atlases—cataloging immune and microbial cell states—are on the horizon. We also anticipate more dual-species atlases, where host cells and associated microbes in the same microenvironment are jointly profiled.

Data integration and modeling are other frontiers. Sophisticated computational frameworks (machine learning, network inference) will be applied to single-cell infection data. There is excitement around predictive modeling: for instance, the *Legionella* study trained a machine learning model on early metabolic changes that predicted infection outcome in single cells with high accuracy (Dramé et al. 2025). In the future, similar predictive models might forecast disease progression or treatment response from a patient's single-cell profile.

Clinically, single-cell insights are being translated into practice. Understanding heterogeneity may improve diagnostics: for example, identifying a rare immune subset that predicts poor prognosis could lead to more targeted therapies. Vaccines may be guided by single-cell data, as Penaranda et al. suggest that dissecting the integrated host–pathogen process at single-cell resolution "will inform development of vaccines with greater productive and protective host immunity" (Penaranda and Hung 2019). Moreover, single-cell screens could identify host factors essential for bacterial survival, suggesting new drug targets.

In conclusion, single-cell techniques are revolutionizing bacterial infection research by providing a detailed, granular view of host and pathogen biology. As one review optimistically states, single-cell RNA-seq is uncovering the "mechanism of infectious diseases" and inspiring novel therapies and interventions (Huang et al. 2021). By revealing the hidden diversity and dynamics of infection in space, time, and cell types, these methods are filling critical gaps left by bulk analyses. Moving forward, the synergy of single-cell genomics, proteomics, imaging, and computation promises to paint a comprehensive picture of infection biology. This integrative understanding will not only advance basic science but also pave the way for precision approaches to diagnose, monitor, and treat bacterial diseases. The next decade is likely to see single-cell analysis becoming an indispensable tool in immunology, microbiology, and infectious disease medicine, translating detailed cellular insights into better outcomes for patients.

References

Campbell IW, Hullahalli K, Waldor MK (2026) Quantifying host-microbe interactions with bacterial lineage tracing. Science (New York, N.Y.) 391(6780):34–40. https://doi.org/10.1126/science.adx5362

Cheng C, Chen W, Jin H, Chen X (2023) A review of single-cell RNA-Seq annotation, integration, and cell-cell communication. Cells 12(15):1970. https://doi.org/10.3390/cells12151970

Davis KM, Isberg RR (2016) Defining heterogeneity within bacterial populations via single cell approaches. Bioessays 38(8):782–790. https://doi.org/10.1002/bies.201500121

Dewangan NK, Mohiuddin SG, Sensenbach S, Karki P, Orman MA (2024) Uncovering bacterial-mammalian cell interactions via single-cell tracking. BMC Biol 22(1):256. https://doi.org/10.1186/s12915-024-02056-z

Dramé M, Garcia-Rodriguez FJ, Ershov D et al (2025) Backtracking metabolic dynamics in single cells predicts bacterial replication in human macrophages. Nat Commun 16:9189. https://doi.org/10.1038/s41467-025-64225-0

Galbusera L, Bellement-Theroue G, Urchueguia A, Julou T, van Nimwegen E (2020) Using fluorescence flow cytometry data for single-cell gene expression analysis in bacteria. PLoS One 15(10):e0240233. https://doi.org/10.1371/journal.pone.0240233

Huang W, Wang D, Yao YF (2021) Understanding the pathogenesis of infectious diseases by single-cell RNA sequencing. Microbial Cell (Graz, Austria) 8(9):208–222. https://doi.org/10.15698/mic2021.09.759

Lei H, Wang C, Wang Y, Wang C (2021) Single-cell RNA-Seq revealed profound immune alteration in the peripheral blood of patients with bacterial infection. Int J Infect Dis 103:527–535. https://doi.org/10.1016/j.ijid.2020.11.205

Li J, Zhang H, Menguy N, Benzerara K, Wang F, Lin X, Chen Z, Pan Y (2017) Single-cell resolution of uncultured Magnetotactic bacteria via fluorescence-coupled electron microscopy. Appl Environ Microbiol 83(12):e00409–e00417. https://doi.org/10.1128/AEM.00409-17

Li N, Chen S, Xu X, Wang H, Zheng P, Fei X, Ke H, Lei Y, Zhou Y, Yang X, Ouyang Y, Xie C, He C, Hu Y, Cao Y, Li Z, Xie Y, Ge Z, Shu X, Lu N et al (2025) Single-cell transcriptomic profiling uncovers cellular complexity and microenvironment in gastric tumorigenesis associated with Helicobacter pylori. J Adv Res 74:471–491. https://doi.org/10.1016/j.jare.2024.10.012

Lloréns-Rico V, Simcock JA, Huys GRB, Raes J (2022) Single-cell approaches in human microbiome research. Cell 185(15):2725–2738. https://doi.org/10.1016/j.cell.2022.06.040

Lötstedt B, Stražar M, Xavier R et al (2024) Spatial host–microbiome sequencing reveals niches in the mouse gut. Nat Biotechnol 42:1394–1403. https://doi.org/10.1038/s41587-023-01988-1

Lu J, Sheng Y, Qian W, Pan M, Zhao X, Ge Q (2023) scRNA-seq data analysis method to improve analysis performance. IET Nanobiotechnol 17(3):246–256. https://doi.org/10.1049/nbt2.12115

Madhu B, Miller BM, Levy M (2023) Single-cell analysis and spatial resolution of the gut microbiome. Front Cell Infect Microbiol 13:1271092. https://doi.org/10.3389/fcimb.2023.1271092

Moran J, Feltham L, Bagnall J, Goldrick M, Lord E, Nettleton C, Spiller DG, Roberts I, Paszek P (2023) Live-cell imaging reveals single-cell and population-level infection strategies of *Listeria monocytogenes* in macrophages. Front Immunol 14:1235675. https://doi.org/10.3389/fimmu.2023.1235675

Nalehua MR, Zaia J (2024) A critical evaluation of ultrasensitive single-cell proteomics strategies. Anal Bioanal Chem 416(9):2359–2369. https://doi.org/10.1007/s00216-024-05171-6

Papalexi E, Satija R (2018) Single-cell RNA sequencing to explore immune cell heterogeneity. Nat Rev Immunol 18(1):35–45. https://doi.org/10.1038/nri.2017.76

Penaranda C, Hung DT (2019) Single-cell RNA sequencing to understand host-pathogen interactions. ACS Infect Dis 5(3):336–344. https://doi.org/10.1021/acsinfecdis.8b00369

Sesen M, Whyte G (2020) Image-based single cell sorting automation in droplet microfluidics. Sci Rep 10:8736. https://doi.org/10.1038/s41598-020-65483-2

Spitzer MH, Nolan GP (2016) Mass cytometry: single cells, many features. Cell 165(4):780–791. https://doi.org/10.1016/j.cell.2016.04.019

Xiao K, Cao Y, Han Z et al (2025) A pan-immune panorama of bacterial pneumonia revealed by a large-scale single-cell transcriptome atlas. Sig Transduct Target Ther 10:5. https://doi.org/10.1038/s41392-024-02093-8

Single-Cell Analysis for Cancer

10

Zheyu Ding and Shensi Shen

Abstract

Single-cell technologies have transformed cancer research by enabling high-resolution characterization of tumor heterogeneity and the microenvironment. These approaches uncover rare malignant cell states, clonal dynamics, and immunosuppressive subsets that shape tumor progression and therapeutic resistance. They have also advanced biomarker discovery, minimal residual disease monitoring, and prediction of immunotherapy response, opening new avenues for patient stratification and precision oncology. Yet, key challenges remain, including scalability, protocol standardization, and data privacy. Emerging innovations, such as spatial multi-omics, organoid-based systems, and artificial intelligence, promise to bridge these gaps and accelerate clinical translation. Collectively, single-cell technologies represent a paradigm shift in oncology, with profound implications for diagnosis, prognosis, and therapy.

Keywords

Single-cell sequencing · Tumor heterogeneity · Tumor microenvironment · Precision oncology · Spatial omics

10.1 Introduction

Cancer remains one of the leading causes of mortality worldwide, placing a substantial burden on global healthcare systems (Maomao et al. 2022; Klein 2021; Perkins et al. 2023). A major source of its complexity lies in its tumor heterogeneity, which encompasses genetic, epigenetic, and phenotypic variations both within a single tumor (intratumoral heterogeneity) and across patients (intertumoral

Z. Ding · S. Shen (✉)
Department of Thoracic Surgery, Institute of Thoracic Oncology, Frontiers Science Center for Disease-related Molecular Network, West China Hospital, Sichuan University, Chengdu, China

V. Singh (ed.), *Advances in Single Cell Analysis*,
https://doi.org/10.1007/978-981-95-9789-5_10

heterogeneity) (Haffner et al. 2021). This heterogeneity shapes tumor progression, therapeutic response, and resistance. Furthermore, the dynamic interactions between malignant tumor cells and components of the tumor microenvironment (TME), including immune, stromal, and vascular cells, further contribute to the adaptive capacity and complexity of cancer (de Visser and Joyce 2023).

Traditional bulk tissue profiling techniques have provided invaluable insights, yet they average signals from thousands to millions of cells (Li et al. 2021). As a result, they may obscure rare populations or unique states that may critically influence tumor progression, metastasis, or therapy resistance. Bulk approaches also fail to resolve cell-type-specific responses or capture the spatial organization of cells within the TME, thereby limiting strategies for effective personalized treatments (Kuksin et al. 2021).

Single-cell technologies have transformed oncology by enabling high-resolution profiling of individual cells, offering unprecedented insights into tumor and TME heterogeneity (Kuksin et al. 2021). These approaches allow researchers to trace lineage relationships, identify resistant subpopulations such as persister cells, and dissect complex cell–cell interactions that drive cancer progression and therapeutic outcomes (Eyler et al. 2020; Baysoy et al. 2023; Russo et al. 2024). As such, single-cell analyses have become central to both fundamental cancer research and translational precision oncology.

10.2 Key Technologies at Single-Cell Level

10.2.1 ScRNA-seq and SnRNA-seq

Single-cell RNA sequencing (scRNA-seq) and single-nucleus RNA sequencing (snRNA-seq) are widely applied approaches for transcriptomic profiling at single-cell resolution. ScRNA-seq is best suited for fresh tumor tissues in which viable, dissociated cells can be isolated (Slyper et al. 2020).

However, many clinical tumor samples are preserved by freezing or fixation, making intact cell isolation challenging. In these contexts, snRNA-seq, which captures RNA from isolated nuclei, provides a more appropriate alternative (Nickason et al. 2025). This method has proven effective for reliable transcriptomic characterization of frozen samples, including solid tumors such as cervical squamous cell carcinoma (Ou et al. 2022), pancreatic ductal adenocarcinoma (Chen et al. 2025), and glioblastoma (Wang et al. 2024).

10.2.2 Multi-omics Single-Cell Technologies

To more fully elucidate the crosstalk between intracellular and intercellular mechanisms, multi-omics single-cell platforms, often described as multimodal omics approaches, have been established (Vandereyken et al. 2023).

Among these, single-cell transcriptomics is the most established and is often combined with other modalities to connect gene expression with phenotypic heterogeneity in an unbiased manner (Baysoy et al. 2023). For example, genome-transcriptome integration, as achieved by methods such as G&T-seq, enables parallel profiling of genomic DNA and mRNA from the same cell, thereby linking genomic alterations to transcriptional states (Macaulay et al. 2015). Similarly, combining transcriptome with epigenome through approaches such as scATAC-seq reveals chromatin accessibility, transcription factor activity, and the regulatory landscape that shape gene expression (Buenrostro et al. 2015). Another powerful strategy is the integration of transcriptome and proteome, exemplified by CITE-seq, which allows simultaneous measurement of unbiased gene expression profiles and multiplexed surface protein markers (Stoeckius et al. 2017). Collectively, these multimodal approaches provide novel insights into the molecular networks that govern cellular heterogeneity and cancer progression.

10.2.3 Spatial Multi-omics Technologies at Single-Cell Level

The single-cell multi-omics approaches described above require dissociation of cells from their native tissue, thereby losing information on the physical interactions that occur within and between cells (Browaeys et al. 2020; Efremova et al. 2020). To address this limitation, emerging spatial multi-omics technologies at the single-cell level aim to map molecular features and cell–cell interactions within intact tissues at a genome-wide scale (Baysoy et al. 2023; Tang et al. 2023).

Among these, spatial transcriptomics is the most advanced, enabling gene expression profile to be directly mapped onto tissue sections while preserving the spatial context of cellular populations. Techniques such as 10x Visium, Slide-seq (Rodriques et al. 2019), and MERFISH (Chen et al. 2015) have become widely adopted for studying the spatial organization of tumors and other biological systems. More recently, spatial transcriptomics has been combined with additional omics layers, giving rise to integrated platforms that profile genome, epigenome, and proteome features in situ. Notable examples include Slide-DNA-seq (Zhao et al. 2022), Spatial ATAC-seq (Deng et al. 2022a), Spatial ATAC-RNA-seq (Zhang et al. 2023), Spatial CUT&Tag (Deng et al. 2022b), SM-Omics (Vickovic et al. 2022), and Spatial CITE-seq (Liu et al. 2023a).

10.2.4 Data Integration Frameworks and Computational Tools

As single-cell multi-omics data become increasingly complex, effective data integration is essential. The primary goal is to achieve robust and sensitive identification of cell types or cell states, thereby facilitating insights into the regulatory mechanisms and functional diversity of cells (Hao et al. 2021). Argelaguet et al. categorized single-cell data integration strategies into three main classes: horizontal, vertical, and diagonal integration, each tailored to different experimental and analytical contexts (Argelaguet et al. 2021).

Accordingly, computational tools for single-cell multi-omics can be broadly divided into matched and unmatched pipelines. Matched pipelines analyze multiple omics layers measured simultaneously from the same cell, exemplified by tools such as Seurat v4 (Hao et al. 2021) and SCHEMA (Singh et al. 2021). In contrast, unmatched pipelines integrate unaligned datasets generated from separate experiments (Baysoy et al. 2023), with representative tools including LIGER (Welch et al. 2019), GLUE (Cao and Gao 2022), and Seurat v5 (Hao et al. 2024).

10.3 Tumor Heterogeneity Uncovered by Single-Cell Technologies

10.3.1 Intratumoral Heterogeneity: Genetic, Epigenetic, and Transcriptional Diversity

Intratumoral heterogeneity is a major driver of cancer complexity and treatment failure, arising from a variety of sources such as genetic mutations, epigenetic modifications, and transcriptional variability (Mathur et al. 2024). Recent developments in single-cell technologies have enabled high-resolution profiling of this heterogeneity, resulting in the creation of single-cell atlases across multiple tumor types. These efforts have provided valuable insights into the nature of intratumoral heterogeneity, uncovered novel regulatory mechanisms in cancer cells, and offered powerful tools for advancing tumor biology.

For example, Mathur et al. combined 3D neuronavigation during surgical resection with scATAC-seq and scRNA-seq to reveal the sources and spatial patterning of intratumoral heterogeneity, thereby identifying potential therapeutic targets to overcome heterogeneity-driven resistance (Mathur et al. 2024). Similarly, by integrating scRNA-seq and scATAC-seq data from human ovarian and endometrial cancers, Regner et al. identified newly emergent regulatory elements in malignant cells that activate hallmark cancer pathways, underscoring the pivotal contribution of intratumoral heterogeneity (Regner et al. 2021). Taken together, these findings highlight the transformative potential of single-cell analysis for dissecting tumor heterogeneity and informing therapeutic strategies.

10.3.2 Tumor Cell State Identification and Plasticity

Single-cell technologies have uncovered rare and transient cell states that remain hidden in bulk analyses (Shaffer et al. 2017; Tombor et al. 2021). These states, including drug-tolerant persister cells (Shen et al. 2019, 2020a; Wang et al. 2025), cancer stem-like cells (Chan et al. 2021), and invasive phenotypes (Li et al. 2023), are key drivers of tumor progression and therapeutic resistance (Pu et al. 2023; Shen et al. 2020b).

In melanoma, for example, Rambow et al. performed scRNA-seq to profile malignant tumor cells from patients with BRAF mutations and identified four distinct drug-tolerant transcriptional states. Among these, a neural crest stem cell-like state emerged as a key driver of therapeutic resistance (Rambow et al. 2018), illustrating how cellular heterogeneity shapes treatment outcomes. Likewise, in small-cell lung carcinoma (SCLC), Chan et al. analyzed tumors from 21 patients using scRNA-seq and revealed biological complexities not captured by bulk-level sequencing. Their findings showed that SCLC—particularly the SCLC-N subtype—exhibits greater heterogeneity than lung adenocarcinomas and displays substantial subtype plasticity, including interconversion between SCLC-A and SCLC-N subtypes (Chan et al. 2021). These studies underscore the dynamic nature of tumor evolution and highlight the power of single-cell approaches to resolve clinically relevant heterogeneity.

10.3.3 Evolutionary Trajectories and Clonal Architecture

Single-cell sequencing has revealed the evolutionary trajectory of tumors, capturing the spatial and temporal progression of genetic and transcriptional changes (Nolan et al. 2023). Tumor clonal evolution is not a linear process but rather a dynamic interplay among genetic alterations, epigenetic modifications, and microenvironmental influences (Eyler et al. 2020).

For instance, Heiser et al. integrated single-cell and spatial transcriptomic data from colorectal cancers to delineate patient-specific tumor evolutionary trajectories, while simultaneously mapping alterations in the TME and clonal architecture. By stratifying tumors based on their evolutionary trajectories, the study sheds light on the coordinated evolution of colorectal cancer and its tumor microenvironment, offering a conceptual framework that may guide the development of more precise targeted therapies (Heiser et al. 2023). In parallel, single-cell-based lineage tracing enables visualization of tumor evolution at clonal resolution, providing a powerful approach to study tumor progression and identify novel opportunities for therapeutic intervention (Yang et al. 2022a).

10.4 Tumor Microenvironment at Single-Cell Resolution

10.4.1 Characterizing Immune Cell Diversity in Tumors

TME comprises a complex mix of immune, stromal, and vascular cells (de Visser and Joyce 2023). The composition and functional state of immune populations within the TME critically influence tumor growth, metastasis, and response to treatment (Chan et al. 2021). Single-cell data allows for high-resolution characterization of immune subsets, providing insights into their activation, differentiation, and roles in tumor progression (Li et al. 2023).

For example, Chen et al. applied scRNA-seq to analyze the TME in nasopharyngeal carcinoma, generating a refined classification of immune cell populations (Chen et al. 2020). Myeloid cells were subdivided into macrophages, monocytes, plasmacytoid dendritic cells (pDCs), and dendritic cells (DCs); T/NK lineages were categorized into $CD8^+$ T cells, conventional $CD4^+$ T cells, regulatory T cells (Tregs), dysfunctional T cells, and natural killer (NK) cells; and B-cell compartments into B cells, germinal center B cells, $FCRL4^+$ memory B cells, and plasma cells (Chen et al. 2020). Notably, this study also identified novel immune subsets, including $CLEC9A^+$ dendritic cells, thereby establishing a paradigm for immune cell classification within the TME based on single-cell technologies.

10.4.2 Tumor-Infiltrating Lymphocytes and Immunosuppressive Cells

Tumor-infiltrating lymphocytes (TILs) are well-established biomarkers of antitumor immunity and can be routinely assessed in clinical practice using standard hematoxylin and eosin staining (Passaro et al. 2024; Shin et al. 2017). Bridging basic research and clinical translation, single-cell technologies have been instrumental in defining the diversity of TIL subpopulations within the TME. For instance, Zheng et al. performed scRNA-seq on 316 patients across 21 tumor types to construct a pan-cancer RNA atlas of T cells, enabling a systematic comparison of their heterogeneity and dynamics (Zheng et al. 2021). This study revealed that distinct T-cell subsets correlate with patient-specific characteristics, highlighting potential determinants of the TME and providing a foundation for the rational design of precision T-cell-based immunotherapies. Similarly, Yang et al. profiled TIL-derived B cells from 649 patients across 19 tumor types using single-cell RNA sequencing, generating a comprehensive pan-cancer B-cell RNA atlas (Yang et al. 2024). Their analysis uncovered two tumor-enriched subpopulations—stress-response memory B cells and tumor-associated atypical B cells—with prognostic potential across cancers, providing a foundation for deeper exploration of B-cell diversity and function in oncology.

Not all immune populations, however, exert antitumor activity. Immunosuppressive subsets, such as Tregs, myeloid-derived suppressor cells, and tumor-associated macrophages, can attenuate immune response and promote tumor progression (Bremnes et al. 2016). Single-cell technologies have accelerated research in this domain as well, enabling detailed dissection of the heterogeneity and functional states of these inhibitory populations. For instance, Alvisi et al. employed scRNA-seq and complementary approaches to show that Tregs are hyperactivated in intrahepatic cholangiocarcinoma, resulting in severely restricted immune cell infiltration and consequently, a profoundly compromised antitumor immune response (Alvisi et al. 2022).

10.4.3 Response and Resistance to Immune Checkpoint Inhibitors

Immune checkpoint inhibitors (ICIs), specifically anti-PD-1 and anti-CTLA-4 antibodies, act by targeting the dysfunctional immune system and promoting CD8-positive T cells killing cancer cells, thereby revolutionizing the landscape of cancer immunotherapy (Carlino et al. 2021; Wang et al. 2023a). However, patient responses to ICI therapy exhibit significant heterogeneity, which leads to inconsistent clinical outcomes (Dall'Olio et al. 2021). Fortunately, single-cell analyses have been increasingly applied to identify predictive biomarkers of response, such as specific immune cell subsets or the expression patterns of immune checkpoint molecules (Zhang et al. 2021; Yang et al. 2022b; Liu et al. 2022, 2025).

For instance, Liu et al. conducted a meta-analysis of published single-cell datasets and revealed that $CXCL13^+$ $CD8^+$ T cells are positively associated with favorable responses to ICI therapy, with treatment further enriching this subset in responsive tumors (Liu et al. 2022). Zhang et al. performed scRNA-seq on patients with advanced triple-negative breast cancer and identified $CXCL13^+$ $CD8^+$ T cells as key mediators of effective responses to anti-PD-1 therapy. They further showed that paclitaxel treatment diminished the abundance of this T-cell subset, thereby impacting clinical outcomes (Zhang et al. 2021). Collectively, these findings underscore the power of single-cell technologies in resolving immune heterogeneity and advancing the development of robust biomarkers to guide precision immunotherapy.

10.4.4 Predicting Immunotherapy Outcomes Using Single-Cell Data

The predictive potential of single-cell analyses in immunotherapy is immense (Liu et al. 2024). By integrating single-cell sequencing with advanced bioinformatics techniques, it is now possible to forecast patient responses based on the cellular composition and functional states of the immune cells in the TME (Hu et al. 2023).

For example, scRNA-seq and TCR-seq analyses of tumor samples from 234 NSCLC patients treated with neoadjuvant chemo-immunotherapy revealed striking differences in the immune landscape between responders and non-responders (Liu et al. 2025). Samples from patients achieving a major pathological response (MPR) displayed elevated levels of $FGFBP2^+$ NK/NK-like T cells, memory B cells, and effector T cells, whereas non-MPR ones were enriched for $CCR8^+$ regulatory T cells (Liu et al. 2025). In another study, Hu et al. applied scRNA-seq to profile tumors from NSCLC patients receiving neoadjuvant PD-1 blockade combined with chemotherapy, identifying $FCRL4^+FCRL5^+$ memory B cells and $CD16^+CX3CR1^+$ monocytes as predictive biomarkers of immunotherapy response (Hu et al. 2023). Together, these findings underscore how single-cell technologies can explore immune heterogeneity and uncover cellular predictors of immunotherapy efficacy.

10.5 Clinical Applications and Translational Insights

10.5.1 Biomarker Discovery and Patient Stratification

The ability to identify rare cell populations and their associated molecular signatures is crucial for developing effective biomarkers (Passaro et al. 2024). Single-cell technologies have facilitated the identification of biomarkers capable of predicting disease progression, therapeutic response, and relapse (Wang et al. 2023b; Li et al. 2024; Wu et al. 2024; Deng et al. 2024; Dong et al. 2024). For example, through an integrative analysis of bulk, single cell, and spatial transcriptomics validated by fluorescent staining, Li et al. identified PSME2 as a pan-cancer biomarker correlated with M1 macrophage infiltration (Li et al. 2024). In osteosarcoma models, PSME2 overexpression notably suppressed tumor cell proliferation, migration, and invasion (Li et al. 2024). These results indicate that PSME2, reflecting M1 macrophage infiltration, could serve both as a prognostic marker and a potential therapeutic target.

Moreover, single-cell profiling can be used to stratify patients based on their tumor's cellular makeup. For instance, a recent large-scale scRNA-seq analysis of normal breast tissue and primary breast tumors showed that, relative to normal epithelial cells, luminal progenitor cells display increased sensitivity to neoadjuvant chemotherapy, PARP inhibitors, and immunotherapy (Gao et al. 2024). Additionally, the study conducted both in vitro and in vivo validation of PLK1 as an LP-subtype-specific therapeutic target, thereby laying the foundation for precise prognosis and patient stratification in breast cancer (Gao et al. 2024).

10.5.2 Monitoring Minimal Residual Disease

Minimal residual disease (MRD) represents the minute fraction of malignant cells that escape eradication following therapy and lie beyond the detection limits of existing imaging and screening approaches (Pantel and Alix-Panabières 2019; Costa et al. 2022). Detecting MRD is closely linked to subsequent cancer recurrence and can provide an early warning period of months before clinical relapse is apparent (Medford et al. 2023). Consequently, the study of MRD carries significant clinical implications for early intervention and improved patient management. Due to their sensitivity, single-cell technologies have become powerful tools for detecting and characterizing MRD.

Because MRD is routinely applied in the clinical management of hematologic malignancies—such as leukemia, lymphoma, and multiple myeloma—single-cell technologies have been more widely investigated in these cancers (Ledergor et al. 2018; Cui et al. 2024; Zhang et al. 2022). For example, Zhang et al. employed scRNA-seq in combination with B-cell receptor sequencing to investigate B-cell acute lymphoblastic leukemia and discovered that inhibition of hypoxia signaling pathways sensitizes leukemic cells to chemotherapy (Zhang et al. 2022). Furthermore, MRD-related studies in solid tumors are equally important. For instance, Lemaitre et al. applied single-cell spatial transcriptomics to hepatocellular carcinoma patients and revealed that PD-L1$^+$ macrophages maintain MRD by activating TGFβ signaling in stem-like cancer cells, indicating that disrupting this interaction could prevent MRD-driven HCC relapse (Lemaitre et al. 2024). Together, these studies illustrate how single-cell technologies are reshaping MRD research and hold the promise to enable earlier detection, deeper mechanistic insight, and more effective intervention strategies across both hematologic and solid tumors.

10.5.3 Single-Cell-Based Precision Oncology

Progress in single-cell sequencing is transforming cancer research and is expected to become a cornerstone of translational oncology (Melnekoff and Laganà 2022). By enabling the profiling of individual tumor cells, single-cell analysis reveals key mutations, gene-expression patterns, and drug sensitivities that are specific to each patient's tumor (Ding et al. 2022; Krishna et al. 2021; Franken et al. 2024; Valdes-Mora et al. 2018). These insights hold substantial promise for precision oncology, guiding the development of personalized treatment regimens.

For instance, Sinha et al. leveraged publicly available bulk and single-cell expression datasets from large-scale drug screens to build a computational pipeline, PERCEPTION, tailored for precision oncology (Sinha et al. 2024). This framework accurately predicted drug responses in cultured cell lines and patient-derived primary tumor cells and was further validated in clinical trials for multiple myeloma and breast cancer. Importantly, it also captured the emergence of resistance in lung cancer patients treated with tyrosine kinase inhibitors. This study represents how single-cell omics can be leveraged to stratify patients, anticipate therapeutic resistance, and inform clinical decision-making.

10.5.4 Single-Cell Profiling in Circulating Tumor Cells

Circulating tumor cells (CTCs) are rare cancer cells that have slough off from the primary tumor into the bloodstream to circulate, and their presence is associated with metastasis (Hu et al. 2023). By using single-cell profiling for CTCs, researchers can track the evolution of metastatic disease, evaluate metastatic potential, identify markers for early detection of relapse, and help clinicians for treatment personalization (Pailler et al. 2019; Lei et al. 2021; Zhu et al. 2018; Zhang et al. 2024; Lin et al. 2021).

For example, Liu et al. performed scRNA-seq on CTCs, primary tumors, and metastatic lesions from patients with pancreatic ductal adenocarcinoma, revealing that platelet-derived RGS18 protects CTCs from NK cell-mediated immune surveillance through engagement of the HLA-E:CD94-NKG2A immune checkpoint axis. This study represents a paradigm for understanding how CTCs evade immune surveillance (Liu et al. 2023b). Similarly, Sun et al. conducted scRNA-seq across CTCs from multiple cancer types and discovered that CTCs primarily evade NK cell-mediated innate immune surveillance through CD155–TIGIT immune checkpoint interactions (Sun et al. 2025). Together, these studies highlight the critical role of single-cell analysis of CTCs in uncovering mechanisms of immune evasion and guiding the development of targeted therapies for metastatic cancer.

10.6 Perspectives

10.6.1 Technical Limitations and Scalability Issues

Despite the transformative potential, single-cell technologies face several technical hurdles that limit their broader application in oncology. A major challenge lies in sample quality, as single cells need to be isolated rapidly and accurately with minimal damage. Mechanical dissociation can compromise cell integrity, while enzymatic digestion used to release single cells from tissue may alter cell viability or transcriptional states (Ding et al. 2020). Furthermore, technical noise introduced during library preparation, amplification, or sequencing can obscure true biological variation, complicating downstream interpretation (Hwang et al. 2018).

Beyond technical precision, scalability represents another critical bottleneck. Current single-cell platforms remain resource-intensive in both cost and computational demand, restricting their application in large patient cohorts or longitudinal clinical trials (Li et al. 2020). The high expenses associated with sequencing depth and multimodal profiling further impede their integration into routine diagnostics. Overcoming these challenges will require innovations in microfluidic technologies, cost-efficient library preparation, and high-throughput automation pipelines that can lower costs while preserving accuracy (Ziegenhain et al. 2017). Without such improvements, the scalability of single-cell technologies in real-world clinical settings will remain a formidable barrier.

10.6.2 Standardization and Reproducibility in Clinical Settings

The clinical application of single-cell technologies remains limited due to the absence of standardized experimental and analytical procedures (Lähnemann et al. 2020). Differences in tissue handling, dissociation protocols, sequencing depth, and data processing pipelines often yield divergent outcomes (Mereu et al. 2020). Such variability can obscure biological signals and reduce reproducibility across laboratories. The absence of universally accepted reference standards makes it difficult to benchmark emerging methods, slowing the pace of clinical adoption.

To overcome these barriers, harmonization of protocols and quality control frameworks is urgently needed (You et al. 2024). Establishing standardized guidelines for tissue collection, preservation, sequencing, and data analysis will ensure comparability across studies and clinical centers. Consistent quality control and regulatory frameworks will be essential for integrating single-cell profiling into precision oncology and for building the trust necessary to apply these tools in clinical decision-making.

10.6.3 Data Privacy and Ethical Concerns in Single-Cell Cancer Data

The generation of high-resolution single-cell datasets raises important challenges in data governance and ethics (Wang et al. 2023c). Because these datasets contain highly detailed genetic information, they pose a risk of patient re-identification. Robust frameworks for data encryption, secure storage, and controlled sharing will therefore be essential to protecting patient privacy (Gürsoy et al. 2020; Tadi et al. 2024).

Beyond privacy, broader ethical considerations must also be addressed. The use of patient-derived samples necessitates transparent and comprehensive informed consent processes, explicitly outlining potential risks of tissue collection, intended use of the data, and policies for data sharing within the research community (Lee et al. 2024). Addressing these issues will require a multidisciplinary effort, uniting oncologists, ethicists, policymakers, and patient advocates to establish ethical and legal frameworks that protect patient rights and sustain public trust while fostering scientific innovation.

10.6.4 Future Perspectives: Organoid Modeling and AI Integration

Future directions in single-cell cancer research will increasingly rely on the convergence of advanced experimental platforms and computational innovations. Organoid models, when integrated with spatial multi-omics technologies, provide unprecedented spatial and functional insights into tumor ecosystems (Passaro et al. 2024; Fukushima et al. 2025). These integrative approaches enhance the precision of biomarker-driven drug development and help overcome challenges posed by intratumoral heterogeneity, ultimately enabling the design of more personalized and effective therapeutic strategies (Passaro et al. 2024).

In parallel, rapid progress in artificial intelligence, particularly with large language models (Ding et al. 2024), has begun to reshape single-cell data analyses. The intersection of single-cell profiling and large language models offers intuitive, user-friendly tools that reduce technical barriers, allowing researchers without extensive computational expertise to explore and interpret complex single-cell datasets with greater efficiency (Ding et al. 2024; Yang et al. 2022c; Hou and Ji 2024).

Looking ahead, single-cell technologies are poised to transform both cancer research and precision oncology (Chan et al. 2021; Li et al. 2023). By integrating state-of-the-art experimental approaches with advanced computational models, these tools will deepen our understanding of tumor heterogeneity, tumor–microenvironment interactions, and therapeutic resistance. Ultimately, the integration of single-cell profiling into clinical practice holds the promise to move oncology from generalized treatment paradigms toward truly individualized strategies that anticipate disease trajectories and optimize therapeutic outcomes.

Competing Interests S. Shen reports personal fees from Agence nationale de la recherche (France), Krebsliga Schweiz (Switzerland), KWF Kankerbestrijding (Netherlands), Austrian Research Funding, and Belgian Foundation against Cancer, and serving as an Associate Editor for *Oncogenesis* (Springer Nature, London, UK); no other disclosures were reported.

References

Alvisi G et al (2022) Multimodal single-cell profiling of intrahepatic cholangiocarcinoma defines hyperactivated Tregs as a potential therapeutic target. J Hepatol 77:1359–1372. https://doi.org/10.1016/j.jhep.2022.05.043

Argelaguet R, Cuomo ASE, Stegle O, Marioni JC (2021) Computational principles and challenges in single-cell data integration. Nat Biotechnol 39:1202–1215. https://doi.org/10.1038/s41587-021-00895-7

Baysoy A, Bai Z, Satija R, Fan R (2023) The technological landscape and applications of single-cell multi-omics. Nat Rev Mol Cell Biol 24:695–713. https://doi.org/10.1038/s41580-023-00615-w

Bremnes RM et al (2016) The role of tumor-infiltrating lymphocytes in development, progression, and prognosis of non-small cell lung cancer. J Thorac Oncol 11:789–800. https://doi.org/10.1016/j.jtho.2016.01.015

Browaeys R, Saelens W, Saeys Y (2020) NicheNet: modeling intercellular communication by linking ligands to target genes. Nat Methods 17:159–162. https://doi.org/10.1038/s41592-019-0667-5

Buenrostro JD et al (2015) Single-cell chromatin accessibility reveals principles of regulatory variation. Nature 523:486–490. https://doi.org/10.1038/nature14590

Cao ZJ, Gao G (2022) Multi-omics single-cell data integration and regulatory inference with graph-linked embedding. Nat Biotechnol 40:1458–1466. https://doi.org/10.1038/s41587-022-01284-4

Carlino MS, Larkin J, Long GV (2021) Immune checkpoint inhibitors in melanoma. Lancet (London, England) 398:1002–1014. https://doi.org/10.1016/s0140-6736(21)01206-x

Chan JM et al (2021) Signatures of plasticity, metastasis, and immunosuppression in an atlas of human small cell lung cancer. Cancer Cell 39:1479–1496.e1418. https://doi.org/10.1016/j.ccell.2021.09.008

Chen KH, Boettiger AN, Moffitt JR, Wang S, Zhuang X (2015) RNA imaging. Spatially resolved, highly multiplexed RNA profiling in single cells. Science 348:aaa6090. https://doi.org/10.1126/science.aaa6090

Chen YP et al (2020) Single-cell transcriptomics reveals regulators underlying immune cell diversity and immune subtypes associated with prognosis in nasopharyngeal carcinoma. Cell Res 30:1024–1042. https://doi.org/10.1038/s41422-020-0374-x

Chen MM et al (2025) Integrated single-cell and spatial transcriptomics uncover distinct cellular subtypes involved in neural invasion in pancreatic cancer. Cancer Cell. https://doi.org/10.1016/j.ccell.2025.06.020

Costa LJ et al (2022) Daratumumab, carfilzomib, lenalidomide, and dexamethasone with minimal residual disease response-adapted therapy in newly diagnosed multiple myeloma. J Clin Oncol 40:2901–2912. https://doi.org/10.1200/jco.21.01935

Cui J et al (2024) Identification of therapy-induced clonal evolution and resistance pathways in minimal residual clones in multiple myeloma through single-cell sequencing. Clin Cancer Res 30:3919–3936. https://doi.org/10.1158/1078-0432.Ccr-24-0545

Dall'Olio FG et al (2021) Tumour burden and efficacy of immune-checkpoint inhibitors. Nat Rev Clin Oncol 19:75–90. https://doi.org/10.1038/s41571-021-00564-3

de Visser KE, Joyce JA (2023) The evolving tumor microenvironment: from cancer initiation to metastatic outgrowth. Cancer Cell 41:374–403. https://doi.org/10.1016/j.ccell.2023.02.016

Deng Y et al (2022a) Spatial profiling of chromatin accessibility in mouse and human tissues. Nature 609:375–383. https://doi.org/10.1038/s41586-022-05094-1

Deng Y et al (2022b) Spatial-CUT&Tag: spatially resolved chromatin modification profiling at the cellular level. Science 375:681–686. https://doi.org/10.1126/science.abg7216
Deng Y et al (2024) Multicellular ecotypes shape progression of lung adenocarcinoma from ground-glass opacity toward advanced stages. Cell Rep Med 5:101489. https://doi.org/10.1016/j.xcrm.2024.101489
Ding S, Chen X, Shen K (2020) Single-cell RNA sequencing in breast cancer: understanding tumor heterogeneity and paving roads to individualized therapy. Cancer Commun (London, England) 40:329–344. https://doi.org/10.1002/cac2.12078
Ding K et al (2022) Single cell heterogeneity and evolution of breast cancer bone metastasis and organoids reveals therapeutic targets for precision medicine. Ann Oncol 33:1085–1088. https://doi.org/10.1016/j.annonc.2022.06.005
Ding Z et al (2024) Exploring the potential of large language model–based chatbots in challenges of ribosome profiling data analysis: a review. Brief Bioinform 26. https://doi.org/10.1093/bib/bbae641
Dong B et al (2024) Plasma proteometabolome in lung cancer: exploring biomarkers through bidirectional Mendelian randomization and colocalization analysis. Hum Mol Genet 33:1688–1696. https://doi.org/10.1093/hmg/ddae110
Efremova M, Vento-Tormo M, Teichmann SA, Vento-Tormo R (2020) CellPhoneDB: inferring cell-cell communication from combined expression of multi-subunit ligand-receptor complexes. Nat Protoc 15:1484–1506. https://doi.org/10.1038/s41596-020-0292-x
Eyler CE et al (2020) Single-cell lineage analysis reveals genetic and epigenetic interplay in glioblastoma drug resistance. Genome Biol 21:174. https://doi.org/10.1186/s13059-020-02085-1
Franken A et al (2024) CD4(+) T cell activation distinguishes response to anti-PD-L1+anti-CTLA4 therapy from anti-PD-L1 monotherapy. Immunity 57:541–558.e547. https://doi.org/10.1016/j.immuni.2024.02.007
Fukushima T et al (2025) An organoid library unveils subtype-specific IGF-1 dependency via a YAP-AP1 axis in human small cell lung cancer. Nat Cancer 6:874–891. https://doi.org/10.1038/s43018-025-00945-y
Gao ZJ et al (2024) Single-cell analyses reveal evolution mimicry during the specification of breast cancer subtype. Theranostics 14:3104–3126. https://doi.org/10.7150/thno.96163
Gürsoy G et al (2020) Data sanitization to reduce private information leakage from functional genomics. Cell 183:905–917.e916. https://doi.org/10.1016/j.cell.2020.09.036
Haffner MC et al (2021) Genomic and phenotypic heterogeneity in prostate cancer. Nat Rev Urol 18:79–92. https://doi.org/10.1038/s41585-020-00400-w
Hao Y et al (2021) Integrated analysis of multimodal single-cell data. Cell 184:3573–3587.e3529. https://doi.org/10.1016/j.cell.2021.04.048
Hao Y et al (2024) Dictionary learning for integrative, multimodal and scalable single-cell analysis. Nat Biotechnol 42:293–304. https://doi.org/10.1038/s41587-023-01767-y
Heiser CN et al (2023) Molecular cartography uncovers evolutionary and microenvironmental dynamics in sporadic colorectal tumors. Cell 186:5620–5637.e5616. https://doi.org/10.1016/j.cell.2023.11.006
Hou W, Ji Z (2024) Assessing GPT-4 for cell type annotation in single-cell RNA-seq analysis. Nat Methods 21:1462–1465. https://doi.org/10.1038/s41592-024-02235-4
Hu J et al (2023) Tumor microenvironment remodeling after neoadjuvant immunotherapy in non-small cell lung cancer revealed by single-cell RNA sequencing. Genome Med 15:14. https://doi.org/10.1186/s13073-023-01164-9
Hwang B, Lee JH, Bang D (2018) Single-cell RNA sequencing technologies and bioinformatics pipelines. Exp Mol Med 50:1–14. https://doi.org/10.1038/s12276-018-0071-8
Klein AP (2021) Pancreatic cancer epidemiology: understanding the role of lifestyle and inherited risk factors. Nat Rev Gastroenterol Hepatol 18:493–502. https://doi.org/10.1038/s41575-021-00457-x
Krishna C et al (2021) Single-cell sequencing links multiregional immune landscapes and tissue-resident T cells in ccRCC to tumor topology and therapy efficacy. Cancer Cell 39:662–677.e666. https://doi.org/10.1016/j.ccell.2021.03.007

Kuksin M et al (2021) Applications of single-cell and bulk RNA sequencing in onco-immunology. Eur J Cancer (Oxford, England: 1990) 149:193–210. https://doi.org/10.1016/j.ejca.2021.03.005
Lähnemann D et al (2020) Eleven grand challenges in single-cell data science. Genome Biol 21:31. https://doi.org/10.1186/s13059-020-1926-6
Ledergor G et al (2018) Single cell dissection of plasma cell heterogeneity in symptomatic and asymptomatic myeloma. Nat Med 24:1867–1876. https://doi.org/10.1038/s41591-018-0269-2
Lee AT, Chang EF, Paredes MF, Nowakowski TJ (2024) Large-scale neurophysiology and single-cell profiling in human neuroscience. Nature 630:587–595. https://doi.org/10.1038/s41586-024-07405-0
Lei Y et al (2021) Applications of single-cell sequencing in cancer research: progress and perspectives. J Hematol Oncol 14:91. https://doi.org/10.1186/s13045-021-01105-2
Lemaitre L et al (2024) Spatial analysis reveals targetable macrophage-mediated mechanisms of immune evasion in hepatocellular carcinoma minimal residual disease. Nat Cancer 5:1534–1556. https://doi.org/10.1038/s43018-024-00828-8
Li B et al (2020) Cumulus provides cloud-based data analysis for large-scale single-cell and single-nucleus RNA-seq. Nat Methods 17:793–798. https://doi.org/10.1038/s41592-020-0905-x
Li Y, Ma L, Wu D, Chen G (2021) Advances in bulk and single-cell multi-omics approaches for systems biology and precision medicine. Brief Bioinform 22. https://doi.org/10.1093/bib/bbab024
Li Y et al (2023) Aurora A kinase inhibition induces accumulation of SCLC tumor cells in mitosis with restored interferon signaling to increase response to PD-L1. Cell Rep Med 4. https://doi.org/10.1016/j.xcrm.2023.101282
Li R et al (2024) PSME2 offers value as a biomarker of M1 macrophage infiltration in pan-cancer and inhibits osteosarcoma malignant phenotypes. Int J Biol Sci 20:1452–1470. https://doi.org/10.7150/ijbs.90226
Lin D et al (2021) Circulating tumor cells: biology and clinical significance. Signal Transduct Target Ther 6:404. https://doi.org/10.1038/s41392-021-00817-8
Liu B, Zhang Y, Wang D, Hu X, Zhang Z (2022) Single-cell meta-analyses reveal responses of tumor-reactive CXCL13(+) T cells to immune-checkpoint blockade. Nat Cancer 3:1123–1136. https://doi.org/10.1038/s43018-022-00433-7
Liu Y et al (2023a) High-plex protein and whole transcriptome co-mapping at cellular resolution with spatial CITE-seq. Nat Biotechnol 41:1405–1409. https://doi.org/10.1038/s41587-023-01676-0
Liu X et al (2023b) Immune checkpoint HLA-E:CD94-NKG2A mediates evasion of circulating tumor cells from NK cell surveillance. Cancer Cell 41:272–287.e279. https://doi.org/10.1016/j.ccell.2023.01.001
Liu H et al (2024) Integrated analysis of single-cell and bulk RNA sequencing data reveals a myeloid cell-related regulon predicting neoadjuvant immunotherapy response across cancers. J Transl Med 22:486. https://doi.org/10.1186/s12967-024-05123-9
Liu Z et al (2025) A single-cell atlas reveals immune heterogeneity in anti-PD-1-treated non-small cell lung cancer. Cell 188:3081–3096.e3019. https://doi.org/10.1016/j.cell.2025.03.018
Macaulay IC et al (2015) G&T-seq: parallel sequencing of single-cell genomes and transcriptomes. Nat Methods 12:519–522. https://doi.org/10.1038/nmeth.3370
Maomao C et al (2022) Current cancer burden in China: epidemiology, etiology, and prevention. Cancer Biol Med 19:1121–1138. https://doi.org/10.20892/j.issn.2095-3941.2022.0231
Mathur R et al (2024) Glioblastoma evolution and heterogeneity from a 3D whole-tumor perspective. Cell 187:446–463.e416. https://doi.org/10.1016/j.cell.2023.12.013
Medford AJ et al (2023) Molecular residual disease in breast cancer: detection and therapeutic interception. Clin Cancer Res 29:4540–4548. https://doi.org/10.1158/1078-0432.Ccr-23-0757
Melnekoff DT, Laganà A (2022) Single-cell sequencing technologies in precision oncology. Adv Exp Med Biol 1361:269–282. https://doi.org/10.1007/978-3-030-91836-1_15
Mereu E et al (2020) Benchmarking single-cell RNA-sequencing protocols for cell atlas projects. Nat Biotechnol 38:747–755. https://doi.org/10.1038/s41587-020-0469-4

Nickason CC et al (2025) A toolkit for single-nucleus characterization of glioblastoma. Methods Mol Biol 2944:227–237. https://doi.org/10.1007/978-1-0716-4654-0_18
Nolan E, Lindeman GJ, Visvader JE (2023) Deciphering breast cancer: from biology to the clinic. Cell 186:1708–1728. https://doi.org/10.1016/j.cell.2023.01.040
Ou Z et al (2022) Single-nucleus RNA sequencing and spatial transcriptomics reveal the immunological microenvironment of cervical squamous cell carcinoma. Adv Sci (Weinheim, Baden-Wurttemberg, Germany) **9**:e2203040. https://doi.org/10.1002/advs.202203040
Pailler E et al (2019) Acquired resistance mutations to ALK inhibitors identified by single circulating tumor cell sequencing in ALK-rearranged non-small-cell lung cancer. Clin Cancer Res 25:6671–6682. https://doi.org/10.1158/1078-0432.Ccr-19-1176
Pantel K, Alix-Panabières C (2019) Liquid biopsy and minimal residual disease—latest advances and implications for cure. Nat Rev Clin Oncol 16:409–424. https://doi.org/10.1038/s41571-019-0187-3
Passaro A et al (2024) Cancer biomarkers: emerging trends and clinical implications for personalized treatment. Cell 187:1617–1635. https://doi.org/10.1016/j.cell.2024.02.041
Perkins RB, Wentzensen N, Guido RS, Schiffman M (2023) Cervical cancer screening: a review. JAMA 330:547–558. https://doi.org/10.1001/jama.2023.13174
Pu Y et al (2023) Drug-tolerant persister cells in cancer: the cutting edges and future directions. Nat Rev Clin Oncol 20:799–813. https://doi.org/10.1038/s41571-023-00815-5
Rambow F et al (2018) Toward minimal residual disease-directed therapy in melanoma. Cell 174:843–855 e819. https://doi.org/10.1016/j.cell.2018.06.025
Regner MJ et al (2021) A multi-omic single-cell landscape of human gynecologic malignancies. Mol Cell 81:4924–4941.e4910. https://doi.org/10.1016/j.molcel.2021.10.013
Rodriques SG et al (2019) Slide-seq: a scalable technology for measuring genome-wide expression at high spatial resolution. Science 363:1463–1467. https://doi.org/10.1126/science.aaw1219
Russo M et al (2024) Cancer drug-tolerant persister cells: from biological questions to clinical opportunities. Nat Rev Cancer 24:694–717. https://doi.org/10.1038/s41568-024-00737-z
Shaffer SM et al (2017) Rare cell variability and drug-induced reprogramming as a mode of cancer drug resistance. Nature 546:431–435. https://doi.org/10.1038/nature22794
Shen S et al (2019) An epitranscriptomic mechanism underlies selective mRNA translation remodelling in melanoma persister cells. Nat Commun 10:5713. https://doi.org/10.1038/s41467-019-13360-6
Shen S et al (2020a) Melanoma Persister cells are tolerant to BRAF/MEK inhibitors via ACOX1-mediated fatty acid oxidation. Cell Rep 33:108421. https://doi.org/10.1016/j.celrep.2020.108421
Shen S, Vagner S, Robert C (2020b) Persistent cancer cells: the deadly survivors. Cell 183:860–874. https://doi.org/10.1016/j.cell.2020.10.027
Shin DS et al (2017) Primary resistance to PD-1 blockade mediated by JAK1/2 mutations. Cancer Discov 7:188–201. https://doi.org/10.1158/2159-8290.Cd-16-1223
Singh R, Hie BL, Narayan A, Berger B (2021) Schema: metric learning enables interpretable synthesis of heterogeneous single-cell modalities. Genome Biol 22:131. https://doi.org/10.1186/s13059-021-02313-2
Sinha S et al (2024) PERCEPTION predicts patient response and resistance to treatment using single-cell transcriptomics of their tumors. Nat Cancer 5:938–952. https://doi.org/10.1038/s43018-024-00756-7
Slyper M et al (2020) A single-cell and single-nucleus RNA-Seq toolbox for fresh and frozen human tumors. Nat Med 26:792–802. https://doi.org/10.1038/s41591-020-0844-1
Stoeckius M et al (2017) Simultaneous epitope and transcriptome measurement in single cells. Nat Methods 14:865–868. https://doi.org/10.1038/nmeth.4380
Sun Y et al (2025) Platelet-mediated circulating tumor cell evasion from natural killer cell killing through immune checkpoint CD155-TIGIT. Hepatology (Baltimore, Md.) 81:791–807. https://doi.org/10.1097/hep.0000000000000934

Tadi AA, Alhadidi D, Rueda L (2024) PPPCT: privacy-preserving framework for parallel clustering transcriptomics data. Comput Biol Med 173:108351. https://doi.org/10.1016/j.compbiomed.2024.108351
Tang X et al (2023) Spatial patterns of the cap-binding complex eIF4F in human melanoma cells. Comput Struct Biotechnol J 21:1157–1168. https://doi.org/10.1016/j.csbj.2023.01.040
Tombor LS et al (2021) Single cell sequencing reveals endothelial plasticity with transient mesenchymal activation after myocardial infarction. Nat Commun 12:681. https://doi.org/10.1038/s41467-021-20905-1
Valdes-Mora F et al (2018) Single-cell transcriptomics in cancer immunobiology: the future of precision oncology. Front Immunol 9:2582. https://doi.org/10.3389/fimmu.2018.02582
Vandereyken K, Sifrim A, Thienpont B, Voet T (2023) Methods and applications for single-cell and spatial multi-omics. Nat Rev Genet 24:494–515. https://doi.org/10.1038/s41576-023-00580-2
Vickovic S et al (2022) SM-omics is an automated platform for high-throughput spatial multiomics. Nat Commun 13:795. https://doi.org/10.1038/s41467-022-28445-y
Wang SJ, Dougan SK, Dougan M (2023a) Immune mechanisms of toxicity from checkpoint inhibitors. Trends Cancer 9:543–553. https://doi.org/10.1016/j.trecan.2023.04.002
Wang Q et al (2023b) Single-cell omics: a new perspective for early detection of pancreatic cancer? Eur J Cancer (Oxford, England : 1990) 190:112940. https://doi.org/10.1016/j.ejca.2023.112940
Wang S et al (2023c) scFed: federated learning for cell type classification with scRNA-seq. Brief Bioinform 25. https://doi.org/10.1093/bib/bbad507
Wang X et al (2024) Single-cell multi-omics sequencing uncovers region-specific plasticity of glioblastoma for complementary therapeutic targeting. Sci Adv 10:eadn4306. https://doi.org/10.1126/sciadv.adn4306
Wang Z et al (2025) Drug-tolerant persister cells in cancer: bridging the gaps between bench and bedside. Nat Commun 16:10048. https://doi.org/10.1038/s41467-025-66376-6
Welch JD et al (2019) Single-cell multi-omic integration compares and contrasts features of brain cell identity. Cell 177:1873–1887.e1817. https://doi.org/10.1016/j.cell.2019.05.006
Wu N et al (2024) MerTK(+) macrophages promote melanoma progression and immunotherapy resistance through AhR-ALKAL1 activation. Sci Adv 10:eado8366. https://doi.org/10.1126/sciadv.ado8366
Yang D et al (2022a) Lineage tracing reveals the phylodynamics, plasticity, and paths of tumor evolution. Cell 185:1905–1923.e1925. https://doi.org/10.1016/j.cell.2022.04.015
Yang L et al (2022b) Single-cell transcriptome analysis revealed a suppressive tumor immune microenvironment in EGFR mutant lung adenocarcinoma. J Immunother Cancer 10. https://doi.org/10.1136/jitc-2021-003534
Yang F et al (2022c) scBERT as a large-scale pretrained deep language model for cell type annotation of single-cell RNA-seq data. Nat Mach Intell 4:852–866. https://doi.org/10.1038/s42256-022-00534-z
Yang Y et al (2024) Pan-cancer single-cell dissection reveals phenotypically distinct B cell subtypes. Cell 187:4790–4811.e4722. https://doi.org/10.1016/j.cell.2024.06.038
You Y et al (2024) Systematic comparison of sequencing-based spatial transcriptomic methods. Nat Methods 21:1743–1754. https://doi.org/10.1038/s41592-024-02325-3
Zhang Y et al (2021) Single-cell analyses reveal key immune cell subsets associated with response to PD-L1 blockade in triple-negative breast cancer. Cancer Cell 39:1578–1593.e1578. https://doi.org/10.1016/j.ccell.2021.09.010
Zhang Y et al (2022) Elucidating minimal residual disease of paediatric B-cell acute lymphoblastic leukaemia by single-cell analysis. Nat Cell Biol 24:242–252. https://doi.org/10.1038/s41556-021-00814-7
Zhang D et al (2023) Spatial epigenome-transcriptome co-profiling of mammalian tissues. Nature 616:113–122. https://doi.org/10.1038/s41586-023-05795-1
Zhang YW, Gvozdenovic A, Aceto N (2024) A molecular voyage: multiomics insights into circulating tumor cells. Cancer Discov 14:920–933. https://doi.org/10.1158/2159-8290.Cd-24-0218
Zhao T et al (2022) Spatial genomics enables multi-modal study of clonal heterogeneity in tissues. Nature 601:85–91. https://doi.org/10.1038/s41586-021-04217-4

Zheng L et al (2021) Pan-cancer single-cell landscape of tumor-infiltrating T cells. Science 374:abe6474. https://doi.org/10.1126/science.abe6474

Zhu Z, Qiu S, Shao K, Hou Y (2018) Progress and challenges of sequencing and analyzing circulating tumor cells. Cell Biol Toxicol 34:405–415. https://doi.org/10.1007/s10565-017-9418-5

Ziegenhain C et al (2017) Comparative analysis of single-cell RNA sequencing methods. Mol Cell 65:631–643.e634. https://doi.org/10.1016/j.molcel.2017.01.023

Unraveling Brain Complexity: The Applications of Single-Cell Technologies in Neurological Disorders

11

Ansh Mehta, Shuvomoy Banerjee, and Anand K. Tiwari

Abstract

Neurological disorders are among the most heterogeneous and complex categories of disease attributable to the complex cellular changes, coupled with an equally complex microenvironment in the brain. It is hard to underline the impact and importance of single-cell technologies (SCTs) such as single-cell RNA sequencing (scRNA-seq), single-nucleus RNA sequencing (snRNA-seq), and single-cell assay for transposase-accessible chromatin using sequencing (scATAC-seq) in neuroscience. SCTs allow the researchers to unbiasedly profile thousands of individual cells and their molecular states in the normal and diseased brain tissues. SCTs also revealed previously unappreciated transcriptional and epigenetic changes and have illuminated the progression and heterogeneity of the disease comprehensively. This book chapter presents an overview of the technological concepts, protocols of single-cell platforms, and the research applications of single-cell approaches in the study of neurological disorders. This chapter covers model systems, such as postmortem human brains, induced pluripotent stem cell (iPSC)-derived brain organoids, and animal models. It also provides a discussion of findings involving Alzheimer's disease, Parkinson's disease, multiple sclerosis, epilepsy, neurodevelopmental disorders, and neuropsychiatric disorders while indicating how a single-cell perspective can provide access to disease-relevant pathways and cell states in the neuropathology that were previously beyond reach. Here we discuss the limitations and challenges, such as sample preparation, data interpretation, and analysis complexities, as challenges of the research category.

A. Mehta · A. K. Tiwari (✉)
Genetics & Developmental Biology Laboratory, School of Biotechnology & Bioengineering, Institute of Advanced Research, Gandhinagar, Gujarat, India
e-mail: anandk.tiwari@iar.ac.in

S. Banerjee
Biochemistry Laboratory, School of Biotechnology & Bioengineering, Institute of Advanced Research, Gandhinagar, Gujarat, India

V. Singh (ed.), *Advances in Single Cell Analysis*,
https://doi.org/10.1007/978-981-95-9789-5_11

Keywords

Single-cell genomics · Multi-omics · Neurodegenerative diseases · Induced pluripotent stem cells · Single-cell RNA sequencing

11.1 Introduction

The complexity of the mammalian brain is reflected in the large diversity of cell types (neurons, astrocytes, oligodendrocytes, microglia, endothelial cells, and pericytes) and the complex networks of these cells that sum into cognition, behavior, and homeostasis (Siletti et al. 2023). Neurological disorders stem from minor, cell-type-specific alterations to heterogeneous tissues from which cellular type heterogeneity brings a considerable challenge to understanding these disorders (Piwecka et al. 2023). Neurodegenerative disorders like Alzheimer's disease, Parkinson's disease, and amyotrophic lateral sclerosis share selective vulnerability of the population of neurons in a complex environment of glial cells and vascular cells; however, common bulk tissue-based measurement, and subsequent averaging of the measurements of millions of cells, erases any subtle cellular events (Piwecka et al. 2023).

A neurological disease is a disease process that exists at a multicellular level (Wang et al. 2022). Alzheimer's disease leads to cognitive impairment due to neuronal loss and subsequent reactive gliosis, while microglial activation is involved in amyloid-β clearance and patterns of synaptic remodeling (Mathys et al. 2024). Parkinson's disease is implicated from loss of dopaminergic neurons in substantia nigra. Important changes occur within astrocytes and microglia that contribute to neuroinflammation and the maintenance of dopamine homeostasis. Multiple sclerosis illustrates a complex relationship among demyelinating plaques, loss of oligodendrocyte progenitors, and invasion of immune cells (Elkjaer et al. 2022). These interdependent cellular processes illustrate the importance of studying mechanisms of disease at a single-cell level, which will allow these neuronal diseases to be targeted at more accurate therapy and biomarker levels (Shi et al. 2025).

Traditional tools such as histopathology, bulk sequencing of RNA, and proteomics can identify a variety of molecular signatures related to neurodegeneration, but cannot provide information about cell-type specificity or rare cell states (Woo and Eyun 2025). Bulk transcriptomics may identify global gene-expression changes but is unable to distinguish whether such changes are in neurons, astrocytes, or invading immune cells (Jovic et al. 2022). In like manner, in situ hybridization and immunohistochemistry give spatial information but are limited to a few markers within an experiment due to spectral and antibody-specificity limitations, excluding broad molecular profiling (Woo and Eyun 2025). As a result, important events like early synaptic dysfunction in preclinical Alzheimer's disease or early activation of microglia in Huntington's disease remain undetected until the disease is advanced, making early diagnosis and intervention attempts difficult (Shi et al. 2025).

Single-cell technologies (SCTs) have emerged within the past decade as game-changing tools that have the ability to overcome these limitations by interrogating single cells' transcriptome, epigenome, and proteome (Liu et al. 2024). Single-cell RNA sequencing (scRNA-seq) has the ability to conduct unbiased profiling of thousands of cells in parallel to find out novel cell types, developmental processes, and stress-response states (Woo and Eyun 2025). Single-nucleus RNA sequencing (snRNA-seq) allows these analyses to be extended to years-preserved postmortem tissue, preserving nuclear RNA of neurons and glia with dissociation-induced artifacts under control (Waag et al. 2023). Single-cell ATAC-seq (scATAC-seq) captures chromatin accessibility landscapes at cell-resolution, defining cell-type-specific regulatory characteristics and transcription factor networks that control disease-associated gene expression (Wang et al. 2025). Furthermore, combined multi-omics modalities integrate scRNA-seq with scATAC-seq or pair transcriptomics and spatial transcriptomics to combine gene-expression data with three-dimensional tissue architecture, further advancing our understanding of cellular microenvironments and intercellular interactions in vivo (Yang et al. 2025).

The applications of SCT in disease research and investigation are numerous. SCTs have first defined cellular hierarchies and found rare cell populations that cause disease. In Alzheimer's disease, prefrontal cortex scRNA-seq found a neuronal subtype enriched for neuronal survival genes that decreases early in pathology, suggesting candidate neuroprotective pathways (Mathys et al. 2024). In Parkinson's disease, snRNA-seq of substantia nigra nuclei revealed that AGTR1-expressing dopaminergic neurons are more susceptible and have activated stress-response pathways, which can be targeted to enhance neuronal resilience (Kamath et al. 2022). Second, SCTs allow for temporal mapping of disease progression. Single-cell atlases of stroke models have temporally mapped dynamic changes in microglial and astrocyte phenotypes between acute and chronic inflammation, providing clues on therapeutic intervention timing (Qiu et al. 2022). Third, SCTs enable precision medicine strategies via cell-type-dependent molecular grouping of individuals (Fig. 11.1). In multiple sclerosis, single-nucleus transcriptome analysis of normal-appearing tissue separated inflammatory subtypes that were associated with lesion burden and therapeutic outcome, with the potential for personalized immunomodulatory therapy (Elkjaer et al. 2022).

Overall, SCTs have opened a new chapter in the study of neurological diseases by interrogating the cellular heterogeneity of the human brain, revealing cell-type-specific disease mechanisms, and allowing dynamic, high-resolution mapping of disease processes. With ongoing developments in SCT platforms—adding enhanced sensitivity, spatial resolution, and integrated multi-omics capabilities—they will likely transform diagnostic strategies and speed the design of targeted therapies against the characteristic cellular landscapes of individual patients.

Single-cell technologies help in separating the different cell populations while all the cells are collected and analyzed together in bulk analysis. While the biomolecules to be analyzed remain the same, i.e., protein markers, RNA, and DNA, SCTs help in understanding the expression patterns of each cell population independently without averaging out the expression profiles of each cell type in the tissue. Researchers are able to identify the cell-specific functions rather than having to understanding the disease manifestation mechanism in the tissue as a whole, allowing for niched analysis.

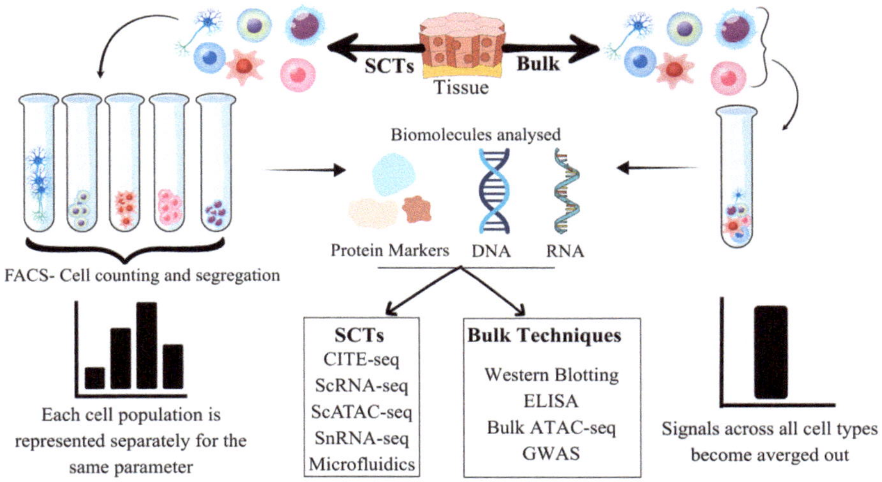

Fig. 11.1 Single-cell technologies and bulk analysis for tissue analysis

11.2 Single-Cell Techniques Used for Analysis of Neurological Disorders

11.2.1 Fluorescence-Activated Cell Sorting (FACS)

FACS has facilitated the assessment of neurological disorders for over four decades. However, categorizing the types of cells in the brain can be difficult because of the overlap in the markers and proteins that exist in many cells.

FACS is routinely used to identify and differentiate between the physiological and morphological phenotypes represented in many cells that have disease pathology. Researchers have been able to recognize and isolate neurons and astrocytes based on marker expression in both fresh and frozen brain tissue (Rubio et al. 2016). Presently, the utility of FACS has evolved from solely sorting GFP-expressing neurons to sorting fluorescent antibody labeled antigen expressing cells. This changes many factors in productivity for many studies with the implementation of fluorescent markers. Markers like NeuN, Fos, and Th1 are routinely used to identify neurons or other cell types, and GFAP, Oligo2, Iba1, GLT1, and CD11b are markers identified as useful to help identify glial cells (Rubio et al. 2016).

FACS enables detailed analysis at the cell-specific level in various neurological diseases. Using this technology allows researchers to distinguish the level of presence of certain biological markers across the different cell types present in the brain, rather than a summed or non-specific analysis. With this specific information researchers can begin to understand how the disease occurs differently in neural versus glial cells, particularly with neurodegenerative disease onset (e.g., Alzheimer's disease). The technology enables the identification of cell types producing disease-specific markers (Tournier et al. 2020).

FACS has also been utilized to isolate plasmacytoid dendritic cells for the study of lupus-prone mice (Liao et al. 2016). Some cells are more likely to be involved in disease pathology and disease manifestation compared to others, and FACS provides a way to objectively identify these patterns. For research access, for example, microglia are so affected by AD pathology that assessing markers they produce is paramount for discriminatory pathogenicity. Additionally, FACS is particularly well-suited for the removal of non-specific cell contamination, then concentrating a specific cell population for removal (Rayaprolu et al. 2020), especially if immune cell populations are conducting a study or experiments of expression, miRNA, and proteomic studies on neurological diseases, including MS. This was key, as FACS could also largely help us organize the heterogeneity in the same cell type observed with similar phenotypic expression (Miedema et al. 2020). FACS even provides a benefit for NSNs. As an example, Fukuda et al. reported that the FACS-ing of ESCs to express sox1 would prevent the risk of a tumor in the mouse brain, which improves augmentations of cell transplantation success for neurodegenerative diseases such as Parkinson's or Alzheimer's (Fukuda et al. 2006).

This method also has inherent drawbacks. Extracellular markers can be analyzed without difficulty whereas binding intracellular antibodies requires fixation and permeability steps that can lead to residual loss of necessary biomolecules (Schwarz et al. 2015). Furthermore, antibody binding can trigger a signaling cascade that can potentially change the functional state of the cell (Liao et al. 2016).

11.2.2 Microarray-Based Single-Cell Transcriptomics

Microarray technology represents an integrated point between high-density probe technologies and cell-based single-cell methods. When applied in the context of neurological disorders, microarray technologies provide researchers with the capacity to identify and evaluate heterogeneity in the nervous system at the population level, identifying and analyzing heterogeneity across a large number of cells all at once. This approach allows researchers to identify targeted gene-expression patterns in a wide variety of cells simultaneously. Commonly, these specific cells are isolated from bulk tissue by using methodologies such as laser capture microdissection (LCM), and then lysed and amplified (to overcome the challenges of low RNA output from individual cells) by T7 RNA polymerase-based amplification. This technique has previously been found useful for analyzing anatomically distinct regions of the brain and identified specific regions (hippocampal formation, neocortex) as areas of distortion under pathogenetic influence (Ginsberg et al. 2006).

Transcription profiling using microarrays has enabled researchers to interrogate the patterns that differentiate dying, or dead, neurons from neurons that survived, directly at a molecular level, identifying vulnerable cell populations in neurodegenerative diseases, such as AD and PD (Boone et al. 2013; Pozojevic and Spielmann 2023). The technique has also been further employed to analyze transcriptomic changes that arise as a result of mutations in the huntingtin protein. Huntington's disease, in fact, was the first neurological disorder studied on a microarray

(Mastrokolias et al. 2015). Using the microarray profiling, Dickson et al. reported that 735 genes were differentially expressed in the hypothalamus of mouse models with HD (Dickson et al. 2022). Using a similar microarray-based analytics, Wang et al. suggested that the ratio of inhibitory to excitatory neurons was increased based on how their geneexpressions compared to each other.

One major limitation of microarray analysis is that it requires pre-designed probes for the analysis of transcripts. This limits it to analysis of known transcripts, as there are probe design considerations that would restrict it for analyses of transcripts that were unknown. Additionally, microarrays have lower levels of sensitivity for low abundance transcripts, because probe-molecule binding would be greatly reduced.

11.2.3 Single-Cell RNA Sequencing

Single-cell RNA sequencing is a transformative technology that has allowed scientists to transform the understanding of neurological disorders. Single-cell RNA sequencing addresses the limitations associated with bulk RNA sequencing approaches.

Using scRNA-seq to study different neurodegenerative diseases and disorders has provided unprecedented insight into the mechanisms of these diseases. With the use of scRNA-seq, researchers were able to identify three molecular subtypes exhibited in Alzheimer's disease based on differences in patterns of Aβ and Tau dysregulation across different cells (Awuah et al. 2023). scRNA-seq approaches to amyotrophic lateral sclerosis have distinguished between vulnerable populations of motor neurons and identified gene-expression strictness or dosage dependencies of genes underlying ALS (Liu et al. 2023). Researchers were also able to identify very distinct, exclusive populations of highly specific glial cell populations with some features of reactive astrocytes and oligodendrocyte precursor cells based on glial molecular markers in human brain samples obtained from patients with temporal lobe epilepsy (Awuah et al. 2023).

The advent of single-cell RNA sequencing (scRNA-seq) technology has enabled the validation of the development of additional in vitro HD models, including human striatal organoids derived from human-induced pluripotent stem cells (Liu et al. 2022). This technology is fundamental to examining how different cell types respond to brain injuries, including traumatic brain injury and spinal cord injury. As the study seeks to compare the difference in expression patterns from cells after injury compared to the adjacent uninjured cells, understanding differences in sensitivity to environmental modifiers, such as temperature, could assist in directing future therapeutic treatment strategies (Albert et al. 2025; Chelyshev and Ermolin 2023).

Although the potential for avenues of analyzation has expanded rapidly with the advancement of this technology, there are drawbacks associated with its use. The methodology requires the cells to be dissociated, which is not an option for frozen tissue samples. Even beyond this issue, the anxiety of the dissociation process can impact the fragile cell populations being studied. Furthermore, the dissociation process can induce cellular stress to the cells being dissociated, leading them into a stress response that can affect the resultant transcriptome analysis. To address this particularity of analysis, methodologies utilizing cold-active proteases are being developed; unfortunately, this methodology has not been experimentally practicable in every dilemma (Wu et al. 2018).

The tissue is collected and subjected to mechanical dissociation like mincing and then immersed in a pool of enzymes such as trypsin, collagenase, and dipase. The dissociated cells are then sorted based on viability, i.e., whether they are dead or living. Dead cells take up dyes like trypan blue while living cells are not able to. The plethora of living cells are then separated and collected using FACS by identifying unique, cell-specific markers. Through microfluidics, each cell is then placed into one droplet of partitioning oil containing a gel bead-in-emulsion GEM. The surface of the GEM consists of an adapter. Among other things, it contains a 10x barcode, a Unique Molecular Identifier and a poly(dT) chain that binds to the polyA region of the mRNA. cDNA is synthesized within the oil droplet and then amplified. The oil partitioning is then broken and the amplified, barcoded, and indexed cDNA from all the cells is pooled together and sequenced. The analysis is then done on the basis of the gene-expression matrix, clustering, and visualization.

11.2.4 Single-Nucleus RNA Sequencing

Single-nucleus RNA sequencing (snRNA-seq) has emerged as a powerful variant offering certain advantages over scRNA-seq (Fig. 11.2). This technique is preferred in situations where whole cells cannot be dissociated easily or when dealing with sensitive and fragile cells like neurons that may get damaged during the process of cell segregation. Moreover, it is ideal to use the technique when the nuclear RNA content is to be analyzed exclusively. SnRNA-seq is specially used for frozen brain tissue samples where intact cells cannot be recovered, but their nuclei can be isolated (Bakken et al. 2018). The technique has been a boon, especially in the case of neurodegenerative disease research, where fresh tissue samples are rarely available, but there is a need for comprehensive molecular analysis of the plethora of heterogeneous cells (Walter et al. 2023).

These advances have facilitated the progress of neurological disease research. Using snRNA-seq, researchers identified 16 subtypes of neurons present in the cerebral cortex itself, in the postmortem brain (Kim et al. 2023). Through analysis of snRNA-seq transcriptomics, variations in both gray matter and white matter were observed in multiple sclerosis brain tissue samples. Even with different compositions of cell types in different samples, the cell-specific gene expression remained quite similar across all the samples (Macnair et al. 2025). SnRNA-seq was also used to confirm that the developmental and neurodegenerative transcriptomic profiles were affected in neurons and glial cells, even in the pre-manifestation stages of Huntington's disease (Carlos Estevez-Fraga et al. 2023). The technique was also used to analyze the distinction between the different stages and treatments of tumors

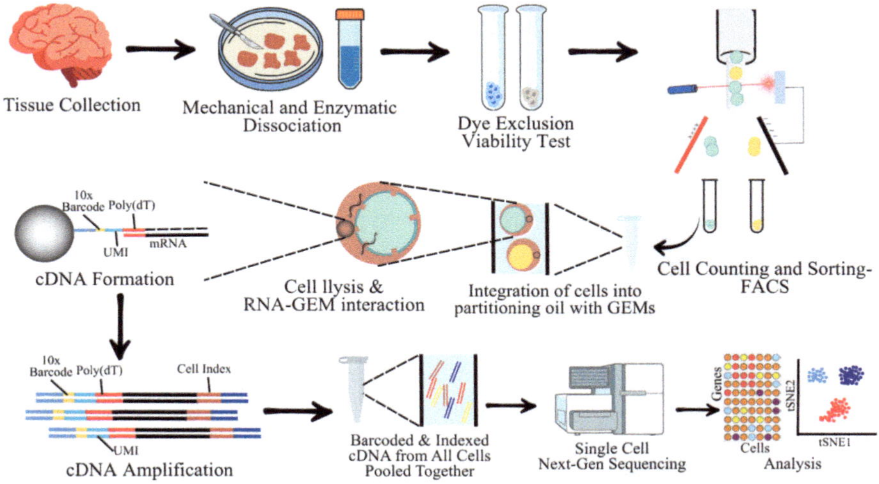

Fig. 11.2 Schematic workflow of ScRNA sequencing

and the corresponding tumor microenvironment in mice models (Borja et al. 2025). The gene-expression changes in the amygdala of postmortem human brain samples affected by autism spectrum disorder were analyzed using snRNA-seq, and it was found that the expression patterns were affected in excitatory neuron populations (Herrero et al. 2020).

While snRNA-seq is one of the most advanced and efficient techniques for gene-expression analysis, it only captures the mRNA present inside the nuclei, leading to lower abundance and reduced sensitivity, especially for those genes that are expressed at low levels. Moreover, RNA present outside the nuclei cannot be analyzed, leaving out essential molecules for analysis, like mature mRNAs, mitochondrial RNAs, and ribosomal RNAs. Moreover, due to the requirement of greater sample tissue to compensate for the losses during nuclear isolation, the technique remains unsuitable for small biopsies and other samples (Deleersnijder et al. 2021).

11.2.5 Single-Cell Assay for Transposase-Accessible Chromatin Using Sequencing (ScATAC-seq)

ScATAC-seq, an epigenomic technology, enables the interpretation of chromatin accessibility at the single-cell level. The technique focuses on the open regions of the chromatin, helping us identify the active enhancers, promoters, and regulatory elements present in the genome. It uses the Tn5 transposase enzyme which inserts adapters into the open regions of the chromatin. The more accessible the DNA region, the more it is tagged. The fragments of each cell are uniquely barcoded to help understand the origin of the tagged DNA. The technique is almost always applied in integration with other techniques, like snRNA-seq. Such integrated multi-omics will be discussed in the next sub-section.

The technique is revolutionary in understanding epigenetic and cis-regulatory elements controlling gene expression in the diseased state but it faces the issue of low coverage. A lot of the DNA material is lost during library preparation due to which only around 3% entries are found to be nonzero compared to >10% of them in techniques like scRNA-seq (Li et al. 2021). The sequence of the transposase may also induce biases where the frequency of transposition of preferred sequences may be higher than the others.

ScATAC-seq was used to identify cis-regulatory elements that contribute to the disease and the genes that they may target in Alzheimer's disease. Oligodendrocyte-associated regulatory elements that contain links to APOE and CLU genes that exhibit a varied chromatin accessibility pattern in the diseased state were identified (Morabito et al. 2021). As the primary analytical technique, scATAC-seq was used to create an epigenetic atlas, which was further utilized to identify SNPs associated with Alzheimer's and Parkinson's disease (Corces et al. 2020).

11.2.6 Integrated Multi-omics

Single-cell multi-omics approaches have revolutionized the analytics of neurological disorders by enabling simultaneous readouts pertaining to varied molecular functions. This helps us understand the complex molecular interactions and networks, providing better insights into disease manifestation and expression mechanisms. Various single-cell methodologies are integrated together. Some examples are sc/snRNA-seq and sc/snATAC-seq; scRNA-seq and DNA methylation study; and sc-Trio-seq and CITE-seq. Such techniques help researchers capture the transcriptomic, proteomic, and epigenetic backgrounds altogether.

113,207 nuclei of control and Parkinson's disease-affected brain samples were analyzed using an integrated scATAC-seq and snRNA-seq approach, which yielded the findings that the dopaminergic neurons, oligodendrocytes, and microglia were tremendously affected and the key genes that were affected in these cells were also identified. Using the same technique, another research group was able to identify 89 SNPs that were associated with PD, and yet another group was able to identify the neuroinflammatory undertakings associated with PD (Choi et al. 2024). Using multi-omics, researchers have been able to find out that transcriptional regulation was altered in de novo loss-of-function variants of genes that are involved in the pathogenesis of Autism Spectrum Disorder (Nassir et al. 2021). Patch-seq analysis was used to analyze the electrophysiological characteristics of brain tissues of people with drug-resistant epilepsy. The study revealed a group of transcriptionally and morphologically altered cortical pyramidal neurons (Ge et al. 2025). For cells undergoing rapid changes and differentiation, like glioblastoma tumors, for instance, these multi-omics techniques provide the best avenues for analytics as well as exploring the therapeutic potential of various treatment options (Ma et al. 2020). Using single-cell multi-omics, Wang et al. discovered a novel phenotype of B cells post cerebral ischemia with enhanced ability for phagocytosis (Wang et al. 2023). Following the induction of spinal cord injuries in mice, the cell types and genes being expressed were tracked from the acute through the chronic phases using snRNA-seq. Various rare healing-oriented cellular types were found along with the expression of regeneration-associated genes being expressed in the neurons (Matson et al. 2022). SnRNA-seq and ATACseq analysis by Lim et al. shows that the downstream targets of OLIG1 and OLIG2 genes, i.e., the OL maturation genes show reduced chromatin accessibility in NueN-negative cells (Lim et al. 2022).

Despite revolutionary capabilities, there are substantial challenges in single-cell multi-omics. As mentioned previously, scATAC-seq exhibits a high number of non-zero entries, creating data sparsity. The integration of techniques is a challenging front due to differences in modalities, features, and differences in the throughput between the technologies. The computational demands are extensive and necessitate sophisticated algorithms, analyzing a variety of biological signals. Nevertheless, ongoing advances to integrate techniques such as scMODAL and improvements in wet-lab protocols continue to expand the scope of multi-omics to analyze neurological conditions, including stroke, traumatic brain injury, multiple sclerosis, and Huntington's disease (Ma et al. 2020; Wang et al. 2025).

11.3 Model Systems for Single-Cell Analysis of Neurological Disorders

Single-cell approaches have transformed our understanding of neurological disease, by providing unparalleled resolution at the single-cell level in molecular terms. Nevertheless, the success of these methods is contingent upon the choice of appropriate model systems, which each has its own advantages and disadvantages to answer specific research questions. In this paragraph, we outline three principal model systems used in single-cell neuroscience research: postmortem human brain tissue; iPSC-derived brain organoids; and animal models; and demonstrate their impact toward our understanding of the pathogenesis of neurological disease.

11.3.1 Postmortem Human Brain

Under postmortem conditions, the study of human brain tissue is the reference standard to examine neurological disease in its intrinsic pathological context, as it offers direct access to tissue-specific and disease-altered neural circuits and cell populations that cannot be simulated in availed models (Padoan et al. 2022). The utilization of brain tissues rejoined through brain banks has also been essential to single-cell RNA sequencing and single-nucleus RNA sequencing projects in many neurological diseases, yielding unique information about human-specific disease mechanisms (Guo et al. 2025).

The primary advantage of postmortem human brain tissue is that it offers researchers, the in vivo representation of disease pathology, and the opportunity to identify spatial alterations in the brain's highly complex, structured, and three-dimensional architecture. Model organisms often provide similar approaches; however, brain tissue obtained from humans preserves species-specific molecular signatures, genetic heterogeneity, and cellular responses that are often lost or poorly defined in non-human models. Postmortem tissue studies have provided pivotal information about the pathogenesis of Alzheimer's disease, with single-nucleus RNA sequencing of more than 191,890 nuclei in prefrontal cortex tissue samples characterizing disease-specific molecular subtypes and cell-type-specific vulnerabilities that directly correlated with clinical dementia severity (Mathys et al. 2019). In the same way, Parkinson's disease studies in substantia nigra tissue samples have determined AGTR1-expressing dopaminergic neurons as the most vulnerable group with selective molecular mechanisms of neuronal damage (Kamath et al. 2022).

Multiple sclerosis studies have especially been aided by postmortem human brain examinations, in which single-nucleus RNA sequencing stratified individuals into different inflammatory subtypes according to transcriptomic patterns in normal-appearing brain areas below imaging thresholds. These studies have demarcated heterogeneous glial responses and delineated seven distinct microglial clusters with discrete functions in demyelination and remyelination, the targets of therapy for neuroinflammation modulation (Yong 2022). In addition, epilepsy research using temporal lobe epilepsy tissue samples has discovered cell-type-specific transcriptional reprogramming in more than 110,000 neuronal transcriptomes, identifying widespread remodeling of GABAergic interneurons and principal neurons and layer-specific dysregulation of glutamate signaling pathways (Pfisterer et al. 2020).

Nonetheless, human postmortem brain tissue has several important limitations to be carefully weighed. The most critical limitation is the postmortem interval (PMI), which progressively degrades RNA integrity as the time between death and tissue preservation increases. It has been illustrated that PMI negatively correlates with RNA Integrity Number (RIN) scores, whereby longer PMI may produce artificially low gene-expression values (White et al. 2018). PMI impacts on single-nucleus transcriptomes were examined and established that as little as three hours of postmortem delay reduces the proportion of genes that are differentially expressed across disease and control samples, indicating that PMI may mask disease-specific transcriptomic signatures (Olney et al. 2025). Neurons and interneurons are similarly very sensitive to postmortem conditions and respond with extreme upregulation of DNA repair, immune response, and stress-pathway genes at extended PMI (Olney et al. 2025).

Heterogeneity of the human population adds extra layers of complexity, as exposure of postmortem brains to various environmental stresses, medications, and comorbidities affects single-cell transcriptomic profiles. Exposure to psychotropic drugs in diseased brains, absent in control tissue, can significantly alter gene-expression patterns and complicate interpretation of disease-specific alterations (Lauterbach 2012). Furthermore, high-quality postmortem tissue remains scarce, especially for rare neurological disorders or specific demographic groups, limiting the scale and statistical power of investigational studies (Danner et al. 2024).

11.3.2 iPSC-Derived Brain Organoids

Human-induced pluripotent stem cell-derived brain organoids are new model systems bridging animal models and human disease, with unprecedented potential to analyze neurological disorders in a reproducible and controlled manner while retaining human-specific cellular features. The three-dimensional multicellular organoids recapitulate essential features of human brain development and pathology, presenting models for mechanistic analysis and therapeutic screening not possible with other model systems (Zhang et al. 2020).

The biggest strength of iPSC-derived brain organoids is that they can model human neurodevelopment and disease progression in real time, following dynamic cellular processes for periods that span weeks. The use of patient iPSC-derived cerebral organoids to investigate autism spectrum disorder has identified augmented generation of GABAergic neurons through heightened expression of FOXG1, showing the capacity of organoids to uncover pathogenic processes in idiopathic conditions where the foundation is not yet understood (Zhang et al. 2020). Likewise, CHD8 mutation-similar organoids, one of the major autism risk genes, have exhibited dysregulated gene expression in differentiation-relevant GABAergic neuron genes, corroborating patient study findings with the benefit of controlled experimental manipulation (Walter et al. 2023).

Timothy syndrome studies depict the singular potential of assembloid technology, with cortical and subpallium spheroids mixed together to produce more intricate brain region interactions. These assembloids have demonstrated disabled interneuron saltation and mobility reversible by L-type calcium channel blockers, giving clear-cut evidence of therapeutic efficacy that would be hard to obtain in other model systems (Zhang et al. 2020). The CHOOSE platform, combining CRISPR-mediated gene disruption with single-cell RNA sequencing, has allowed pooled loss-of-function screening in cerebral organoids, which demonstrates that disruption of 36 high-risk autism genes impacts cell fate specification with specific vulnerability in dorsal intermediate progenitors and upper-layer excitatory neurons (Costamagna et al. 2021).

Schizophrenia research has used organoids to model genetic and idiopathic schizophrenia. DISC1-mutant organoids have shown morphological abnormalities and diminished proliferation reversed by modulation of the WNT pathway, while patient-derived organoids without canonical gene mutations have exhibited abnormal cortical organization associated with nuclear FGFR1 activity. These findings illustrate the means by which organoids can identify cell processes for disease variants and expose therapeutic possibilities through pharmacological manipulations (Costamagna et al. 2021).

Although they hold enormous potential, iPSC-derived brain organoids have gigantic limitations that limit their utility for some research studies. Reproducibility is likely the largest limitation with heterogeneity due to several causes such as iPSC line variation, reprogramming efficiency, differentiation protocols, and intrinsic stochastic processes of self-organization (Zhang et al. 2020). Genetic and epigenetic changes introduced by reprogramming can have varying results in different clones leading to various phenotypes that are especially challenging to treat diseases of heterogeneous penetrance and severity (Urrestizala-Arenaza et al. 2024).

Limitation of vascularization in the majority of organoid protocols is a core deficiency, as dual SMAD inhibition suppresses mesodermal cell lineage commitment, precluding endothelial cell and pericyte development. It limits nutrient and oxygen exchange to 200–400 micrometers depth, leading to necrotic cores and stress responses of cells that can complicate disease-relevant phenotypes. Insufficiency of microvascularization is especially limiting in research on neurodegenerative disease where blood-brain barrier disruption and neurovascular interactions are central to pathogenesis (Urrestizala-Arenaza et al. 2024).

Maturation is a key challenge since extended culture durations are needed for brain organoids to achieve phases pertinent for the majority of neurological disorders. Organoid maturation time for experiments is adding increasing experimental cost and decreasing throughput, whereas methods trying to speed up maturation through pharmacological intervention threaten to disrupt inherent development programs with potentially undesirable impacts on physiological validity (Qian et al. 2019).

11.3.3 Animal Models

Animal models have been the key reagents for neurosciences research, providing the benefits of single-cell accessibility through genetic tractability, experiment condition control through standardization, and disease development modeling capability over lifetimes. The conjunction of single-cell RNA sequencing with conventional animal model systems has provided neurological disease access unprecedentedly while facilitating controlled experimental manipulation unavailable in human studies.

Mouse models are the most prevalent animal system for single-cell neuroscientific applications, sharing about 80% of human individuals' genetic homology and with access to advanced genetic manipulation techniques (Choi et al. 2021). Alzheimer's disease studies with mouse models have demonstrated species-specific and conserved molecular patterns, and single-cell characterization of 5xFAD mice showed microglial activation status and astrocytic responses that mirror human pathology and uncover mouse-specific protective responses (Keren-Shaul et al. 2017). Mouse models of Parkinson's disease have gained important information regarding dopaminergic neuron vulnerability through single-cell RNA sequencing experiments. PINK1 knockout mouse research has revealed evidence for early indications of mitochondrial pathology preceding neuronal death, and α-synuclein overexpression models have revealed cell-type-specificity of response to protein aggregate stress (Oliveras-Salvá et al. 2014; Sun et al. 2018). Sample access at specific time points and ability to manage environmental variables have allowed for precise elucidation of mechanisms of disease progression not feasible in human investigations.

Zebrafish models are increasingly being emphasized in research on neurological disorders because they provide unparalleled advantages for working with single cells. Optical transparency of zebrafish larvae allows real-time monitoring of cell behavior throughout development and disease, while fecundity and generation time provide capabilities for large-scale genetic screens (Neziri et al. 2024). New investigations in zebrafish epilepsy models have used single-cell RNA sequencing to determine neuronal subtypes with seizure-associated transcriptional modifications, demonstrating conserved epileptogenesis mechanisms across vertebrate species (Choi et al. 2021).

Tractability of the zebrafish genome has been especially useful for the study of rare neurological conditions. The use of SCN1A mutant zebrafish to study Dravet syndrome, a severe form of epilepsy, has integrated single-cell RNA sequencing with behavioral studies to define dysfunction in GABAergic interneurons and therapeutic strategies (Tiraboschi et al. 2020). Comparable investigation of intellectual disability syndromes in zebrafish models has uncovered developmental pathways of cognitive impairment directly applicable to human pathology (Choi et al. 2021).

Drosophila models are of particular value for investigating general mechanisms of neurodegeneration because they have advanced genetic tools and a well-characterized nervous system. Mass single-cell RNA sequencing experiments in Drosophila Alzheimer's disease models have identified conserved stress response pathways and defined the manner in which amyloid-β toxicity impacts different populations of neurons. Mass genetic screens in Drosophila have given access to screen for modifier genes that influence the severity of the disease, providing insights into protective mechanisms that can be translated into the therapeutic arena (Lau et al. 2025).

Yet, animal models also have essential limitations which should be understood when working with single-cell data. The differences in brain organization and anatomy among species, gene-expression patterns, and disease susceptibility can restrict the translational relevance of observations (Zhao and Bhattacharyya 2018). Mouse models of Alzheimer's disease are usually a failure at modeling the full range of human pathology, i.e., neurofibrillary tangle formation and widespread loss of neurons seen in human patients (De Plano et al. 2024). These disparities also hold the potential to uncover mechanisms in mice that cannot be translated for human therapeutics.

The short-duration courses of disease development in animal models, as convenient as they are for purposes of experimentation, will not mimic the decades-long course of human neurodegenerative disease. The sudden initiation of disease in animal models can activate cellular stress response and compensation pathways that are qualitatively distinct from those observed in long-term human disease development, potentially leading to a misinterpretation of disease pathogenic mechanisms (Pound and Ritskes-Hoitinga 2018).

Standardization of the environment in laboratory animal facilities and experimentation control can be unable to reflect the heterogeneity and richness of environmental exposures that human disease development experiences. The absence of comorbidities, drug exposures, and lifestyle factors that define the human neurological diseases presents a major limitation in the translation of results from animal models to clinical use (Pound and Ritskes-Hoitinga 2018).

Internationally, all model systems have certain strengths and limitations for single-cell neurosciences studies, and what one needs is optimal for certain individual research needs. Postmortem human brains provide authentic disease pathology but are plagued with technical complications and experimenter control constraints. iPSC-based organoids provide human-specific controlled modeling capabilities but do not have full physiological sophistication. Animal models provide advanced experiment manipulations and controlled conditions but can have constrained translational applicability. Convergent integration of data from numerous model systems is the strongest approach to further elucidate neurological illness and establish effective therapeutic strategies.

11.4 Conclusion

Single-cell technologies have dramatically changed our understanding of neurological disorders by revealing the brain's cellular and molecular complexities in a way that was unimaginable. Procedures like scRNA-seq, snRNA-seq, and scATAC-seq have enabled researchers to identify the most vulnerable cells, pinpoint gene-expression changes in a single-cell type, and track the development of a disease over time—all these being outside the realm of capability of conventional bulk analysis. In summary, single-cell approaches applied to alterative diseases (e.g., Alzheimer's, Parkinson's, and epilepsy) discovered some of the mechanistic disease features that had been hiding behind a curtain, adding further pieces of evidence for the specificity of neuronal loss. Postmortem human tissue, iPSC-derived organoids, and animal models all have their limitations (albeit provide perspectives that complement each other and advance our knowledge), but no one system provides more than a partial understanding. In addition, recent multi-omics and spatial approaches are beginning to connect gene expression to a given location of the brain and together represent different cells types to offer yet an additional important layer of understanding. There remain some practical challenges: technical problems in sample preparation, difficulty in managing large datasets, shortage of high-quality human tissue, and the requirement for more efficient computational tools; nevertheless, these issues are being addressed one at a time. As the field advances in terms of speed and sensitivity of the methods, single-cell technologies will probably be even more effective in identifying novel drug targets and ultimately facilitating the implementation of personalized treatment plans based on the cellular profile unique to each patient. Essentially, single-cell technologies have revealed an entirely new dimension of how neurological diseases originate and develop. With the continuous development of the tools and their use in different patient cohorts, these means will be the main factor in a radical change of the way we diagnose, understand, and eventually treat neurological disorders.

Acknowledgments The authors are thankful to the Puri Foundation for Education in India for Infrastructure support at IAR, Gandhinagar. The IAR reference No. from the Registrar's office is duly acknowledged.

Conflict of Interest The authors of this manuscript declare that there is no conflict of interest.

References

Albert V, Faruq M, Verma S, Sardana V, Choudhary K, Agrawal D (2025) Single-cell RNA sequencing of brain tissue samples from severe traumatic brain injury patients: a protocol for cellular heterogeneity and transcriptional alteration study. Indian J Neurotrauma 22(2):198–204. https://doi.org/10.1055/s-0045-1809063

Awuah WA, Ahluwalia A, Ghosh S, Roy S, Tan JK, Adebusoye FT, Ferreira T, Bharadwaj HR, Shet V, Kundu M, Weng Yee AL, Abdul-Rahman T, Atallah O (2023) The molecular landscape of neurological disorders: insights from single-cell RNA sequencing in neurology and neurosurgery. Eur J Med Res 28:529. https://doi.org/10.1186/s40001-023-01504-w

Bakken TE, Hodge RD, Miller JA, Yao Z, Nguyen TN, Aevermann B, Barkan E, Bertagnolli D, Casper T, Dee N, Garren E, Goldy J, Graybuck LT, Kroll M, Lasken RS, Lathia K, Parry S, Rimorin C, Scheuermann RH et al (2018) Single-nucleus and single-cell transcriptomes compared in matched cortical cell types. PLoS One 13(12):e0209648. https://doi.org/10.1371/journal.pone.0209648

Boone DR, Sell SL, Hellmich HL (2013) Laser capture microdissection of enriched populations of neurons or single neurons for gene expression analysis after traumatic brain injury. J Vis Exp 74:50308. https://doi.org/10.3791/50308

Borja M, Tran V, Ding Y, Granados A, McGeever E, Kim Y, Detweiler A, Mekonen H, Paul S, Pisco AO, Neff NF, Tabernero A (2025) Single-nucleus RNA sequencing reveals a preclinical model for the most common subtype of glioblastoma. Commun Biol 8(1):1–19. https://doi.org/10.1038/s42003-025-08092-x

Chelyshev YA, Ermolin IL (2023) RNA sequencing and spatial transcriptomics in traumatic spinal cord injury (review). Sovrem Tekhnologii Med 15(6):75–86. https://doi.org/10.17691/stm2023.15.6.08

Choi T, Choi T, Lee Y, Choe S, Kim C (2021) Zebrafish as an animal model for biomedical research. Exp Mol Med 53(3):310–317. https://doi.org/10.1038/s12276-021-00571-5

Choi H, Kim H, Chung H, Lee S, Kim J (2024) Application of computational algorithms for single-cell RNA-seq and ATAC-seq in neurodegenerative diseases. Brief Funct Genomics 24:elae044. https://doi.org/10.1093/bfgp/elae04

Corces MR, Shcherbina A, Kundu S, Gloudemans MJ, Frésard L, Granja JM, Louie BH, Eulalio T, Shams S, Bagdatli ST, Mumbach MR, Liu B, Montine KS, Greenleaf WJ, Kundaje A, Montgomery SB, Chang HY, Montine TJ (2020) Single-cell epigenomic analyses implicate candidate causal variants at inherited risk loci for Alzheimer's and Parkinson's diseases. Nat Genet 52(11):1158–1168. https://doi.org/10.1038/s41588-020-00721-x

Costamagna G, Comi GP, Corti S (2021) Advancing drug discovery for neurological disorders using iPSC-derived neural organoids. Int J Mol Sci 22(5):2659. https://doi.org/10.3390/ijms22052659

Danner B, Gonzalez AD, Corbett WC, Alhneif M, Etemadmoghadam S, Parker-Garza J, Flanagan ME (2024) Brain banking in the United States and Europe: importance, challenges, and future trends. J Neuropathol Exp Neurol 83(4):219. https://doi.org/10.1093/jnen/nlae014

De Plano LM, Saitta A, Oddo S, Caccamo A (2024) Navigating Alzheimer's disease mouse models: age-related pathology and cognitive deficits. Biomolecules 14(11):1405. https://doi.org/10.3390/biom14111405

Deleersnijder D, Callemeyn J, Arijs I, Naesens M, Van Craenenbroeck AH, Lambrechts D, Sprangers B (2021) Current methodological challenges of single-cell and single-nucleus RNA-sequencing in glomerular diseases. J Am Soc Nephrol 32(8):1838–1852. https://doi.org/10.1681/ASN.2021020157

Dickson E, Dwijesha AS, Andersson N, Lundh S, Björkqvist M, Petersén Å, Soylu-Kucharz R (2022) Microarray profiling of hypothalamic gene expression changes in Huntington's disease mouse models. Front Neurosci 16:1027269. https://doi.org/10.3389/fnins.2022.1027269

Elkjaer ML, Röttger R, Baumbach J, Illes Z (2022) A systematic review of tissue and single cell transcriptome/proteome studies of the brain in multiple sclerosis. Front Immunol 13:761225. https://doi.org/10.3389/fimmu.2022.761225

Estevez-Fraga C, Altmann A, Parker CS, Rachael IS, Costa B, Chen Z, Manzoni C, Zarkali A, Durr A, Roos RAC, Landwehrmeyer B, Leavitt BR, Rees G, Tabrizi SJ, McColgan P (2023) Genetic topography and cortical cell loss in Huntington's disease link development and neurodegeneration. Brain 146(11):4532–4546. https://doi.org/10.1093/brain/awad275

Fukuda H, Takahashi J, Watanabe K, Hayashi H, Morizane A, Koyanagi M, Sasai Y, Hashimoto N (2006) Fluorescence-activated cell sorting-based purification of embryonic stem cell-derived neural precursors averts tumor formation after transplantation. Stem Cells (Dayton, Ohio) 24(3):763–771. https://doi.org/10.1634/stemcells.2005-0137

Ge Q, Yang J, Huang F, Dai X, Chen C, Guo J, Wang M, Zhu M, Shao Y, Xia Y, Zhou Y, Peng J, Deng S, Shi J, Hu Y, Zhang H, Wang Y, Wang X, Li M et al (2025) Multimodal single-cell analyses reveal molecular markers of neuronal senescence in human drug-resistant epilepsy. J Clin Invest 135(5):e188942. https://doi.org/10.1172/JCI188942

Ginsberg SD, Che S, Counts SE, Mufson EJ (2006) Single cell gene expression profiling in Alzheimer's disease. NeuroRx 3(3):302–318. https://doi.org/10.1016/j.nurx.2006.05.007

Guo Y, Ma J, Ma X, Jiang C, Ye K, Xu J, Yang X, Li S, Zhang J, Huang Y, Huang H, Zhao X (2025) Single-nucleus RNA sequencing revealed ribosomal genes expression defines the cellular state of oligodendrocytes and precursor cells in post-mortem ischemia mouse brains. Int J Biol Macromol 329:147708. https://doi.org/10.1016/j.ijbiomac.2025.147708

Herrero MJ, Velmeshev D, Hernandez-Pineda D, Sethi S, Sorrells S, Banerjee P, Sullivan C, Gupta AR, Kriegstein AR, Corbin JG (2020) Identification of amygdala-expressed genes associated with autism spectrum disorder. Mol Autism 11(1):39. https://doi.org/10.1186/s13229-020-00346-1

Jovic D, Liang X, Zeng H, Lin L, Xu F, Luo Y (2022) Single-cell RNA sequencing technologies and applications: a brief overview. Clin Transl Med 12(3):e694. https://doi.org/10.1002/ctm2.694

Kamath T, Abdulraouf A, Burris SJ, Langlieb J, Gazestani V, Nadaf NM, Balderrama K, Vanderburg C, Macosko EZ (2022) Single-cell genomic profiling of human dopamine neurons identifies a population that selectively degenerates in Parkinson's disease. Nat Neurosci 25(5):588–595. https://doi.org/10.1038/s41593-022-01061-1

Keren-Shaul H, Spinrad A, Weiner A, Matcovitch-Natan O, Dvir-Szternfeld R, Ulland TK, David E, Baruch K, Lara-Astaiso D, Toth B, Itzkovitz S, Colonna M, Schwartz M, Amit I (2017) A unique microglia type associated with restricting development of Alzheimer's disease. Cell 169(7):1276–1290.e17. https://doi.org/10.1016/j.cell.2017.05.018

Kim N, Kang H, Jo A, Yoo A, Lee O (2023) Perspectives on single-nucleus RNA sequencing in different cell types and tissues. J Pathol Transl Med 57(1):52. https://doi.org/10.4132/jptm.2022.12.19

Lau F, Binacchi R, Brugnara S, Cumplido-Mayoral A, Savino SD, Khan I, Orso A, Sartori S, Bellosta P, Carl M, Poggi L, Provenzano G (2025) Using single-cell RNA sequencing with drosophila, zebrafish, and mouse models for studying Alzheimer's and Parkinson's disease. Neuroscience 573:505–517. https://doi.org/10.1016/j.neuroscience.2025.03.042

Lauterbach EC (2012) Psychotropic drug effects on gene transcriptomics relevant to Parkinson's disease. Prog Neuro-Psychopharmacol Biol Psychiatry 38(2):107–115. https://doi.org/10.1016/j.pnpbp.2012.03.011

Li Z, Kuppe C, Ziegler S, Cheng M, Kabgani N, Menzel S, Zenke M, Kramann R, Costa IG (2021) Chromatin-accessibility estimation from single-cell ATAC-seq data with scOpen. Nat Commun 12(1):1–14. https://doi.org/10.1038/s41467-021-26530-2
Liao X, Makris M, Luo XM (2016) Fluorescence-activated cell sorting for purification of plasmacytoid dendritic cells from the mouse bone marrow. J Vis Exp 117:54641. https://doi.org/10.3791/54641
Lim RG, Wu J, Gold MP, Reidling JC, Tang G, Adam M, Dansu DK, Park H, Casaccia P, Miramontes R, Lau A, Hickman RA, Khan F, Paryani F, Tang A, Ofori K, Miyoshi E, Michael N et al (2022) Huntington disease oligodendrocyte maturation deficits revealed by single-nucleus RNAseq are rescued by thiamine-biotin supplementation. Nat Commun 13(1):1–23. https://doi.org/10.1038/s41467-022-35388-x
Liu C, Fu Z, Wu S, Wang X, Zhang S, Chu C, Hong Y, Wu W, Chen S, Jiang Y, Wu Y, Song Y, Liu Y, Guo X (2022) Mitochondrial HSF1 triggers mitochondrial dysfunction and neurodegeneration in Huntington's disease. EMBO Mol Med 14(7):e15851. https://doi.org/10.15252/emmm.202215851
Liu H, Guan L, Deng M, Bolund L, Kristiansen K, Zhang J, Luo Y, Zhang Z (2023) Integrative genetic and single cell RNA sequencing analysis provides new clues to the amyotrophic lateral sclerosis neurodegeneration. Front Neurosci 17:1116087. https://doi.org/10.3389/fnins.2023.1116087
Liu X, Peng T, Xu M, Lin S, Hu B, Chu T, Liu B, Xu Y, Ding W, Li L, Cao C, Wu P (2024) Spatial multi-omics: deciphering technological landscape of integration of multi-omics and its applications. J Hematol Oncol 17:72. https://doi.org/10.1186/s13045-024-01596-9
Ma A, McDermaid A, Xu J, Chang Y, Ma Q (2020) Integrative methods and practical challenges for single-cell multi-omics. Trends Biotechnol 38(9):1007. https://doi.org/10.1016/j.tibtech.2020.02.013
Macnair W, Calini D, Agirre E, Bryois J, Jäkel S, Smith RS, Kukanja P, Stokar-Regenscheit N, Ott V, Foo LC, Collin L, Schippling S, Urich E, Nutma E, Marzin M, Ansaloni F, Amor S, Magliozzi R, Heidari E, Robinson MD et al (2025) snRNA-seq stratifies multiple sclerosis patients into distinct white matter glial responses. Neuron 113(3):396–410.e9. https://doi.org/10.1016/j.neuron.2024.11.016
Mastrokolias A, Ariyurek Y, Goeman JJ, Van Duijn E, Roos RA, Van Der Mast RC, Van Ommen GB, Den Dunnen JT, t Hoen PA, van Roon-Mom W (2015) Huntington's disease biomarker progression profile identified by transcriptome sequencing in peripheral blood. Eur J Hum Genet 23(10):1349–1356. https://doi.org/10.1038/ejhg.2014.281
Mathys H, Peng Z, Gao F, Mohammadi S, Young JZ, Menon M, He L, Abdurrob F, Jiang X, Martorell AJ, Ransohoff RM, Hafler BP, Bennett DA, Kellis M, Tsai L (2019) Single-cell transcriptomic analysis of Alzheimer's disease. Nature 570(7761):332–337. https://doi.org/10.1038/s41586-019-1195-2
Mathys H, Boix CA, Akay LA, Xia Z, Ng AP, Jiang X, Abdelhady G, Galani K, Mantero J, Band N, James BT, Babu S, Louderback K, Prokopenko D, Tanzi RE, Bennett DA, Tsai L, Kellis M (2024) Single-cell multiregion dissection of Alzheimer's disease. Nature 632(8026):858–868. https://doi.org/10.1038/s41586-024-07606-7
Matson KJ, Russ DE, Kathe C, Hua I, Maric D, Ding Y, Krynitsky J, Pursley R, Sathyamurthy A, Squair JW, Levi BP, Courtine G, Levine AJ (2022) Single cell atlas of spinal cord injury in mice reveals a pro-regenerative signature in spinocerebellar neurons. Nat Commun 13(1):1–16. https://doi.org/10.1038/s41467-022-33184-1
Miedema A, Wijering MHC, Eggen BJL, Kooistra SM (2020) High-resolution transcriptomic and proteomic profiling of heterogeneity of brain-derived microglia in multiple sclerosis. Front Mol Neurosci 13:583811. https://doi.org/10.3389/fnmol.2020.583811
Morabito S, Miyoshi E, Michael N, Shahin S, Martini AC, Head E, Silva J, Leavey K, Perez-Rosendahl M, Swarup V (2021) Single-nucleus chromatin accessibility and transcriptomic characterization of Alzheimer's disease. Nat Genet 53(8):1143. https://doi.org/10.1038/s41588-021-00894-z

Nassir N, Bankapur A, Samara B, Ali A, Ahmed A, Inuwa IM, Zarrei M, Safizadeh Shabestari SA, AlBanna A, Howe JL, Berdiev BK, Scherer SW, Woodbury-Smith M, Uddin M (2021) Single-cell transcriptome identifies molecular subtype of autism spectrum disorder impacted by de novo loss-of-function variants regulating glial cells. Hum Genomics 15:68. https://doi.org/10.1186/s40246-021-00368-7

Neziri S, Köseoğlu AE, Köseoğlu GD, Özgültekin B, Özgentürk NÖ (2024) Animal models in neuroscience with alternative approaches: evolutionary, biomedical, and ethical perspectives. Animal Model Exp Med 7(6):868. https://doi.org/10.1002/ame2.12487

Oliveras-Salvá M, Macchi F, Coessens V, Deleersnijder A, Gérard M, Van der Perren A, Van den Haute C, Baekelandt V (2014) Alpha-synuclein-induced neurodegeneration is exacerbated in PINK1 knockout mice. Neurobiol Aging 35(11):2625–2636. https://doi.org/10.1016/j.neurobiolaging.2014.04.032

Olney KC, Gibson KA, Cadiz MP, Rahimzadeh N, Swarup V, Fryer JD (2025) Postmortem interval leads to loss of disease-specific signatures in brain tissue. ENeuro 12(3):ENEURO.0505-24.2025. https://doi.org/10.1523/ENEURO.0505-24.2025

Padoan CS, Garcia LF, Crespo KC, Longaray VK, Martini M, Contessa JC, Kapczinski F, De Oliveira FH, Goldim JR, Magalhães P (2022) A qualitative study exploring the process of postmortem brain tissue donation after suicide. Sci Rep 12(1):1–15. https://doi.org/10.1038/s41598-022-08729-5

Pfisterer U, Petukhov V, Demharter S, Meichsner J, Thompson JJ, Batiuk MY, Vasistha NA, Thakur A, Mikkelsen J, Adorjan I, Pinborg LH, Pers TH, Von Engelhardt J, Kharchenko PV, Khodosevich K (2020) Identification of epilepsy-associated neuronal subtypes and gene expression underlying epileptogenesis. Nat Commun 11(1):1–19. https://doi.org/10.1038/s41467-020-18752-7

Piwecka M, Rajewsky N, Rybak-Wolf A (2023) Single-cell and spatial transcriptomics: deciphering brain complexity in health and disease. Nat Rev Neurol 19(6):346–362. https://doi.org/10.1038/s41582-023-00809-y

Pound P, Ritskes-Hoitinga M (2018) Is it possible to overcome issues of external validity in preclinical animal research? Why most animal models are bound to fail. J Transl Med 16(1):304. https://doi.org/10.1186/s12967-018-1678-1

Pozojevic J, Spielmann M (2023) Single-cell sequencing in neurodegenerative disorders. Mol Diagn Ther 27(5):553. https://doi.org/10.1007/s40291-023-00668-9

Qian X, Song H, Ming G (2019) Brain organoids: advances, applications and challenges. Development 146(8):dev166074. https://doi.org/10.1242/dev.166074

Qiu M, Zong J-b, He Q-w, Liu Y-x, Wan Y, Li M, Zhou Y-f, Jie-hong W, Bo H (2022) Cell heterogeneity uncovered by single-cell RNA sequencing offers potential therapeutic targets for ischemic stroke. Aging Dis 13(5):1436–1454. https://doi.org/10.14336/AD.2022.0212

Rayaprolu S, Gao T, Xiao H, Ramesha S, Weinstock LD, Shah J, Duong DM, Dammer EB, Webster JA Jr, Lah JJ, Wood LB, Betarbet R, Levey AI, Seyfried NT, Rangaraju S (2020) Flow-cytometric microglial sorting coupled with quantitative proteomics identifies moesin as a highly-abundant microglial protein with relevance to Alzheimer's disease. Mol Neurodegener 15(1):28. https://doi.org/10.1186/s13024-020-00377-5

Rubio FJ, Li X, Liu R, Cimbro R, Hope BT (2016) Fluorescence activated cell sorting (FACS) and gene expression analysis of Fos-expressing neurons from fresh and frozen rat brain tissue. J Vis Exp 114:54358. https://doi.org/10.3791/54358

Shi W, Zhang Z, Xu X, Tian Y, Feng L, Huang X, Du Y, Li Z (2025) Single-cell and spatial transcriptomics integration: new frontiers in tumor microenvironment and cellular communication. Front Immunol 16:1649468. https://doi.org/10.3389/fimmu.2025.1649468

Siletti K, Hodge R, Albiach AM, Lee KW, Ding L, Hu L, Lönnerberg P, Bakken T, Casper T, Clark M, Dee N, Gloe J, Hirschstein D, Shapovalova NV, Keene CD, Nyhus J, Tung H, Yanny AM, Arenas E et al (2023) Transcriptomic diversity of cell types across the adult human brain. Science. https://doi.org/10.1126/science.add7046

Sun L, Shen R, Agnihotri SK, Chen Y, Huang Z, Büeler H (2018) Lack of PINK1 alters glia innate immune responses and enhances inflammation-induced, nitric oxide-mediated neuron death. Sci Rep 8(1):1–16. https://doi.org/10.1038/s41598-017-18786-w

Tiraboschi E, Martina S, van der Ent W, Grzyb K, Gawel K, Cordero-Maldonado ML, Poovathingal SK, Heintz S, Satheesh SV, Brattespe J, Xu J, Suster M, Skupin A, Esguerra CV (2020) New insights into the early mechanisms of epileptogenesis in a zebrafish model of Dravet syndrome. Epilepsia 61(3):549–560. https://doi.org/10.1111/epi.16456

Tournier BB, Tsartsalis S, Ceyzériat K, Medina Z, Fraser BH, Grégoire C, Kövari E, Millet P (2020) Fluorescence-activated cell sorting to reveal the cell origin of radioligand binding. J Cereb Blood Flow Metab. https://doi.org/10.1177/0271678X19860408

Urrestizala-Arenaza N, Cerchio S, Cavaliere F, Magliaro C (2024) Limitations of human brain organoids to study neurodegenerative diseases: a manual to survive. Front Cell Neurosci 18:1419526. https://doi.org/10.3389/fncel.2024.1419526

Walter TJ, Suter RK, Ayad NG (2023) An overview of human single-cell RNA sequencing studies in neurobiological disease. Neurobiol Dis 184:106201. https://doi.org/10.1016/j.nbd.2023.106201

Waag R, Bohacek J (2023) Single-Nucleus RNA-Sequencing in Brain Tissue. Curr Protoc. Nov;3(11):e919. https://doi.org/10.1002/cpz1.919. PMID: 37987152

Wang Y, LeDue JM, Murphy TH (2022) Multiscale imaging informs translational mouse modeling of neurological disease. Neuron 110(22):3688–3710. https://doi.org/10.1016/j.neuron.2022.09.006

Wang R, Li H, Ling C, Zhang X, Lu J, Luan W, Zhang J, Shi L (2023) A novel phenotype of B cells associated with enhanced phagocytic capability and chemotactic function after ischemic stroke. Neural Regen Res 18(11):2413. https://doi.org/10.4103/1673-5374.371365

Wang G, Zhao J, Lin Y, Liu T, Zhao Y, Zhao H (2025) ScMODAL: a general deep learning framework for comprehensive single-cell multi-omics data alignment with feature links. Nat Commun 16(1):1–13. https://doi.org/10.1038/s41467-025-60333-z

White K, Yang P, Li L, Farshori A, Medina AE, Zielke HR (2018) Effect of postmortem interval and years in storage on RNA quality of tissue at a repository of the NIH NeuroBioBank. Biopreserv Biobank 16(2):148–157. https://doi.org/10.1089/bio.2017.0099

Woo H, Eyun S (2025) Applications and techniques of single-cell RNA sequencing across diverse species. Brief Bioinform 26(4). https://doi.org/10.1093/bib/bbaf354

Wu H, Kirita Y, Donnelly EL, Humphreys BD (2018) Advantages of single-nucleus over single-cell RNA sequencing of adult kidney: rare cell types and novel cell states revealed in fibrosis. J Am Soc Nephrol 30(1):23. https://doi.org/10.1681/ASN.2018090912

Yang P, Jin K, Yao Y, Jin L, Shao X, Li C, Lu X, Fan X (2025) Spatial integration of multi-omics single-cell data with SIMO. Nat Commun 16(1):1–15. https://doi.org/10.1038/s41467-025-56523-4

Yong VW (2022) Microglia in multiple sclerosis: protectors turn destroyers. Neuron 110(21):3534–3548. https://doi.org/10.1016/j.neuron.2022.06.023

Zhang DY, Song H, Ming G (2020) Modeling neurological disorders using brain organoids. Semin Cell Dev Biol 111:4. https://doi.org/10.1016/j.semcdb.2020.05.026

Zhao X, Bhattacharyya A (2018) Human models are needed for studying human neurodevelopmental disorders. Am J Hum Genet 103(6):829. https://doi.org/10.1016/j.ajhg.2018.10.009

Single-Cell Analysis for Cardiac Diseases 12

Kirtan Dave and Saloni Gautam

Abstract

Cardiovascular diseases (CVDs) remain the foremost cause of mortality worldwide, driven by a complex interaction of genetic, molecular, and cellular events that shape cardiac injury and repair. Recent progress in single-cell RNA sequencing (scRNA-seq), together with bulk RNA-seq and advanced in silico computational analysis, has greatly expanded our ability to characterize the cellular landscape of the heart with unprecedented resolution. This chapter examines multi-omics datasets to demonstrate how different cardiac cell populations can drive vascular dysfunction, fibrosis, inflammation, and disease progression. Our joint analysis has identified fibroblasts, endothelial cells, and macrophages as significant mediators of cardiac remodeling. Actively recruited fibroblasts mediate extracellular matrix (ECM) deposition and scarring, which facilitates further remodeling, while the endothelial-to-mesenchymal transition (EndMT) is a driver of microvascular loss and ischemia. Immune-cell profiling demonstrates a shift from pro-inflammatory (M1) to reparative (M2) macrophage phenotypes, reflecting dynamic immune adaptation during chronic injury. Pathway enrichment highlights the involvement of NFκB, PI3K–Akt, and TGF-β signaling pathways, accompanied by metabolic downregulation in more advanced disease states. Using cell–cell communication and ligand–receptor interaction modeling, we further show how cross-talk between endothelial and immune compartments drives inflammation, angiogenesis, and structural remodeling. Integrating scRNA-seq data with GTEx cardiac transcriptomics and KEGG/GSEA pathway analyses reveals several druggable molecular targets linked to fibroblast activa-

K. Dave (✉)
Parul Institute of Applied Sciences, Parul University, Vadodara, Gujarat, India

Bioinformatics Laboratory, Research and Development Cell, Parul University, Vadodara, Gujarat, India
e-mail: kirtan.dave30360@paruluniversity.ac.in

S. Gautam
Parul Institute of Applied Sciences, Parul University, Vadodara, Gujarat, India

V. Singh (ed.), *Advances in Single Cell Analysis*,
https://doi.org/10.1007/978-981-95-9789-5_12

tion, cytokine regulation, and vascular inflammation. Together, these findings provide a comprehensive systems-level view of cardiac remodeling and underscore the translational value of single-cell and in silico approaches for biomarker discovery and therapeutic development in cardiovascular disease.

Keywords

Cardiovascular disease · Single-cell RNA sequencing · In silico analysis · Transcriptomics · Cardiac remodeling

12.1 Introduction

Cardiovascular diseases (CVDs), which comprise coronary artery disease, stroke, heart failure, rheumatic heart disease, and peripheral arterial disease, remains the leading cause of death and disability in the world (Martin et al. 2025). Recent updates from global surveillance and statistics indicate that CVDs account for more deaths than any other disease category and are contributing to high costs for health systems and society (Chong et al. 2025a). The burden of cardiovascular disease (CVD) is both substantial and changing. The absolute number of people with cardiovascular diseases and the absolute number of deaths from CVD, have all risen over the past few decades as populations are aging and expanding, even with modest declines in some age-standardized rates in high-income settings (Lindstrom et al. 2022). This signifies an increasing number of patients requiring long-term care, secondary prevention and chronic disease management in all settings, including both high-income and low- and middle-income countries (Chong et al. 2025a). The individual lifetime risk and population burden of CVD are influenced by a small number of predominantly modifiable risk factors: raised blood pressure, dysglycemia/diabetes, raised body mass index, unhealthy lipid levels, and smoking (Global Cardiovascular Risk Consortium et al. 2025). Two large consortium analyses published in 2025 illustrate how not having or modifying these typical risk factors leads to a significantly lower lifetime risk of CVD events and longer life expectancy, demonstrating prevention as the single most influential tool to reduce future CVD burden (Bleakley et al. 2013).

Future populations and health may see common populations and health effects related to risk factors and indications of disease. The contemporary modeling studies and reviews all have two linked interpretive messages; the first is that trajectories vary by region, with many of our high-income countries showing improvements in terms of age-standardized mortality and DALYs, whereas many of the poorest regions (low- and middle-income countries) may see the largest absolute increase in the incidence of deaths related to CVD attributable to demographic change and rising metabolic risk. Second, future population health will be shaped by the "syndemic" interaction of multiple modifiable risks—risks that cluster in space across groups and potentially reinforce one another—and therefore coordinated multisectoral prevention strategies are thought to be required (Chong et al. 2025a, b).

International and regional policy reports also highlight the immense economic and social impacts of sustained CVD burden: productivity losses, catastrophic health expenditures for households, and increasing stress on health services, which are already stretched on other priority agendas (Gheorghe et al. 2018). Therefore, international agencies and professional organizations recommend integrating initiatives aimed at preventing at a population level (policy and environment) and screening and managing risk factors at the primary-care level and equitably providing access to acute and chronic cardiovascular care (Chauhan and Aeri 2013).

Worldwide and regional policy documentation also reiterates the significant economic and social impacts of the ongoing burden of CVD: lost productivity, catastrophic health expenditures to households, and increased burden on health services already burdened by other priorities (Pereira and Pereira 2020). As a result, global and professional organizations are advocating for "integrated strategies" that include population-level prevention (policy and environment), detection and management of risk at the primary-care level, and equitable access to acute and chronic cardiovascular services (Schwalm et al. 2016).

12.1.1 Limitations of Conventional Bulk Transcriptomics in Cardiac Research

Traditional bulk transcriptomics (bulk RNA-seq and microarrays) has been invaluable for identifying gene-level alterations in cardiac disease, but important conceptual and technical shortcomings limit its capacity to resolve the complicated biology of the heart (Palmer et al. 2024). Firstly, bulk methods average RNA signals across all cells in the sample, masking cell-type-specific expression and blurring contributions from rare or spatially restricted populations. In the heart an organ with a heterogeneous cellular composition that dynamically changes in disease (cardiomyocytes, fibroblasts, endothelial cells, immune cells, pericytes) averaging can therefore lead investigators to erroneous conclusions about whether cell types drive transcriptional alterations. Single-cell and spatial experiments have repeatedly portrayed disease-associated signatures that are occult or confounded in bulk profiles (Ralph Patrick et al. 2024).

Secondly, bulk gene expression changes may reflect changes in cell composition rather than true per-cell transcriptional regulation. For instance, following myocardial infarction, fibrosis or immune infiltration increases the proportion of fibroblasts or immune cells respectively; if bulk gene expression was increased, the population dysregulation would be difficult to infer since it is merely tracking relative changes in cell composition (Zhang et al. 2024). In an attempt to correct for this, deconvolution methods have been applied to bulk tissue samples; however, it can be difficult to accurately deconvolute, particularly when considering the quality of the reference atlas used for deconvolutions, as well as unknown cell states that may exist in diseased tissue (He et al. 2025).

Thirdly, bulk methods do not capture spatial context—where gene expression is occurring relative to the tissue architecture—and yet, these spatial relationships are the essence of cardiac pathophysiology and repair (Nguyen et al. 2025). As spatial transcriptomics and imaging-based molecular profiling capture micro-niches and gradients of gene expression that bulk assays are unable to identify, mechanistic understanding and therapeutic target prioritization are limited (Wünnemann et al. 2025).

Fourthly, bulk profiling is not very sensitive to detect rare cell states and transient programs (activated fibroblast subtypes, antigen presenting cell types, stressed cardiomyocyte subpopulations) that may be biologically important despite low abundance. These rare but functionally important populations are often detected with single-cell or high-resolution spatial methods, indicating the absence of data in bulk datasets (Tan et al. 2023; Lázár et al. 2025).

Fifthly, despite the advantages of transcriptomics data, the limits of the techniques and the analysis limit the inference of causality in networks and regulatory programs at a cell-type level from transcriptomics bulk data only, due to the inherent breakdown of batch effects, variations in RNA quality (especially from fibrotic or post-mortem tissue), and the inability to analyze regulatory interactions at single-cell resolution. However, integrative analyses that couple data from bulk, single-cell transcriptomics and spatial transcriptomics can combat these drawbacks, but are typically unable to remediate them entirely (Tzec-Interián et al. 2025; Tang et al. 2025).

12.1.2 Emergence of Single-Cell Technologies in Cardiology

In the past ten years, single-cell and spatial molecular methods have revolutionized cardiac research by advancing the field from tissue-level averages to cell-resolved maps of the normal and pathological heart. While bulk assays traditionally report a single aggregate signal from thousands to millions of mixed cell types, methods such as single-cell RNA sequencing (scRNA-seq), single-nucleus RNA sequencing (snRNA-seq), long-read single-cell isoform profiling, and spatial transcriptomics can now identify discrete cell populations, rare cellular states, and the anatomical locales of the transcriptional programs that underpin injury, repair, and remodeling (Palmer et al. 2024). These platforms have initiated a conceptual shift: the heart is no longer a collection of similar cardiomyocytes and a homogenous stroma but a dynamic mosaic of cell types, with constantly shifting proportions, states, and neighborhood relationships that change with age, stress, and disease (Paik et al. 2020).

Applying single-cell research methods to human and model-system hearts has illuminated many impactful advances. For example, large-scale atlasing efforts have defined the primary cellular compartments of the adult human ventricle and have similarly defined novel cell subtypes (e.g., fibroblast activation states, stressed cardiomyocyte subpopulations, and diverse endothelial phenotypes, as well as complex immune repertoires, by region and disease status). The combination of such studies has identified cell types that either expand or change into a pathogenic program after myocardial infarction, pressure overload or heart failure, and tiled transcription factors, and ligand–receptor pairings that show cell-type-specific expansion associate with cardiac pathology, or might be therapeutic targets, either explicitly or by association (Xiao et al. 2025; Forte et al. 2018).

In addition to cell identity, recent single-cell studies have uncovered the role of alternative splicing and isoform diversity in the heart. Long-read single-nucleus methods have resolved splicing isoform landscapes at near single-cell resolution, identifying isoform shifts that strike alongside disease, serve as buffers to, or reshape core cellular programs—findings from short-read bulk methods cannot currently identify. Indeed, transcript isoform atlases of this resolution enable new mechanistic hypotheses about post-transcriptional regulation in cardiomyopathy and heart failure (Pan et al. 2025). Importantly, the combination of single-cell data with spatial transcriptomics and in situ hybridization now provides cellular identity while maintaining tissue architecture. Utilizing spatial multi-omics studies of myocardial infarct and chronic heart disease, researchers have mapped micro-niches (e.g., the infarct core and border zone), demonstrated spatial gradients of stress and fibrotic programs, and observed localized immune–stromal interactions that determine whether the tissue repairs or maladaptively remodels. The spatially anchored information is crucial for understanding cell–cell communication, and ultimately developing a prioritized set of interventional targets that will operate in anatomical microenvironments (Wünnemann et al. 2025).

Methodological advances also support interpretability and translational potential. Integration frameworks and reference atlases enable a high level of cell annotation confidence and cross-study comparison, whereas ligand–receptor and trajectory analysis allows for inference of intercellular signaling and dynamic state transitions. More broadly, there are still challenges: biases in sampling cardiac tissue in humans (particularly viable myocardium), biases from dissociation, harmonizing data across platforms, and promoting a connection from descriptive atlases of cell populations to causal elements of biology and therapy (Jiang et al. 2024).

12.1.3 Epidemiological Trends and Transcriptomic Innovations in Cardiovascular Research

Cardiovascular diseases (CVDs) still account for the most deaths and disability globally, with causative factors contributing to nearly 19.4 million deaths each year and a large percentage of global disease burden (Mensah et al. 2023). Despite gains in prevention and treatment, the burden of CVD remains high, especially in low- to middle-income countries where healthcare inequities continue to exist (Shi et al. 2025a). Traditional epidemiological and clinical strategies have helped yield significant knowledge of the risk factors of CVD and distribution of disease, but epidemiology and clinical knowledge cannot fully illuminate the molecular and cellular heterogeneity that characterizes disease initiation and progression. Bridging these scales from population burden of CVD to the mechanisms of cell specificity has thus become a central aim of contemporary cardiovascular research (Yang et al. 2025). The main intent of this chapter is to present a consolidated overview of the worldwide and regional burden of cardiovascular diseases and address how recent advances in molecular and transcriptomic approaches and technologies represent a change in the field moving forward. The context begins with updated epidemiological information from Global Burden of Disease (GBD 2025) analyses and reports from the World Health Organization (WHO), which reflect a positive decline in age-standardized mortality in some regions of the world and a universal increase in absolute numbers of cases attributable to aging, urbanization, and lifestyle changes (Schorr et al. 2021). Following this, the chapter discusses how bulk transcriptomic approaches have progressed our understanding of biological processes like cardiac remodeling, fibrosis, inflammation, and metabolic reprogramming over the last 20 years, while also discussing the limitations of the technique to address cell-type-specific or spatially resolved molecular phenomena (Xie et al. 2024).

To address these challenges, single-cell and spatial transcriptomic technologies have emerged and transformed the way we can investigate an emerging landscape of cardiac biology. These next-generation tools have provided researchers with the capability to evaluate gene expression at the resolution of individual cells, discovering rare cardiac cell states, dynamic transcriptional programs, and significant intercellular communication networks, largely for the first time (Palmer et al. 2024; Wünnemann et al. 2025). For instance, recent single-cell and spatial studies showed that cardiac tissues are defined by various fibroblast activation states, specialized endothelial and immune cell populations, and unique stressed-cardiomyocyte subtypes that are crucial for cardiac repair and pathological remodeling (Pan et al. 2025). By combining molecular-level discoveries with population-level disease information, this chapter illustrates how, together, epidemiology, transcriptomics, and systems biology programs can lead to a more comprehensive and actionable understanding of cardiovascular disease mechanisms.

This chapter therefore covers the epidemiological foundations of CVDs, to the molecular insights revealed using high-resolution omics technologies. It summarizes a growing body of work published in clinical research and cardiovascular journals, like The Lancet, The New England Journal of Medicine, Circulation, Nature Cardiovascular Research, and JACC, between the years 2023 and 2025, to create a narrative that illustrates the continuum from burden of disease to cellular complexity to technology (Schmitz et al. 2019). The chapter aims to act as a single point of reference for clinicians, researchers, and students, emphasizing how modern approaches to illness from the global one to the single cell are joining to catalyze the next stage of precision cardiology and translational cardiovascular science (Fig. 12.1).

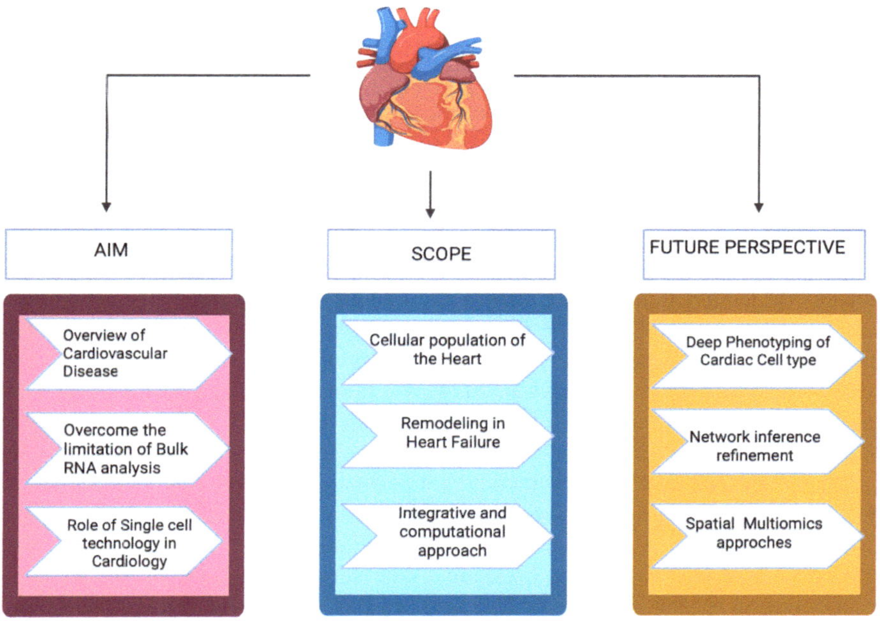

Fig. 12.1 Aim, scope, and future directions in cardiac systems biology (Forte et al. 2018; Zeng et al. 2025a)

12.2 Cellular Landscape of the Healthy and Diseased Heart

12.2.1 Cardiac Cell Populations: Cardiomyocytes, Fibroblasts, Endothelial, and Immune Cells

The mammalian heart is a diverse organ both structurally and functionally, and the cellular composition and intercellular interactions are essential for homeostasis and disease. Cardiomyocytes, the contractile muscle cells, have been the focus of cardiac biology for many years because they are responsible for the pumping function of the heart and are the primary drivers of systolic performance. However, recent single-cell and spatial studies highlight that there are multiple subtypes and stress-responsive states, rather than a single uniform population of cardiomyocytes: subsets of ventricular cardiomyocytes demonstrate distinct transcriptional programs related to contraction, metabolism, and responsiveness to stress, and rare "stressed" cardiomyocyte states appear following ischemia or heart failure (Palmer et al. 2024). It is essential to distinguish between different cardiomyocytes states, as they may have important implications for both vulnerability to injury and the relative ability to recover. Fibroblasts are referred to as the primary stromal cells of the heart, which facilitate the creation of the extracellular matrix (ECM), the creation of scar tissue, and structural remodeling of the heart. Recent studies in single-cell atlases have revealed a remarkable diversity and heterogeneity of cardiac fibroblast phenotypes, including resting populations, activated myofibroblasts, and specific sub-clusters associated with disease which upregulates ECM genes, chemokines, and pro-fibrotic transcriptional programs (Amrute et al. 2023). Interestingly, recent studies also identified heart-failure-specific fibroblast populations that contribute to pathological remodeling via specific defined signaling axes, suggesting that fibroblasts are cell-type-specific functional therapeutic targets rather than merely passive producers of scar tissue (Komuro et al. 2025). Endothelial cells, which form the lining of coronary vessels and the microvasculature, are now viewed as being specialized and regionally heterogeneous in the heart. Multiple single-cell and spatial multimeric studies demonstrate endothelial cellular heterogeneity describing subsets of endothelial cells that are specialized in barrier function, angiogenesis, and immune signaling. Vascular niches are also emerging as areas of localized age and pathological signaling (Palmer et al. 2024; Wünnemann et al. 2025). Given that the microvasculature is responsible for both oxygen delivery and trafficking of inflammatory cells, any alterations in endothelial cell dysfunction or phenotypic duplicity will have devastating effects on myocardial perfusion, inflammation, and repair. In both healthy and diseased myocardium, immune cells represent a contextual and dynamic compartment of myocardial cells. Resident macrophages, dendritic cells, and small lymphocyte subsets maintain tissue survey and facilitate homeostatic remodeling in tissue; after injury (e.g., myocardial infarction), there is also rapid recruitment of circulating neutrophils and monocyte-derived macrophages, which will coordinate clearance of debris, facilitate resolution of inflammation, and subsequently activate fibroblasts (Wünnemann et al. 2025; Amrute et al. 2023). Single-cell profiling has identified pro-inflammatory and reparative immune states, while spatial mapping shows how infiltration of immune cells will often take anatomically constrained routes (e.g., from endocardium to infarct core) highlighting the role of spatial context in interpreting immune–stromal interactions. Overall, modern single-cell and spatial studies have redefined the heart as a dynamic cellular ecosystem. Each major cardiac cell type—including cardiomyocytes, fibroblasts, endothelial cells, and immune cells—comprises multiple functionally relevant subtypes whose composition and interactions are altered during aging, metabolic stress, ischemia, and chronic heart failure. The understanding of these cell-resolved architectures, together with spatial organization, will be essential for identifying specific mechanisms of disease to design tailored therapies that extend beyond cardiomyocyte-centric approaches (Palmer et al. 2024; Wünnemann et al. 2025; Amrute et al. 2023) (Table 12.1).

12.2.2 Cellular Heterogeneity in the Normal Myocardium

A healthy myocardium is not a uniform mass of contractile tissue; rather, it is a complex, spatially organized ecosystem of heterogeneous cell types and states whose interactions underlie normal function and resilience. Early single-cell and single-nucleus atlases showed that the adult human heart is composed of multiple major cell compartments—cardiomyocytes, fibroblasts, endothelial cells, pericytes,

Table 12.1 Association and role of different cardiac cells in different cardiac diseases

Cardiac disease	Associated cell types Cellular state	Mode of action	Reference
Acute myocardial infarction (MI)	Neutrophils $CCR2^+$ Macrophage Activated fibroblasts Endocardial/Endothelial cells	Sterile inflammation; *endocardial* entry routes for myeloid cells; IL-1β/CCL2 axes; border-zone fibroblast activation; spatially distinct repair niches	Kuppe et al. (2022)
Chronic heart failure (HFrEF/HF)	Stress-responsive cardiomyocytes Fibroblast Macrophages	Fetal-gene reprogramming in CMs; TGF-β/IL-11 signaling; immune–fibroblast neighborhoods sustaining fibrosis	Amrute et al. (2023)
Metabolic cardiac disease (HFpEF)	Endothelial and microvascular cells Pro-inflammatory monocytes Perivascular fibroblasts	Systemic inflammation–microvascular dysfunction axis; endothelial activation; adiposity/metabolic cues	Kneuer et al. (2025)
Dilated cardiomyopathy (DCM)	Remodeled cardiomyocytes ECM-producing fibroblasts Resident infiltrating *immune* cells	Sarcomere and cytoskeletal stress programs; fibrosis networks; genetic risk loci inform cell-type targets	Koenig et al. (2022)
Hypertrophic cardiomyopathy (HCM)	Hypertrophied cardiomyocytes Fibroblast	Stress-response transcriptional programs; conduction/action-potential remodeling; fibrosis	Lu et al. (2024)
Atrial fibrillation (AF)	Mesenchymal activated endothelial Atrial fibroblasts T cell	Endothelial plasticity → TGF-β-driven fibroblast activation; ion-handling/stress genes in atrial CMs	Zeng et al. (2025b)
Myocarditis	Cytotoxic T cells Inflammatory macrophage Stressed cardiomyocytes	IFN-γ-dominated programs; IL16/IL18 signatures; immune–stromal circuits	Maatz et al. (2025)
Post-MI and multi-etiology cardiac fibrosis	Specialized Fibroblast Perivascular Macrophage Pericytes	Matrifibrocyte programs; IL-1β from $CCR2^+$ macrophages to FBs; secretome remodeling	Patrick et al. (2024)

smooth muscle cells, and a host of resident immune cells—which are progressively divided into transcriptionally and functionally distinct subpopulations (Litviňuková et al. 2020). More recent studies at higher resolution have extended and elaborated upon this vision with an emphasis on regional specialization (atria versus ventricle, endocardium versus epicardium), as well as developmental and age-related shifts and a more nuanced interpretation of cell states importantly for physiological adaptation. Cardiomyocytes, the electrically and mechanically active cells responsible for contraction, are heterogeneous populations of cells. For example, ventricular myocytes demonstrate gradients of gene programs that correspond to the myocyte's anatomical location and workload for the handling of ions, energy metabolism, and contractile isoforms. Single-cell and single-nucleus profiling efforts have identified cardiomyocyte subpopulations with differing metabolic profiles and stress-responsive transcriptional signatures in even non-pathologic hearts, suggesting there is an intrinsic division of labor among myocytes and differential susceptibility to stressors (Palmer et al. 2024). These transcriptional differences can correlate with differences in sarcomeric gene isoform expression, calcium handling machinery, and mitochondrial gene programs that ultimately modify contractile kinetics and efficiency. Cardiac fibroblasts, once thought to be a fairly uniform stromal compartment, are now understood to be a mosaic of specialized subtypes. The normal myocardium includes, among others, homeostatic fibroblasts that coordinate the maintenance of the extracellular matrix (ECM); perivascular fibroblasts that are associated with the coronary microvasculature; and transcriptionally distinct populations that provide niche-supporting functions for cardiomyocytes and endothelial cells (Kanemaru et al. 2023). This inherent heterogeneity is critical, because certain fibroblast subtypes are prepared to engage distinct repair or fibrotic programs following injury, and their regionalized distribution leads to a differential pattern of ECM deposition and tissue stiffness after tissue stress. There is also substantial diversity in endothelial cells between vascular beds and anatomical regions of the heart. The coronary artery endothelium, capillary endothelium, and endocardium each have distinct gene profiles/functionality in terms of angiogenesis, barrier, immune, and metabolic exchange (Palmer et al. 2024). Spatially resolved studies have delineated endothelial specialization to microanatomical niches, and also suggested that endothelial cells retain organotypic transcriptional "memories" that influence multimodal leukocyte recruitment and regional metabolic supply. The resident immune cells in the healthy heart, namely macrophages, dendritic cells, and a small number of lymphocytes, carry out housekeeping tasks: removing debris, providing trophic support, and regulating fibroblast activity. Single-cell profiling has shown that resident macrophages ordinarily express anti-inflammatory or tissue-supporting programs, and resident macrophages occupy anatomically specialized niches where they engage in ongoing interactions with stromal and parenchymal cells to maintain homeostasis (Amrute et al. 2023). In addition to discrete cell classes, recent applications using long-read transcripts and spatial transcriptomics have highlighted the important roles of isoform diversity and local neighborhoods in normal cardiac biology. Alternative splicing produces repertoires of isoforms that vary by region and cell type, producing another layer of functional specialization

(Pan et al. 2025). Furthermore, spatial mulatomic maps have shown that the function of a particular cell is closely associated with its neighbors and the microenvironment in which it resides: gradients of metabolism, local ECM composition, and proximity to vessels or conduction tissue create micro-niches that condition cellular phenotype (Wünnemann et al. 2025; Kanemaru et al. 2023). Overall, the normal myocardium is a highly structured assembly of heterogeneous cell populations and states that are organized across anatomical axes and micro-niches. Understanding this baseline complexity is critical for interpreting alterations associated with disease, creating targeted treatments, and understanding why specific cardiac regions or cell types display selective vulnerability in different cardiac pathologies (Palmer et al. 2024; Litviňuková et al. 2020).

12.2.3 Remodeling and Cellular Reprogramming in Heart Failure

Heart failure (HF) represents the end stage of a myriad of cardiovascular diseases and is clinically defined by significant structural, cellular, and molecular remodeling of the myocardium. Whereas hemodynamic overload, ischemia, and metabolic change initiate this remodeling process, symptomatic heart failure progression is driven primarily by more complex defects in cardiomyocyte function, intercellular communication, and tissue architecture (Forte et al. 2018). In the last five years, single-cell and spatial transcriptomic studies have provided unprecedented in vivo insight into how different populations of cardiac-related cells respond and adapt in heart failure, including novel transcriptional programs, cell–cell signaling, and maladaptive reprogramming (Wünnemann et al. 2025). Cardiomyocyte dysfunction and reprogramming lies at the heart of remodeling in heart failure. Cardiomyocytes undergo hypertrophy, sarcomeric reorganization, metabolic switching away from fatty acid oxidation, toward glycolysis, and partial dedifferentiation mimicking fetal gene expression programs. While these adaptations are initially beneficial, they ultimately undermine contractile efficiency and exacerbate chamber dilation and wall stress (Dhalla et al. 2024). Emerging studies using single-nucleus sequencing have identified distinct populations of cardiomyocytes that exhibit transcriptional signatures associated with stress, including induction of NPPB, MYH7, and ACTA1, along with activation of pathways associated with unfolded protein response and oxidative stress (Palmer et al. 2024). This "stressed" or dedifferentiated state is variable throughout the myocardium, with spatial studies demonstrating local areas of metabolic inadequacy, and mitochondrial dysfunction, that accompanies fibrotic and ischemic regions (Wünnemann et al. 2025).

A further hallmark of cardiac remodeling also includes the activation and expansion of fibroblasts. Persistent injury and/or chronic stress promote resident fibroblasts to acquire myofibroblast properties that are marked by the expression of high levels of extracellular matrix (ECM) genes like COL1A1, POSTN, and FN1. Myofibroblasts are needed to maintain cardiac structure; however, sustained activation can result in maladaptive fibrosis, greater tissue stiffness, and impaired contractile synchrony with surrounding cells (Kanemaru et al. 2023). Single-cell and spatial multi-omics techniques have demonstrated that there are not homogenous effects of fibroblast activation in heart failure but that there are various fibroblast subtypes present, including inflammatory fibroblasts, matrifibrocytes, and pro-angiogenic fibroblast populations (Komuro et al. 2025). These fibroblast subtypes have ample cross-talk to immune cells and endothelial cells through paracrine signaling using TGF-β, CXCL12, and IL-11 pathways that perpetuate the chronic remodeling process while simultaneously blocking its resolution. Reprogramming also involves endothelial cells and immune populations during heart failure. Endothelial-to-mesenchymal transition (EndoMT) promotes the accumulation of fibroblast-like cells, and impaired angiogenic signaling is associated with capillary density and perfusion deficits (Palmer et al. 2024). Alongside these inflammatory processes, chronic inflammation alters the immune environment in the myocardium, as persistent activation of macrophages and T cells leads to the release of pro-fibrotic cytokines (i.e., TNF-α, IL-1β, IL-6) (Amrute et al. 2023). Importantly, single-cell atlases of failing human hearts reveal the affinity for immune–stromal interactions to establish self-perpetuating feedback loops which support pathological remodeling and promote poor ventricular geometry (Wünnemann et al. 2025). The cross-evidence also points to cellular reprogramming events occurring in heart failure. Under extreme stress, certain non-myocyte populations, such as fibroblasts and endothelial cells, display partial transcriptional reprogramming toward cardiomyocyte-like gene expression states. The reprogramming process has been termed "stress-induced pseudo-differentiation" (Li et al. 2024). While the stress-induced pseudo-differentiation is insufficient for regeneration, it does point to plasticity of the adult heart and the fact that latent regenerative potential may be utilized therapeutically. There have been advances in reprogramming methodologies, with direct programming efforts using transcription factors (e.g., Gata4, Mef2c, Tbx5) and small molecules clustered together or alone to reactivate developmental gene networks and induce cardiomyocyte-like cells from fibroblasts in experimental models (Shi et al. 2025b). In summary, the remodeling of the failing heart is a complex and dynamic process that includes maladaptive cellular responses and reprogramming phenomena across multiple cardiac lineages. Cardiomyocytes show metabolic and structural dedifferentiation; fibroblasts have diverged to pro-fibrotic and inflammatory lineages; endothelial cells lose their angiogenic competence; and immune cells sustain chronic inflammation. Integrative single-cell and spatial multi-omics studies are now painting a picture of the cross-talk between these compartments, leading to a definition of heart failure as not simply a single-cell disorder, but rather a networked failure of intercellular homeostasis (Wünnemann et al. 2025; Komuro et al. 2025). This emerging cellular framework will allow for precision therapies targeting pathogenic cell states and facilitating adaptive remodeling, as opposed to irreversible remodeling.

12.2.4 Immune Cell Infiltration and Inflammatory Signaling in Cardiac Injury

Following myocardial injury such as acute infarction, ischemia–reperfusion, and pressure overload, the heart launches a rapid and coordinated immune response that profoundly influences healing, remodeling, and eventual functional outcome. Immune cell infiltration into the injured myocardium occurs in defined temporal and spatial phases, beginning with neutrophils and monocyte-derived macrophages in the early hours and days, followed by adaptive immune cells (T and B lymphocytes) and resident macrophage reprogramming in subacute to chronic phases (Wünnemann et al. 2025). The modes of infiltration and activation are not uniform but depend substantially on anatomical micro-zones (infarct core, border zone, endocardial surface) and tissue context—a fact recently revealed by spatial multi-omics mapping of post-infarct hearts (Wünnemann et al. 2025). During the initial stage post injury, there is leakage of damage-associated molecular patterns (DAMPs) from dying cardiomyocytes and the damaged extracellular matrix, as well as production of reactive oxygen species and cytokines (e.g., IL-1β, TNF-α) that initiate endothelial activation and upregulation of adhesion molecules (VCAM-1, ICAM-1) and chemokines (e.g., CCL2) that recruit circulating monocytes into the injured myocardium. A seminal study in spatial transcriptomics of murine infarction revealed an unanticipated (to the authors, at least) route, with $CCR2^{+}$ monocytes/macrophages entering via the endocardium in the infarct zone, at least partially mediated by von Willebrand factor (vWF) production by endocardial endothelial cells. This demonstrates that immune infiltration is anatomically constrained, and dynamic (Wang et al. 2025).

Once they have entered the tissue, monocytes differentiate into the various macrophage phenotypes. At first, a pro-inflammatory ("M1-like") phenotype dominates, producing matrix metalloproteinases, reactive nitrogen species, and pro-fibrotic cytokines; then, in the reparative phase, macrophages adopt anti-inflammatory or tissue-remodeling ("M2-like") profiles that promote angiogenesis, fibroblast activation, and, in the case of myocardial lesions, scar maturation (Wang et al. 2025). Recent single-cell sequencing studies in heart failure have also identified macrophage subpopulations (e.g., $TREM2^{+}$, $SPP1^{+}$) to be associated with fibrosis, reactive oxygen species generation, and poorer outcomes (Zeng et al. 2025a). Combined, this work shows that macrophage heterogeneity is functionally relevant and not simply a description. Besides macrophages, neutrophils quickly infiltrate and exacerbate the tissue injury process via protease and ROS release; it is crucial that neutrophils clear and allow the tissue to transition into the repair phase (Zeng et al. 2025a). Lymphocytes, particularly $CD4^{+}$ T helper 1 (Th1) and Th17 cells, also become important in cardiac remodeling by promoting monocyte recruitment and fibroblast activation by IFN-γ, IL-17, and CXCL9/10 (Wang et al. 2025). B cells and regulatory T cells (Tregs) are increasingly described as post-injury inflammatory modulators, but their functional roles in the human cardiac environment are less well defined.

The interaction between immune cells and non-immune cell types in the tissues of the heart is critical for pathological remodeling. Recent findings by Maillard et al. show that infiltrated macrophages produce TGF-β1 and IL-11 that drive resident fibroblasts to transdifferentiate into myofibroblasts, which promote an increased extracellular matrix deposition, and therefore added stiffness and degradation of mechanical coupling (Wang et al. 2025). Spatial multi-omics data also show that macrophage–fibroblast neighborhoods in the border zone are very different compared to remote myocardium with an increased complexity of immune signaling networks (Wünnemann et al. 2025). Finally, prolonged or adaptive immune activation (i.e., sustained infiltration of $CCR2^{+}$ monocyte) is associated with worse ventricular dilation and worse ejection fraction in prospective human studies (Zeng et al. 2025a). Clinically, a clearer and more balanced understanding of cardiac inflammation is emerging. Rather than being viewed solely as harmful, inflammation in the heart is now recognized as a "sterile" yet essential process that supports tissue repair in the aftermath of injury but can also become destructive when excessive or prolonged. In the early phase, inflammatory cells clear dead tissue and trigger repair pathways; however, if this response persists, it can drive fibrosis, scar formation, and long-term ventricular dysfunction. This realization has shifted therapeutic goals from broadly suppressing inflammation to fine-tuning it. Strategies such as IL-1β blockade and CCR2 inhibition are under active investigation, aiming to curb harmful immune activation while preserving its regenerative benefits (Wang et al. 2025). The growing application of single-cell and spatial transcriptomics has added unprecedented precision, helping researchers identify which immune cell subsets act at specific times and locations in the injured heart. These insights offer a path toward targeted immunomodulation, where treatment could be timed and tailored to guide inflammation toward healing rather than harm.

In conclusion, immune cell infiltration and inflammatory signaling in cardiac injury are far from random events. They form a highly coordinated, spatially organized, and time-sensitive network of cellular interactions that determine whether the heart heals or progresses toward failure. Recent advances in single-cell and spatial multi-omics technologies have illuminated this complexity, revealing how specific immune cells communicate with cardiomyocytes, fibroblasts, and endothelial cells to influence recovery. By decoding these intricate patterns, researchers are beginning to transform our understanding of post-injury inflammation from a chaotic cascade to an orchestrated biological program and, in doing so, are opening new avenues for precision therapies that can modulate immune responses with the same sophistication as the heart's own healing mechanisms.

12.2.5 Stromal and Vascular Dynamics in Cardiac Repair

The suspension of cardiac repair after injury, whether from MI or ischemia–reperfusion injury, involves a tightly regulated engagement of stromal and vascular compartments. These two compartments—cardiac fibroblasts, pericytes, endothelial cells, and smooth muscle cells—coordinate inflammation, ECM remodeling, neovascularization, and scar maturation (Forte et al. 2018). Advances in single-cell and spatial transcriptomics have changed our view of the spatially heterogeneous and temporally regulated dynamics of stromal and vascular compartments in a way that's central to healing versus maladaptive remodeling (Wünnemann et al. 2025). Cardiac fibroblasts are a central element in the stroma's response to injury. In the early stages following MI, fibroblasts transition from quiescent ECM-maintaining cells to highly proliferative, secretory myofibroblasts in response to transforming growth factor-beta (TGF-β), interleukin-11 (IL-11), and platelet-derived growth factor (PDGF) signaling (Komuro et al. 2025). These myofibroblasts deposit structural proteins such as collagen I and III to form the provisional matrix that sustains the infarcted area and protects against ventricular rupture. However, prolonged activation of fibroblasts may result in excess ECM deposition, pathological fibrosis, and diastolic dysfunction. Single-cell studies have discovered a striking diversity in fibroblast subtypes pro-inflammatory, matrifibrocyte, and pro-angiogenic lineages each in their own niches with distinct contributions to repair or fibrosis (Amrute et al. 2023). In addition to fibroblast activity, vascular remodeling and angiogenesis are essential for re-establishing perfusion and oxygen delivery to injured myocardium. Endothelial cells that are located adjacent to the infarct border zone can activate hypoxia-inducible genes such as VEGFA and ANGPT2, helping to reorganize into sprouting angiogenic clonesters, which can facilitate capillary growth (Palmer et al. 2024). Single-cell transcriptomic maps have identified specialized groups of endothelial subtypes (tip, stalk, and venous-like) as partners in angiogenic stimulus, while interacting with the fibroblasts and pericytes to stabilize newly formed vessels (Wünnemann et al. 2025; Litviňuková et al. 2020). While these adaptation mechanisms occur, the regenerative angiogenic response in the adult human heart remains paralleled, contributing to chronic hypoxia and scarring maturation processes.

While historically categorized as "structural" components of blood vessels, pericytes and smooth muscle cells are now recognized as dynamic contributors to repair. During the acute inflammation phase, pericytes detach from microvessels and release both pro-angiogenic and immunomodulatory signals, and may even temporarily behave like fibroblast-like phenotypes that assist in ECM production and stabilization of the tissue (Li et al. 2024). Vascular smooth muscle cells (VSMCs), on the other hand, remodel arterioles to aid in the recovery of perfusion, but their chronic activation may also promote microvascular rarefaction. Notably, spatial transcriptomics have demonstrated that the co-localization of pericyte and endothelial cells within the peri-infarct zone predicted successful maturation of blood vessels, pointing to the spatiotemporal importance of coordinated and reciprocal stroma–vascular interactions for long-term cardiac recovery (Wünnemann et al. 2025). Importantly, not all stroma–vascular interactions are limited to structural repair. For example, fibroblasts release angiogenic mediators, including VEGF and angiopoietins, while endothelial cells release paracrine factors (e.g., nitric oxide, endothelin) to modulate fibroblast activation. This reciprocal communication establishes a regulatory feedback loop to determine whether healing occurs adaptively or progresses to maladaptive fibrosis (Komuro et al. 2025). Recent spatial multi-omics have visualized in situ these cross-signaling networks, and have shown fibroblast–endothelial clusters are present and act as a "repair hubs" during the transition from inflammation to regeneration (Wünnemann et al. 2025). New evidence is also emerging to suggest that targeted interference with these stromal and vascular dynamics can ultimately improve outcomes after myocardial injury. For instance, experimental therapies directed at promoting angiogenesis—VEGF delivery, endothelial progenitor cell therapy, and exosomal microRNAs—are being tested in conjunction with interventions targeting fibroblast activation to promote balanced repair. This integrative platform corresponds to the larger concept of precision cardiac regeneration with interventions tailored to manipulate beneficial stromal–vascular signaling while mitigating excessive fibrosis. Overall, cardiac repair is not merely a regenerative scar formation but rather the dynamic interaction of stromal and vascular networks. Fibroblasts, endothelial cells, pericytes, and smooth muscle cells work in unison to reconstruct and remodel myocardium in response to spatiotemporally coordinated signaling cues. Recent advancements in single-cell and spatial multi-omics technologies have yielded unparalleled insights into these dynamic processes, elucidating how local microenvironmental signals influence the transition toward effective repair, ineffective repair, or the progression to active fibrosis—the critical divergence in fibroproliferative tissue pathology. Understanding and translation of these stromal–vascular dynamics to therapeutic strategies could revolutionize cardiac regeneration and repair and limit the progression to failure following myocardial injury (Fig. 12.2).

12.3 Integrative and Computational Approaches

The rapid emergence of single-cell, spatial, epigenomic, and proteomic data in cardiovascular research necessitates integrative and computational frameworks to realize their full potential. In cardiac biology and disease, the use and integration of bulk versus single-cell transcriptomics, modeling of cell–cell communication networks, application of machine learning/AI strategies, linking scRNA-seq to chromatin and proteomic assays, and mapping of spatial architecture are integral or central, rather than optional, approaches to derive mechanistic insights and therapeutic hypotheses from high-dimensional data. The sub-sections below provide an overview of current approaches, frameworks, and challenges around these different but intersecting frontiers (Fig. 12.3).

12.3.1 Integration of Bulk and Single-Cell Transcriptomic Data

Bulk RNA-sequencing (RNA-seq) of whole heart tissue or sorted populations is still a low-cost and widely available tool, but it is fundamentally a measure that averages over heterogeneous cell states. In contrast, single-cell or single-nucleus RNA-seq has the ability to resolve cell-type-specific expressions although it usually lacks sample subsets and the clinical context. As such, integrating bulk and single-cell data present a unique opportunity to obtain scale and resolution. Several methods

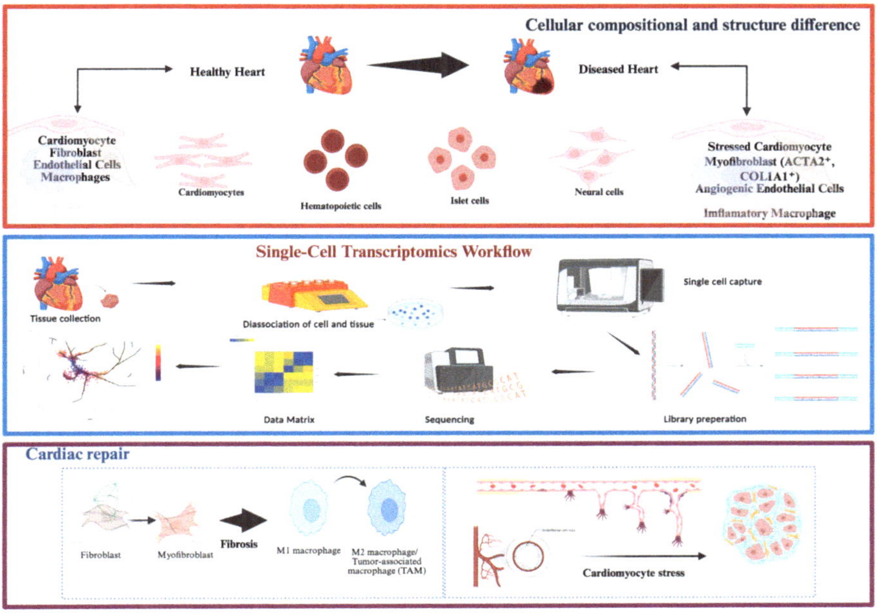

Fig. 12.2 Cellular composition, methodology for single-cell transcriptomics, and cellular repair

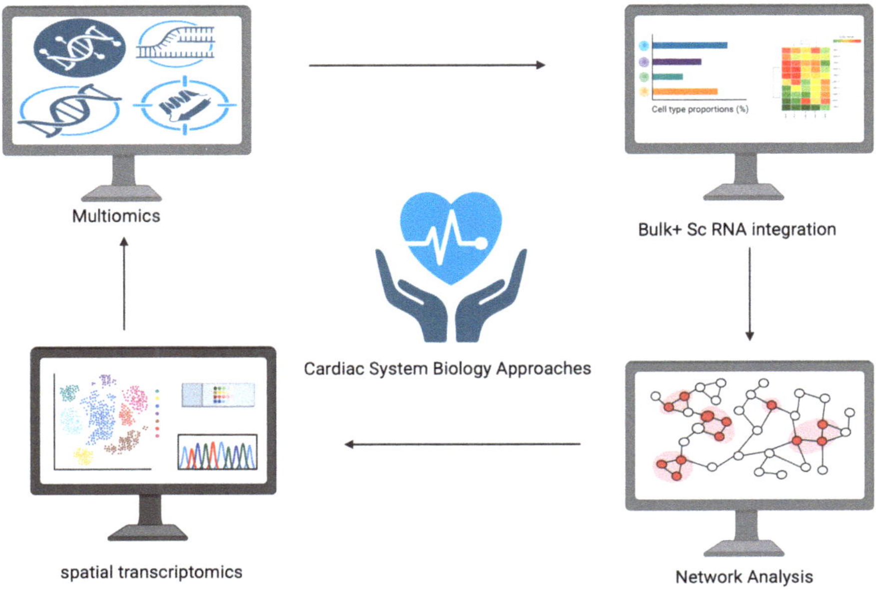

Fig. 12.3 Computational approaches in cardiovascular disease

have recently emerged that deconvolve bulk RNA-seq samples using single-cell reference atlases. These approaches infer cell-type proportions and map bulk differential expression signatures onto specific cellular programs, addressing this key challenge. For instance, one informative study used a machine learning model that integrated bulk and single-cell RNA-seq and determined oxidative stress-related candidate biomarkers (LUM and PCOLCE2) in cardiac tissue. A researcher (Li et al. 2025) used similar integration by examining snRNA-seq and bulk RNA sequencing datasets from patients with atrial fibrillation and identified lactate-metabolism genes associated with T-cell changes, achieving a diagnostic AUC of 0.909 (Long et al. 2025). In cardiovascular disease, these tools allow for linking large patient cohorts (bulk data) with mechanistic insights into cell types (single-cell data) to reveal biomarker discovery, stratification, and hypothesis generation. Major challenges include batch effects, variability with tissue dissociation, and capture efficiency of cell types, and the need for robust algorithms that align bulk and single-cell multimodalities with compositional and activation trends.

12.3.2 Cell–Cell Communication and Network Analysis

It is imperative to understand how cardiac cell types cardiomyocytes, fibroblasts, endothelial cells, and immune cells interact to understand tissue behavior in health and disease. Single-cell data allow us to infer ligand–receptor pairs, signaling pathways, and cellular neighborhoods, and there are many network-analysis tools available (e.g., CellChat, NicheNet, ICELLNET). For example, during cardiac remodeling, mapping the cross-talk among stromal, immune, and endothelial networks has revealed paracrine axes mediating fibrosis and repair. Once a network is mapped, it can be analyzed computationally to identify "hub" cell types or signaling molecules that may inform potential therapeutic targets. In addition, spatial transcriptomics can also provide context to these network inferences. There are fewer recent examples using cardiovascular-focused networks; however, the cell–cell communication modeling paradigm continues to grow exponentially. Nevertheless, in applying these tools, one must acknowledge potential biases (e.g., dropouts, ambient RNA, variability in capture of secreted ligands) and validate the predicted inferences experimentally (Paik et al. 2020).

12.3.3 Machine Learning and AI Applications in Cardiac scRNA-seq

The use of machine learning (ML) and artificial intelligence (AI)-based methods is becoming increasingly common in single-cell RNA sequencing (scRNA-seq) tasks such as cell-type identification, trajectory inference, state-mapping, gene-signature discovery, and disease-subtype stratification. In a recent example of a specific method of artificial intelligence, Nguyen et al. (2023) employed autoencoder, sparse-modeling, and semi-supervised learning approaches to studies of pig and mouse cardiac scRNA-seq data, and demonstrated a substantial improvement in the identification of cardiomyocyte subpopulations and cell-state transitions, finding AI methods significantly outperformed standard clustering/trajectory tools. In a broader review, Das et al. (2025) provided evidence of the increased use of ML and bioinformatics pipelines applied to scRNA-seq across diseases, specifically tracking the rise and use of denoising, embedding, and classification algorithms in deep learning architectures. In cardiac research, ML applications include, but are not limited to, identifying rare cell states associated with heart failure, predicting cell-type composition from clinical samples, and linking cell–cell profiles with imaging or functional phenotypes. Still, challenges remain, especially in terms of practical applications such as interpretability of models, overfitting in small datasets, and reproducibility/generalizability across platforms. Notwithstanding these challenges, we see ML/AI as an essential component of modern cardiac single-cell analytics.

12.3.4 Multi-omics Integration: Linking scRNA-seq with ATAC-seq and Proteomics

Single-cell transcriptomics assesses RNA expression, but the future state and regulatory state are also determined by chromatin accessibility mechanisms (ATAC-seq), DNA methylation, protein abundance/modifications, metabolic states, and spatial context. Integrative multi-omics aims to illuminate how these layers interact. Tan et al. (2023) summarized data integration across tissues, including cardiovascular applications, and proposed that the integration of transcriptome + chromatin state + surface proteome is critical to determine regulatory networks. Tang et al. (2025) more recently recognized potential for integrating scRNA-seq, scATAC-seq, and proteomics in the mammalian heart, pointing to the intrinsic value of joint clustering, modality alignment, and trajectory modeling. The integration of these methods in cardiovascular research directly connects disease-related chromatin accessibility hotspots to cell-type-specific gene expression and the identification of functional protein biomarkers. For example, integration of chromatin accessibility and expression variables may uncover regulatory elements specific to fibroblast subtype enabling pathological activation. However, barriers to data integration exist, such as data sparsity (especially in ATAC and proteomics at single-cell level), high computational demand, and the need for more pipelines and associated cross-modal annotations.

12.3.5 Spatial Transcriptomics for Cardiac Tissue Architecture

While single-cell technologies involve physical and biological dissociation of the cells (thereby losing spatial context), spatial transcriptomics (ST) retains the information of location, and allows the transcriptomic profiles to be visualized and mapped back to the architecture of the tissue. The application of spatial methodologies within cardiac tissues offers researchers the ability to identify gradients and

Table 12.2 Translational potential of computational frameworks in cardiac multi-omics

Computational Approach	Data Source	Application	References
Bulk RNA + single-cell RNA integration	Bulk RNA-seq from patient cohorts + single-cell atlases	Detect oxidative-stress biomarkers and cell-type-specific therapeutic targets in heart failure	Li et al. (2025)
Cell–cell communication and network modeling	scRNA-seq ± spatial datasets from infarcted myocardium	Identify ligand–receptor hubs (e.g., TGF-β, CXCL12) regulating fibrosis; nominate paracrine therapeutic targets	Amrute et al. (2023)
Machine learning and AI pipeline	sc/snRNA-seq + clinical metadata	Automated cell-type annotation, patient stratification, early prediction of heart-failure progression	Nguyen et al. (2023), Das et al. (2025)
Multi-omics integration	scRNA-seq, scATAC-seq, single-cell proteomics	Define enhancer–gene links and transcription-factor hierarchies controlling fibroblast activation	Tan et al. (2023), Tang et al. (2025)
Spatial transcriptomics	Human heart tissue	Map fibroblast-activation and immune-infiltration gradients for precision cardiac pathology	He et al. (2025), Wünnemann et al. (2025)

regions of importance including the infarct, border zone, perivascular niches, pathways, immune infiltration, and microenvironmental interactions. He et al. (2025) developed a new approach, SEU-TCA, in maths modeling and transfer-component analysis to enhance visualization of scRNA-seq datasets over spatial spots (like ST spot maps) and applied this technique to data on human heart development. Molla Desta et al. (2025) reviewed the recent advances in single-cell and spatial methodologies in and through a range of tissues including cardiac. Within the cardiac field, spatial transcriptomics revealed for example that macrophage infiltration within the zone of the infarct is happening along vascular routes and correlating with fibroblast activation in adjacent regions. The advantage while combining spatial transcriptomics with the single-cell datasets is it offers "cellular neighborhoods" and signaling niches that offer cellular states that drive outcomes. The challenge of the computational aspect remains in deconvolving the spots, aligning the modalities, and interpreting spatial patterns of gene-signatures with respect to histology and function (Table 12.2).

References

Amrute JM, Luo X, Penna V, Bredemeyer A, Yamawaki T, Yang S et al (2023) Targeting immune-fibroblast crosstalk in myocardial infarction and cardiac fibrosis. Res Square:rs-3

Bleakley C, Millar A, Harbinson M, McVeigh GE (2013) Lifetime risk of cardiovascular disease: the next generation in risk prediction. Can J Cardiol 29(2):147–150

Chauhan S, Aeri BT (2013) Prevalence of cardiovascular disease in India and it is economic impact-a review. Int J Sci Res Publ 3(10):1–5

Chong B, Jayabaskaran J, Jauhari SM, Chan SP, Goh R, Kueh MTW et al (2025a) Global burden of cardiovascular diseases: projections from 2025 to 2050. Eur J Prev Cardiol 32(11):1001–1015

Chong B, Jayabaskaran J, Jauhari S et al (2025b) The global syndemic of modifiable cardiovascular risk factors projected from 2025 to 2050. JACC 86(3):165–177. https://doi.org/10.1016/j.jacc.2025.04.061

Das K, Das R, Sinnarasan VSP, Paul D, Sheikh MMR, Venkatesan A (2025) Machine learning, bioinformatics and scRNA-seq analysis of hub genes for co-diagnosis of COVID-19 and systemic lupus erythematosus. In Silico Res Biomed:100075

Dhalla NS, Mota KO, Elimban V, Shah AK, de Vasconcelos CM, Bhullar SK (2024) Role of vasoactive hormone-induced signal transduction in cardiac hypertrophy and heart failure. Cells 13(10):856

Forte E, Furtado MB, Rosenthal N (2018) The interstitium in cardiac repair: role of the immune–stromal cell interplay. Nat Rev Cardiol 15(10):601–616

Gheorghe A, Griffiths U, Murphy A, Legido-Quigley H, Lamptey P, Perel P (2018) The economic burden of cardiovascular disease and hypertension in low-and middle-income countries: a systematic review. BMC Public Health 18(1):975

Global Cardiovascular Risk Consortium, Magnussen C, Alegre-Diaz J, Al-Nasser LA, Amouyel P, Aviles-Santa L, Bakker SJL, Ballantyne CM, Bernabé-Ortiz A, Bobak M, Boffetta P, Brenner H, Brunström M, Can G, Carrillo-Larco RM, Checkley W, Dallongeville J, De Bacquer D, de Gaetano G, de Lemos JA et al (2025) Global effect of cardiovascular risk factors on lifetime estimates. N Engl J Med 393(2):125–138. https://doi.org/10.1056/NEJMoa2415879

He J, Yang Y, Jiang R et al (2025) Integration of single-cell and spatial transcriptomics by SEU-TCA reveals the spatial origin of early cardiac progenitors. Genome Biol 26:158. https://doi.org/10.1186/s13059-025-03633-3

Jiang T, Jin X, Gao Y et al (2024) CardioAtlas: deciphering the single-cell transcriptome landscape in cardiovascular tissues and diseases. Biomark Res 12:149. https://doi.org/10.1186/s40364-024-00696-5

Kanemaru K, Kubota S, Wada Y, Fukuda K, Ieda M (2023) Diverse fibroblast subtypes maintain the homeostatic architecture and injury response of the heart. Nat Cardiovasc Res 2(4):320–335. https://doi.org/10.1038/s44161-023-00234-9

Schmitz B, Niehues H, Lenders M, Thorwesten L, Klose A, Krüger M, Brand E & Brand S-M (2019) Effects of high-intensity interval training on microvascular glycocalyx and associated microRNAs. Am J Physiol –Heart Circ Physiol. https://journals.physiology.org/doi/abs/10.1152/ajpheart.00751.2018

Kneuer JM, Müller M, Erbe S, Kokot KE, Rosch S, Müller-Kozarez I et al (2025) Circulating immune cell signature analysis in HFpEF across species. Circ Res 137(5):682–698

Koenig AL, Shchukina I, Amrute J et al (2022) Single-cell transcriptomics reveals cell-type-specific diversification in human heart failure. Nat Cardiovasc Res 1:263–280. https://doi.org/10.1038/s44161-022-00028-6

Komuro J, Hashimoto H, Katsuki T, Kusumoto D, Katoh M, Ko T et al (2025) Heart failure-specific cardiac fibroblasts contribute to cardiac dysfunction via the MYC–CXCL1–CXCR2 axis. Nat Cardiovasc Res:1–17

Kuppe C, Ramirez Flores RO, Li Z, Hayat S, Levinson RT, Liao X, Hannani MT, Tanevski J, Wünnemann F, Nagai JS, Halder M, Schumacher D, Menzel S, Schäfer G, Hoeft K, Cheng M, Ziegler S, Zhang X, Peisker F, Kaesler N et al (2022) Spatial multi-omic map of human myocardial infarction. Nature 608(7924):766–777. https://doi.org/10.1038/s41586-022-05060-x

Lázár E, Mauron R, Andrusivová Ž et al (2025) Spatiotemporal gene expression and cellular dynamics of the developing human heart. Nat Genet. https://doi.org/10.1038/s41588-025-02352-6

Li B, Zheng L, Yang J, Qu L (2024) Targeting oncogene-induced cellular plasticity for tumor therapy. Adv Biotechnol 2(3):24

Li C, Hao R, Li C, Liu L, Ding Z (2025) Integration of single-cell and bulk RNA sequencing data using machine learning identifies oxidative stress-related genes LUM and PCOLCE2 as potential biomarkers for heart failure. Int J Biol Macromol 300:140793

Lindstrom M, DeCleene N, Dorsey H, Fuster V, Johnson CO, LeGrand KE et al (2022) Global burden of cardiovascular diseases and risks collaboration, 1990-2021. J Am Coll Cardiol 80(25):2372–2425

Litviňuková M, Talavera-López C, Maatz H, Reichart D, Worth CL, Lindberg EL et al (2020) Cells of the adult human heart. Nature 588(7838):466–472

Long X, Li J, Zhang Y, Zhang Z (2025) Integrated single-cell and bulk transcriptome analysis reveal lactate metabolism-related signature and T cell alteration in atrial fibrillation. Front Cell Dev Biol 13:1644702. https://doi.org/10.3389/fcell.2025.1644702

Lu J, Ren J, Liu J, Lu M, Cui Y, Liao Y, Zhou Y, Gao Y, Tang F, Wang J, Wang S, Wen L, Song L (2024) High-resolution single-cell transcriptomic survey of cardiomyocytes from patients with hypertrophic cardiomyopathy. Cell Prolif 57(3):*e13557*. https://doi.org/10.1111/cpr.13557

Maatz H, Lindberg EL, Adami E et al (2025) The cellular and molecular cardiac tissue responses in human inflammatory cardiomyopathies after SARS-CoV-2 infection and COVID-19 vaccination. Nat Cardiovasc Res 4:330–345. https://doi.org/10.1038/s44161-025-00612-6

Martin SS, Aday AW, Allen NB, Almarzooq ZI, Anderson CAM, Arora P, Avery CL, Baker-Smith CM, Bansal N, Beaton AZ, Commodore-Mensah Y, Currie ME, Elkind MSV, Fan W, Generoso G, Gibbs BB, Heard DG, Hiremath S, Johansen MC, Kazi DS et al (2025) 2025 Heart Disease and Stroke Statistics: a report of US and global data from the American Heart Association. Circulation 151(8):e41–e660. https://doi.org/10.1161/CIR.0000000000001303

Mensah GA, Fuster V, Murray CJ, Roth GA, Global Burden of Cardiovascular Diseases and Risks Collaborators (2023) Global burden of cardiovascular diseases and risks, 1990-2022. J Am Coll Cardiol 82(25):2350–2473

Molla Desta G, Birhanu AG (2025) Advancements in single-cell RNA sequencing and spatial transcriptomics: transforming biomedical research. Acta Biochim Pol 72:13922. https://doi.org/10.3389/abp.2025.13922

Nguyen T, Wei Y, Nakada Y et al (2023) Analysis of cardiac single-cell RNA-sequencing data can be improved by the use of artificial-intelligence-based tools. Sci Rep 13:6821. https://doi.org/10.1038/s41598-023-32293-1

Nguyen Q, Tung LW, Lin B, Sivakumar R, Sar F, Singhera G, Wang Y, Parker J, Le Bihan S, Singh A, MV Rossi F, Collins C, Bashir J, Laksman Z (2025) Spatial transcriptomics in human cardiac tissue. Int J Mol Sci 26(3):995. https://doi.org/10.3390/ijms26030995

Paik DT, Cho S, Tian L, Chang HY, Wu JC (2020) Single-cell RNA sequencing in cardiovascular development, disease and medicine. Nat Rev Cardiol 17(8):457–473

Palmer JA, Rosenthal N, Teichmann SA, Litvinukova M (2024) Revisiting cardiac biology in the era of single cell and spatial omics. Circ Res 134(12):1681–1702

Pan T, Lu L, Youker K, Shiau CK, Wang M, Lin HY, Nguyen A, He Y, Tong E, Zhu P, Ranka R, Yan Y, Sinha A, Bharat A, Eagar T, Wilcox J, Bhimaraj A, Gao R (2025) Single-cell splicing isoform atlas of the adult human heart and heart failure. Circulation. https://doi.org/10.1161/CIRCULATIONAHA.125.074959. Advance online publication

Patrick R, Janbandhu V, Tallapragada V, Tan SSM, McKinna EE, Contreras O, Ghazanfar S, Humphreys DT, Murray NJ, Tran YTH, Hume RD, Chong JJH, Harvey RP (2024) Integration mapping of cardiac fibroblast single-cell transcriptomes elucidates cellular principles of fibrosis in diverse pathologies. Sci Adv 10(25):eadk8501. https://doi.org/10.1126/sciadv.adk8501

Pereira E, Pereira H (2020) Socioeconomic impact of cardiovascular disease. Revista Portuguesa De Cardiologia (English Edition) 39(5):253–254

Ralph Patrick et al (2024) Integration mapping of cardiac fibroblast single-cell transcriptomes elucidates cellular principles of fibrosis in diverse pathologies. Sci Adv 10:eadk8501. https://doi.org/10.1126/sciadv.adk8501

Schorr EN, Gepner AD, Dolansky MA, Forman DE, Park LG, Petersen KS et al (2021) Harnessing mobile health technology for secondary cardiovascular disease prevention in older adults: a scientific statement from the American Heart Association. Circ Cardiovasc Qual Outcomes 14(5):e000103

Schwalm JD, McKee M, Huffman MD, Yusuf S (2016) Resource effective strategies to prevent and treat cardiovascular disease. Circulation 133(8):742–755

Shi H, Xia Y, Cheng Y, Liang P, Cheng M, Zhang B et al (2025a) Global burden of ischaemic heart disease from 2022 to 2050: projections of incidence, prevalence, deaths, and disability-adjusted life years. Eur Heart J 11:355

Shi Y, Zhang L, Chen M, Calle EA, Langer R (2025b) Regenerative engineering: evolution and its modern significance. In: Regenerative engineering and translational medicine, pp 1–13

Tan WLW, Seow WQ, Zhang A, Rhee S, Wong WH, Greenleaf WJ, Wu JC (2023) Current and future perspectives of single-cell multi-omics technologies in cardiovascular research. Nat Cardiovasc Res 2(1):20–34. https://doi.org/10.1038/s44161-022-00205-7

Tang Z, Alrumaihi F, Alwanian WM, Alharbi HO, Allemailem KS, Alissa M et al (2025) The future of cardiology: integrating single-cell transcriptomics with multi-omics for enhanced cardiac disease insights. Curr Probl Cardiol 50:103005

Tzec-Interián JA, González-Padilla D, Góngora-Castillo EB (2025) Bioinformatics perspectives on transcriptomics: a comprehensive review of bulk and single-cell RNA sequencing analyses. Quant Biol 13(2):e78

Wang X, Chen L, Wei J, Zheng H, Zhou N, Xu X et al (2025) The immune system in cardiovascular diseases: from basic mechanisms to therapeutic implications. Signal Transduct Target Ther 10(1):166

Wünnemann F, Sicklinger F, Bestak K et al (2025) Spatial multiomics of acute myocardial infarction reveals immune cell infiltration through the endocardium. Nat Cardiovasc Res 4:1345–1362. https://doi.org/10.1038/s44161-025-00717-y

Xiao X, Wu W, Mao Q et al (2025) Single-cell transcriptomic profiling reveals cell type heterogeneity between HFpEF and HFrEF. Commun Biol 8:1436. https://doi.org/10.1038/s42003-025-08827-w

Xie Z, Xie H, Xie C, Yang S, Feng Y, Su Z et al (2024) A combined analysis of bulk RNA-seq and scRNA-seq was performed to investigate the molecular mechanisms associated with the occurrence of myocardial infarction. BMC Genomics 25(1):921

Yang L, Zheng B, Gong Y (2025) Global, regional and national burden of ischemic heart disease and its attributable risk factors from 1990 to 2021: a systematic analysis of the Global Burden of Disease study 2021. BMC Cardiovasc Disord 25:625. https://doi.org/10.1186/s12872-025-05022-x

Zeng W, Li J, Liu W, Shi B, Zhao R, Wang Y (2025a) Immune cell dynamics in cardiac diseases: insights from single-cell sequencing. J Cell Mol Med 29(20):e70846

Zeng Z, Wang X, Pei J, Wang P, Li J, Li F, Wang J, Zhao Y, Yu C, Liu H, Zheng Z (2025b) Endothelial plasticity in atrial fibrosis by integrating single-cell sequencing and genetic lineage tracing. Eur Heart J 46(35):3470–3482. https://doi.org/10.1093/eurheartj/ehaf372

Zhang M, Zhang X, Niu J et al (2024) Integrated analysis of single-cell RNA sequencing and bulk RNA data reveals gene regulatory networks and targets in dilated cardiomyopathy. Sci Rep 14:13942. https://doi.org/10.1038/s41598-024-64693-2

Single-Cell Analysis in Precision Medicine

13

Hafiza Areeba Khan, Syed Damin Abbas Hamdani,
Mustafeez Mujtaba Babar, and Jayakumar Rajadas

Abstract

Precision medicine envisions a future where health care is tailored to the unique molecular identity of each individual. However, for decades, a fundamental biological reality has obscured this approach: the profound cellular heterogeneity within every patient. Traditional bulk analytical methods, which average molecular signals across millions of cells, paint a blurred, low-resolution picture of disease, masking the rare but critical cell populations that drive pathogenesis, therapy resistance, and relapse. This chapter charts the revolution that is enabled by single-cell analysis (SCA), a suite of revolutionary technologies that finally brings this cellular mosaic into sharp focus. The core principles of SCA, includ-

H. A. Khan
Department of Biosciences, COMSATS Institute of Information Technology, Islamabad, Pakistan

Shifa College of Pharmaceutical Sciences, Shifa Tameer-e-Millat University, Islamabad, Pakistan

S. D. A. Hamdani
Shifa College of Pharmaceutical Sciences, Shifa Tameer-e-Millat University, Islamabad, Pakistan

M. M. Babar (✉)
Shifa College of Pharmaceutical Sciences, Shifa Tameer-e-Millat University, Islamabad, Pakistan

Advanced Drug Delivery and Regenerative Biomaterials Laboratory of Cardiovascular Institute, Stanford University School of Medicine, Stanford University, Palo Alto, CA, USA
e-mail: mustafeez.babar@fulbrightmail.org

J. Rajadas (✉)
Advanced Drug Delivery and Regenerative Biomaterials Laboratory of Cardiovascular Institute, Stanford University School of Medicine, Stanford University, Palo Alto, CA, USA
e-mail: jayraja@stanford.edu

V. Singh (ed.), *Advances in Single Cell Analysis*,
https://doi.org/10.1007/978-981-95-9789-5_13

ing the detection and barcoding of molecular contents from individual cells, have evolved into high-throughput platforms for genomics, transcriptomics, epigenomics and proteomics. The chapter details how this immense data is tamed by sophisticated computational biology and artificial intelligence approaches, which translates complex datasets into discoverable patterns of cell types, states, and trajectories. Significantly, the important clinical impacts of this technology, from predicting cancer relapse by tracking clonal evolution to identifying the specific neuronal subtypes vulnerable in Alzheimer's disease, have been reported. Still, this approach is not without its challenges. Moreover, the technical, economical, and ethical barriers that must be resolved for rightful integration into clinical practice have been discussed. In conclusion, single-cell analysis goes beyond simply providing new tools. It fundamentally reframes the understanding of biology, affirming that cellular heterogeneity is the central idea that must be decoded. By continuing to unravel this intricate topic, the idea of providing the right treatment to a specific patient at the right time is steadily transitioning from an ideal approach into an achievable reality.

Keywords

Precision medicine · Targeted therapy · Cellular analysis · PCA · Targeted therapies · Pharmacotherapy

13.1 Introduction

Precision medicine has emerged not as a buzzword, but as a transformative framework reshaping modern health care. At its core is the profound idea that each individual is molecularly unique, a complex interplay of genetic blueprint, environmental exposures, and life history. The idea is to move beyond the "one-size-fits-all" model by simplifying and tailoring diagnostic and therapeutic treatments to an individual's specific biological identity. The main objective is to deliver patient-specific treatments at the right time (Wang and Wang 2023). This approach extends far beyond genetic testing alone; it represents an integrative approach that synthesizes data from multi-omics to inform clinical decision-making. The recent developments in technology and bioinformatics have transformed this vision into an increasingly practical component of routine clinical care (Li et al. 2021; Molla and Bitew 2024).

The phrase "precision medicine" may be modern, but its conceptual roots are ancient. For modern precision medicine to realize its full potential, it must achieve two interdependent objectives: the accurate molecular classification of diseases into distinct subtypes and the concurrent development of targeted therapies for each identified subgroup (Weldy and Ashley 2021). Traditional diagnostic techniques often obscure critical variability, grouping biologically diverse patients under a single label. The advent of advanced genomic tools has revealed that many diseases, particularly cancers and autoimmune disorders, are not independent entities but collections of molecularly distinct subtypes. This refined classification enables clinicians to stratify patients and select interventions based on the underlying drivers of their disease. However, accurate subtyping is only valuable if paired with effective, targeted therapeutics, highlighting the translational challenge of bridging deep biological discovery with clinical application (Lähnemann et al. 2020).

The success of this classification hinges on understanding a fundamental biological reality: cellular heterogeneity. Even within a single individual, no two cells are genetically or functionally identical. This diversity exists at multiple levels: between different organs, within a specific tissue or tumour, and even among cells of the same type. This heterogeneity is a primary architect of treatment response variability and a significant obstacle to traditional diagnostic methods. Historically, biomedical research relied heavily on bulk analysis, which is the technique that averages the molecular output of thousands or millions of cells. While invaluable for identifying common signals, this approach operates under the assumption of cellular uniformity. It effectively obscures rare but critical cell subpopulations, such as cancer stem cells responsible for relapse or rare immune cell clones driving autoimmune pathology. It provides a blurred, low-resolution image where intricate cellular details are lost. Cellular heterogeneity is not merely noise; it is the central idea of disease pathogenesis and progression. In oncology, a tumour is not a homogeneous mass but a complex network comprising malignant cells, immune cells, fibroblasts, and vasculature. Critically, the genetic landscape of the cancer cells themselves can be wildly diverse, a phenomenon known as clonal heterogeneity. This diversity creates a formidable challenge: a therapy that effectively kills the dominant cancer cell clone may accidentally select for and allow the expansion of a rare, resistant subclone, leading to therapeutic failure and relapse (Lähnemann et al. 2020).

Similarly, in neurobiology, neurological diseases, for example, Parkinson's or Alzheimer's, likely initiate in specific, vulnerable neuronal subtypes, which are impossible to identify using bulk approaches as they homogenize entire brain regions. The same principle applies to immunology, where the coordinated response to a pathogen is orchestrated by highly specialized and rare immune cell states that are invisible in bulk assays. Therefore, focusing on this cellular heterogeneity is not an optional luxury but an absolute necessity for the advancement of precision medicine. It is the key to unlocking truly personalized therapeutic strategies. By moving from a bulk-level understanding to a high-resolution, single-cell view, we can identify the specific cellular drivers of an individual's disease, predict mechanisms of resistance, and develop interventions that target the entire dysfunctional physiology. This transition from population level averages to the intricate details of a patient's cellular mosaic and represents the next frontier in personalized health care. Figure 13.1 illustrates how single-cell analysis addresses cellular heterogeneity to enable personalized therapeutic strategies, creating a continuous learning system that refines clinical decisions based on real-world outcomes (Inayatullah et al. 2025).

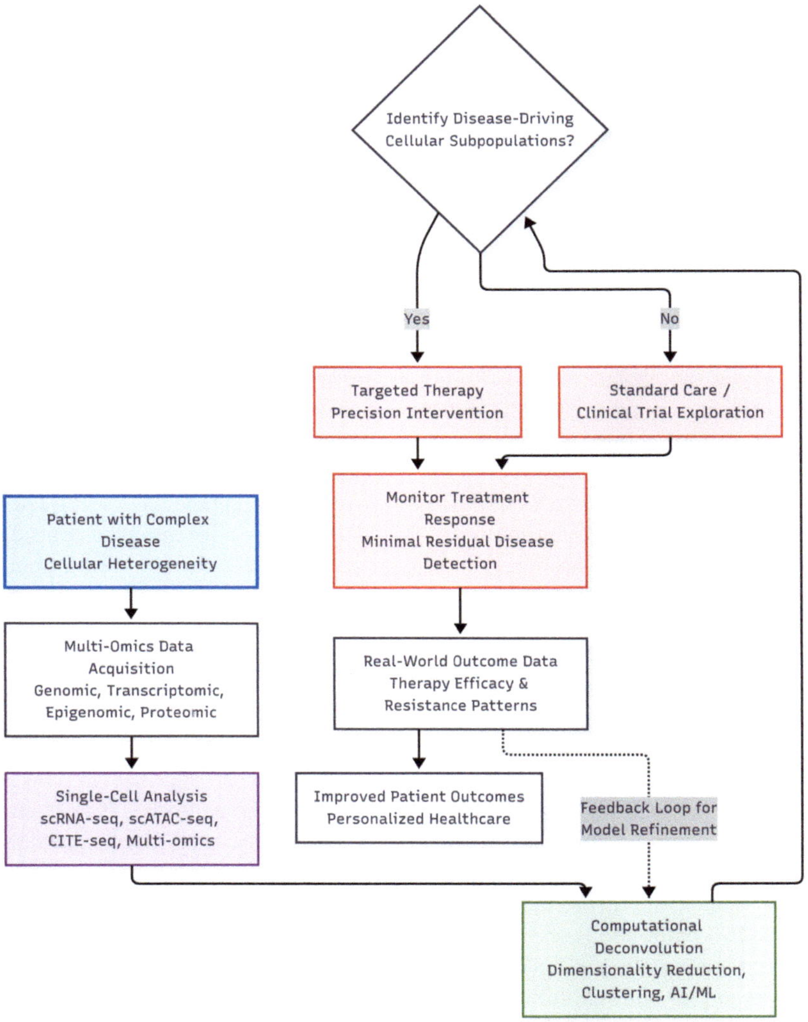

Fig. 13.1 The single-cell powered precision medicine cycle

13.2 Bulk Population Analysis

The promise of precision medicine, as outlined in the introduction, is predicated on recognizing and deciphering cellular heterogeneity to guide clinical decisions. For decades, however, the foundational tools of molecular biology have operated in direct opposition to this principle. Bulk population analysis has been the backbone of biomedical research, yielding monumental discoveries like the assembly of DNA, the human genome sequence, and the identification of major driver mutations in cancer (Hulsen et al. 2019; Li et al. 2021). Yet, for the goals of precision medicine, this approach is fundamentally limited because it deliberately averages signals across thousands to millions of cells, effectively painting a picture of a "typical" or "average" cell that, in reality, may not exist (Lähnemann et al. 2020). While providing a crucial understanding of biology, bulk methods act as a barrier, obscuring the very cellular diversity that precision medicine seeks to target.

The principle of bulk analysis is direct: a tissue sample, whether a tumour biopsy, a blood draw, or a section of an organ, is homogenized into slurry. The genetic material (DNA or RNA) or proteins are then extracted in bulk from this mixture and analysed using techniques like microarray hybridization, bulk RNA sequencing (RNA-seq), or Western blotting (He et al. 2022). The output is a single, aggregate measurement for each gene or protein, representing the average expression level across the entire population of cells in the sample. This averaging is fundamentally incompatible with the principles of precision medicine. It operates on the assumption of relative homogeneity, which, as established, is biologically invalid. The measured value is a composite signal, a weighted average dominated by the most abundant cell types, while signals from rare but functionally critical subpopulations are diluted beyond detection. It is analogous to trying to understand the intricate workings of a complex machine by listening to its overall hum from a distance, rather than examining the function of each individual gear and spring (Lähnemann et al. 2020).

This reliance on bulk analysis presents several intractable problems for implementing true precision medicine. The most significant limitation is the masking of rare cell populations. Many disease mechanisms are driven by small, specialized subsets of cells. In oncology, cancer stem cells (CSCs) or therapy-resistant clones often constitute a tiny fraction (<1–5%) of the total tumour mass but are solely responsible for tumour initiation, metastasis, and relapse due to their innate self-renewal capabilities and enhanced DNA repair mechanisms (Kang et al. 2022; Wegmann et al. 2024). In a bulk RNA-seq experiment, the unique transcriptional profile of these rare cells such as high expression of stemness markers like *ALDH1A1* or *CD44* is statistically masked out by the signal from the bulk, differentiated tumour cells. Consequently, a therapeutic strategy designed from bulk data might effectively target the dominant, proliferating population, for example with a chemotherapeutic agent but would fail to eradicate the quiescent, drug-resistant CSC pool, inevitably leading to disease recurrence. This principle extends beyond cancer; in autoimmune disorders, rare auto reactive T-cell or B-cell clones can drive pathology, and in neurological conditions, specific vulnerable neuronal subtypes initiate degeneration, all of which remain invisible to bulk assays (Ren et al. 2024). Furthermore, bulk methods lead to the intermingling of expression signals, a problem formally known as the "feature attribution problem." A bulk transcriptome profile does not distinguish between which cells express which genes. For example, a high expression level of an immune checkpoint gene like *PD-L1* in a tumour sample is extremely ambiguous. It could mean that every cancer cell expresses a moderate amount, making them universally susceptible to anti-PD-1/PD-L1 therapy or a small, aggressive subset of cancer cells expresses an extremely high amount, suggesting a focused resistance mechanism. The third option is that the signal comes entirely from tumour-infiltrating macrophages or other immune cells, not from the cancer cells at all, indicating a different immune context (Quek et al. 2024). This ambiguity has direct and costly therapeutic consequences. A patient might be prescribed an expensive immune checkpoint inhibitor based on a high bulk *PD-L1* signal that originates solely from tumour-associated macrophages, a scenario known to be associated with poor response, thereby rendering the therapy ineffective and delaying appropriate care (Gao et al. 2022).

Bulk analysis also demonstrates a profound inability to resolve cellular ecosystems and cellular states. Cancers are not static entities but complex, dynamic ecosystems. Within these systems, malignant cells co-evolve and continuously communicate with the dedicated immune, stromal, and vascular cells through direct contact (juxtacrine) and secreted signals (paracrine). While flow cytometry can identify broad cell lineages, bulk omics cannot reconstruct the intricate communication networks that dictate disease progression and therapy response (Dezem et al. 2024; Quek et al. 2024). For instance, it cannot determine if cytotoxic T-cells are actively engaging with cancer cells (via *MHC-I* and *TCR* interactions) or are functionally exhausted and suppressed by regulatory T-cells (Tregs) expressing *CTLA-4* or secreting *TGF-β*. Understanding these specific interactions is critical for developing rational combination therapies, such as pairing a targeted drug with a specific immune checkpoint blocker. Moreover, bulk sequencing fails to capture continuous transitions, such as the epithelial-to-mesenchymal transition (EMT) in cancer metastasis, where cells exist on a spectrum of states rather than in discrete categories. The average signal from a population undergoing EMT is a nonsensical "middle" state that does not accurately represent any individual cell, misleading our understanding of the underlying biology. Additionally, there is a loss of information on allelic expression and splice variants. Bulk RNA-seq measures the total expression level of a gene, but it cannot determine if the two alleles of a gene are expressed equally (biallelic) or if one is silenced (monoallelic), a phenomenon important in imprinting and X-chromosome inactivation. Furthermore, it provides an average of splice variants (isoforms) across the cell population. If two different cell types express different isoforms of the same gene, the bulk data will show a messy, combined profile that obscures the cell-type-specific splicing patterns crucial for understanding functional protein diversity (Ge et al. 2025).

Finally, bulk genomics leads to an oversimplification of tumour evolution and clonal architecture as represented in Fig. 13.2. Bulk whole-exome or whole-genome sequencing can only infer the presence of different subclones through complex computational deconvolution of variant allele frequencies (VAFs), which is an indirect and often imprecise inference. This method struggles to accurately reconstruct the phylogeny of a tumour when subclones are rare or have similar VAFs, and it is completely unable to resolve polygenomic populations where multiple subclones are present in the same sample (Ren et al. 2024). Crucially, without physically isolating single cells, it is impossible to definitively know which mutations co-occur on the same DNA molecule (in *cis*) a fundamental requirement for understanding true clonal relationships and evolutionary trajectories (Naldini et al. 2023; Wegmann et al. 2024). This limitation directly impedes the precision medicine goal of

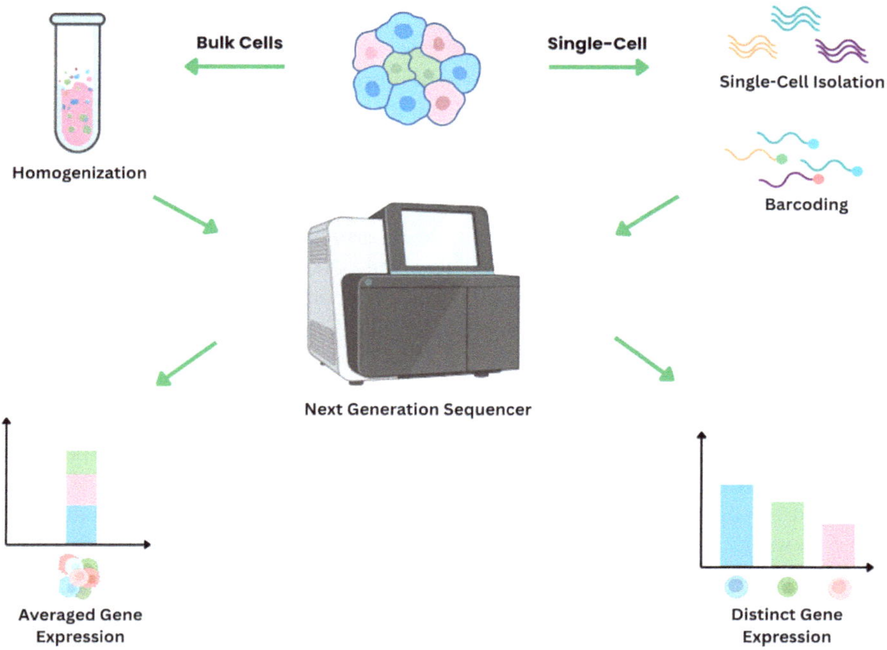

Fig. 13.2 Fundamental differences in bulk and single-cell RNA sequencing approaches

predicting disease course and designing therapies that target a founder clone or proactively block common evolutionary pathways to resistance. For example, the order of acquisition of mutations like *TP53* and *PIK3CA* can influence therapeutic vulnerability, information that is lost in a bulk sequencing assay.

13.3 Principles of Single-Cell Analysis

The limitations of bulk analysis created an imperative for technologies capable of dissecting cellular heterogeneity. This need is now met by the rapid advancement in single-cell analysis (SCA), a suite of revolutionary techniques found on a core principle: to capture and interrogate the molecular makeup of individual cells. This shift from a population-average view to a high-resolution, cell-by-cell perspective is fundamentally transforming our comprehension of biology and disease, providing the specific tools required to comprehend the goals of precision medicine (Inayatullah et al. 2025; Lähnemann et al. 2020; Lim et al. 2024). The underlying principle of SCA is that meaningful biological information, from rare disease-driving cells to subtle transitional states, is lost in averaging and can only be recovered by preserving the identity of each cell throughout the analytical process (Flynn et al. 2023; Jovic et al. 2022).

The journey to modern SCA began with critical historical milestones. The first proof-of-concept studies emerged in the late 2000s with the adaptation of next-generation sequencing to single cells, a feat requiring monumental efforts to amplify the minute quantities of DNA or RNA from a single cell. However, the field truly entered its modern era around 2015–2016 with the commercialization of high-throughput microfluidic platforms, such as droplet-based systems (e.g., 10x Genomics) (He et al. 2022; Jovic et al. 2022). These technologies enabled the parallel profiling of thousands of cells in a single experimental run, moving SCA from a specialized, low-throughput art to a scalable, reproducible science. This scalability was the key that unlocked the potential to move beyond rare discoveries and systematically catalogue cellular diversity in health and disease (Inayatullah et al. 2025; Lim et al. 2024).

The technical workflow of single-cell analysis, particularly for single-cell RNA sequencing (scRNA-seq), follows a generalized pipeline that reflects its core principles. As shown in Fig. 13.1, it begins with single-cell isolation, where live cells are dissociated from a tissue sample and individually captured. This is most commonly achieved through microfluidic devices that encapsulate single cells into nanolitre-sized droplets alongside barcoded beads, or using microwell plates (Jovic et al. 2022). The critical next step is cell-specific barcoding, where every molecule (e.g., mRNA transcript) from a single cell is tagged with a unique cellular barcode. This allows all subsequent sequencing data to be computationally demultiplexed, assigning millions of reads back to the individual cell of origin (Ge et al. 2025). Finally, the genetic material undergoes cDNA synthesis and amplification via in vitro transcription (IVT) or polymerase chain reaction (PCR) to generate sufficient material for sequencing. This entire process is designed to minimize technical noise and batch effects while maximizing the capture of biological signal from each cell (Lotfollahi et al. 2024). The power of SCA is embodied in the diverse omics approaches that can now be applied at single-cell resolution, each shedding light on a different layer of cellular regulation. Among these, the Single-Cell Genomics is focused on the DNA sequence within specific cells. It is indispensable for mapping clonal heterogeneity in cancers, enabling the direct reconstruction of tumour phylogenies by identifying which mutations co-occur in the same cell (Naldini et al. 2023; Wegmann et al. 2024). It can reveal mosaicism in developmental disorders and precisely chart the evolutionary trajectories of pathogens. Techniques like scDNA-seq are crucial for understanding the genetic diversity that bulk sequencing could only infer indirectly (Gao et al. 2022; Kang et al. 2022). Alternatively, Single-Cell Transcriptomics (scRNA-seq) is the most broadly implemented single-cell omics method. It measures the entire set of RNA transcripts (the transcriptome) in each cell, defining its functional state and identity at a specific moment (Ren et al. 2024; Woo and Eyun 2025). ScRNA-seq is the primary tool for discovering novel cell types and states, mapping differentiation trajectories, and understanding how cells respond to stimuli in a heterogeneous population. It can identify the rare immune cell clones activated by a vaccine or the specific neuronal subtypes vulnerable in neurodegeneration (Lim et al. 2024; Quek et al. 2024).

Single-Cell Epigenomics is a combination of techniques that probe the regulatory landscape that controls the expression of genes without changing the sequence of DNA itself. Methods like scATAC-seq (Assay for Transposase-Accessible Chromatin with sequencing) map open chromatin regions, revealing the active regulatory elements in each cell. This allows researchers to understand the regulatory logic behind cellular heterogeneity as to why two genetically identical cells can have vastly different phenotypes by inferring transcription factor activity and charting gene regulatory networks (Dezem et al. 2024; Flynn et al. 2023; Inayatullah et al. 2025). In Single-Cell Proteomics, the sequencing-based methods infer protein abundance from mRNA levels; single-cell proteomics measures proteins directly. Technologies like mass cytometry (CyTOF) and high-dimensional flow cytometry use antibody-based tagging to quantify dozens of proteins simultaneously at single-cell resolution. This is vital for immunology and cancer research, providing functional readouts on cell signalling, phosphorylation status, and surface marker expression that are often not perfectly correlated with transcript levels (Lim et al. 2024; Naldini et al. 2023; Wegmann et al. 2024).

The most recent frontier is multimodal omics, which represents the next evolutionary step in SCA. These technologies capture multiple layers of molecular information from the same cell. For instance, CITE-seq and REAP-seq simultaneously measure transcriptome and surface protein expression, while scNMT-seq can profile the genome, transcriptome, and methylome from one cell. This integrated approach moves beyond correlation to establish direct causal links between different molecular layers, offering an unprecedentedly comprehensive view of cellular identity and function. By moving from a blurred, population-level average to a sharp, multi-dimensional view of each individual cell, these principles of single-cell analysis provide the necessary resolution to finally decode the cellular mosaics of human biology and disease, directly addressing the heterogeneity that is central to precision medicine (Molla and Bitew 2024; Wang and Wang 2023).

13.4 Technological Platforms and Methodologies

The principles of single-cell analysis are dependent on the technological platforms that transform the theoretical ability to study individual cells into a practical reality. The evolution of these methodologies from painstaking, manual protocols to automated, high-throughput workflows has been the critical driver propelling the field from niche applications to mainstream biomedical research. This technological revolution has provided the tools necessary to directly address the limitations of bulk analysis, enabling the precise deconvolution of cellular heterogeneity that is central to precision medicine (Li et al. 2021; Lim et al. 2024). The landscape of single-cell technologies is vast and rapidly evolving, but can be broadly categorized into platforms designed to interrogate specific molecular layers: the genome, transcriptome, epigenome, proteome, and, most recently, multiple modalities simultaneously from the same cell (Flynn et al. 2023; Ge et al. 2025).

13.4.1 Single-Cell RNA Sequencing Platforms

ScRNA-seq is the most commonly adopted single-cell technology, and its development has been driven by microfluidic innovations. Droplet-based methods (e.g., 10x Genomics Chromium platform) dominate large-scale atlas projects. These systems isolate individual cells in nanolitre droplets with unique barcoded beads attached to them, enabling the parallel processing of thousands to tens of thousands of cells per run in a highly cost-effective manner. This high throughput is ideal for comprehensively profiling complex tissues and discovering rare cell populations (Jovic et al. 2022). Alternatively, microwell-based platforms (e.g., BD Rhapsody™) use arrays of nano well chips to capture cells and barcode beads, offering similar scalability with flexibility in sample input and recovery. For the highest sensitivity, particularly for detecting lowly expressed genes or analysing cells with minimal RNA content, plate-based methods (e.g., SMART-Seq2) remain valuable (He et al. 2022). A few of the common sc-RNA sequencing platforms have been summarized in Table 13.1. These methods isolate individual cells into 96- or 384-well plates, providing full-length transcript coverage and superior sensitivity, although at a lower throughput and higher cost per cell. Each platform makes a trade-off between the number of cells profiled and the complexity of information obtained per cell, a key

Table 13.1 Key characteristics of common single-cell RNA sequencing platforms

Technology platform	Examples	Throughput (Cells per Run)	Key strength	Key limitation	Ideal use case
Droplet-Based	10x Genomics Chromium	10,000	High-throughput, cost-effective per cell	Lower sequencing depth per cell	Large-scale cell atlas generation, discovering rare populations
Microwell-Based	BD Rhapsody, ICELL8	10,000	Flexible cell loading, sample multiplexing	Platform-specific chemistry	Profiling precious samples, mixed species experiments
Plate-Based	SMART-Seq2, SMART-Seq v4	384	Full-length transcript coverage, high sensitivity	Low throughput, high cost per cell	Detailed analysis of few cells, alternative splicing studies
Spatial Transcriptomics	10x Visium, NanoString GeoMx	1–5000 spots	Preserves spatial location information	Lower single-cell resolution (may capture multiple cells per spot)	Understanding tissue architecture and cell-cell interactions

consideration in experimental design (Lim et al. 2024). Recent innovations include single-nucleus RNA sequencing (snRNA-seq) for challenging samples like frozen tissues and spatially resolved transcriptomics (e.g., 10x Visium) that preserve crucial architectural context within tissues (Dezem et al. 2024). Platform selection involves balancing throughput, sensitivity, and biological context.

13.4.2 Single-Cell Genomic Methodologies

Interrogating the genome at single-cell resolution is technically challenging because of the difficulty of amplifying a single diploid genome without introducing biases and artifacts. Whole-genome amplification (WGA) methods, such as Multiple Annealing and Looping-Based Amplification Cycles (MALBAC) and Multiple Displacement Amplification (MDA), are used prior to sequencing to generate sufficient material. scDNA-seq is primarily used to resolve clonal architecture in cancers, identify subclonal mutations, and trace lineage relationships in development (Kang et al. 2022; Ren et al. 2024). A powerful extension is Single-Cell Multi-Omics, which allows for the simultaneous sequencing of DNA and RNA from the same cell. Technologies like DR-Seq and G&T-Seq can reveal how genomic alterations (e.g., mutations, copy number variations) directly impact the transcriptional state within that very same cell, providing a direct link between genotype and phenotype (Inayatullah et al. 2025; Li et al. 2021). This is invaluable for understanding the functional consequences of genetic heterogeneity within a tumour, moving beyond mere cataloguing of mutations to seeing their immediate cellular effects.

13.4.3 Single-Cell Epigenomic Methodologies

Understanding gene regulation requires mapping the epigenetic landscape that controls chromatin accessibility and packaging. Single-cell Assay for Transposase-Accessible Chromatin with sequencing (scATAC-seq) is the most common technique to profile the epigenome. It uses a hyperactive Tn5 transposase enzyme to add sequencing adapters into open, nucleosome-free regions of the genome, which are hallmarks of active regulatory elements (enhancers, promoters). By mapping these regions in thousands of single cells, researchers can infer cell-type-specific gene regulatory programs and transcription factor binding motifs, revealing the regulatory logic behind cellular identity and fate decisions (Lähnemann et al. 2020). Other methods include scDNA methylation sequencing (e.g., scBS-seq, scRRBS), which directly maps methylated cytosine bases, providing another critical layer of epigenetic information that is often dysregulated in cancer and developmental disorders (Flynn et al. 2023; Ge et al. 2025). The incorporation of epigenomic data with transcriptomics helps to explain why genes are expressed differentially across a heterogeneous cell population.

13.4.4 Single-Cell Proteomic Methodologies

While transcriptomics infers protein abundance, direct measurement of proteins provides crucial information on post-translational modifications, signalling activity, and surface marker expression that often correlate poorly with mRNA levels. Mass Cytometry (CyTOF—Cytometry by Time-Of-Flight) uses heavy metal-isotope-tagged antibodies instead of fluorophores to measure over 40 proteins simultaneously by time-of-flight mass spectrometry, overcoming the spectral overlap limitations of traditional flow cytometry (Lim et al. 2024). It is exceptionally powerful for deep immunophenotyping of complex tissues like blood and tumour microenvironments. A more recent innovation is single-cell multiplexed proteomics via sequencing, including CITE-seq (Cellular Indexing of Transcriptomes and Epitopes by Sequencing) and REAP-seq. These methods use oligonucleotide-tagged antibodies that allow for the concurrent measurement of dozens of surface proteins alongside the whole transcriptome from the exact same single cells, seamlessly integrating protein and RNA data (Flynn et al. 2023; Molla and Bitew 2024).

13.4.5 Integrated Multi-omics Methodologies

One of the prime achievements of single-cell technology is the ability to capture several molecular modalities from the same individual cell. This moves beyond correlation to establish direct causal links between different layers of regulation. CITE-seq and REAP-seq are prime examples of integrated transcriptome-proteome analysis (Flynn et al. 2023). For linking genotype to phenotype, DNA-genotype and RNA-phenotype co-assay methods like G&T-seq (Genome and Transcriptome sequencing) enable the parallel sequencing of the genome and transcriptome from the same cell, crucial for connecting specific mutations to their functional transcriptional consequences in cancer (Inayatullah et al. 2025; Li et al. 2021). The most complex integrations aim to pair the epigenome with the transcriptome. Techniques like scNMT-seq (single-cell Nucleosome, Methylation, and Transcription sequencing) provide a trifecta of data on chromatin accessibility, DNA methylation, and transcriptome from a single cell, offering a detailed view of the gene regulatory logic (Ge et al. 2025). These multi-omics technologies are rapidly maturing and represent the future of single-cell analysis, promising a truly holistic understanding of cellular function and dysfunction in human disease (Molla and Bitew 2024). The choice of platform is therefore dictated by the specific biological question, balancing the need for depth of information, scale of cell numbers, and integration across molecular layers to build a complete picture of the cellular state.

13.5 Computational Support for Single-Cell Analysis

The powerful technological platforms detailed in the previous sections generate an immense and complex dataset. A single scRNA-seq experiment can easily produce a matrix of millions of data points representing gene expression counting across tens of thousands of cells. This scale and complexity render manual analysis impossible, making sophisticated computational biology not merely an auxiliary tool but the very engine that translates raw sequencing data into biological insight. The role of computation is to manage quality-control, normalize, interpret, and visualize these high-dimensional datasets, transforming them into a comprehensible map of cellular states and relationships. Without these computational frameworks, the promise of single-cell technologies would remain entirely unrealized (Ge et al. 2025; Lähnemann et al. 2020).

In today's research and development landscape, computational analysis is seamlessly integrated into the single-cell workflow. The process begins with primary data analysis in which raw sequencing reads (FASTQ files) are processed through pipelines like Cell Ranger (10x Genomics) or STARsolo to perform sample demultiplexing, alignment to a reference genome, and quantification of gene expression counts per cell, resulting in a count matrix (Flynn et al. 2023; He et al. 2022). This is followed by critical quality control (QC) and filtering, where computational algorithms identify and remove low-quality cells (e.g., those with high mitochondrial gene content indicating apoptosis, or too few detected genes) and doublets (two cells mistakenly captured as one), which can severely distort downstream analysis (Lim et al. 2024). Subsequent steps include normalization to remove technical variations (e.g., sequencing depth) and feature selection to identify the most biologically variable genes for further analysis. This rigorous computational preprocessing ensures that the data reflects true biological heterogeneity rather than technical artifacts, forming an essential foundation for all subsequent discovery (Ge et al. 2025).

The heart of single-cell analysis lies in applying core computational principles to identify patterns within the high-dimensional data. Dimensionality reduction is perhaps the most crucial first step for visualization and exploration. Techniques like Principal Component Analysis (PCA) linearly transform the data to create new axes (principal components) that capture the greatest amount of variance. However, nonlinear methods are often more powerful for capturing complex cellular manifolds. Uniform Manifold Approximation and Projection (UMAP) and t-Distributed Stochastic Neighbour Embedding (t-SNE) are widely used to project high-dimensional data into two or three dimensions, creating intuitive maps where similar cells are positioned near each other, allowing researchers to visually identify potential cell clusters and transitions (Flynn et al. 2023; Ge et al. 2025). Clustering algorithms then systematically partition cells into distinct groups based on the similarity of their gene expression profiles. Methods like Louvain and Leiden clustering operate on a graph of cells built from their high-dimensional distances, effectively identifying discrete cell types and states without prior biological knowledge (He et al. 2022). These unsupervised methods are fundamental for novel cell type discovery, such as identifying new immune cell subtypes or rare cancer stem cell populations that were previously obscured in bulk data (Kang et al. 2022).

Beyond simple classification, computational methods enable the inference of biological trajectories. Algorithms like PAGA and Monocle3 can reconstruct the underlying continuum of cellular processes, such as differentiation pathways or the cell cycle, by ordering cells along a pseudo time trajectory based on their transcriptional progression. This allows researchers to move beyond a static snapshot to a dynamic model of how cells transition from one state to another, which is critical for understanding, development, and disease progression (Ge et al. 2025; Inayatullah et al. 2025). While traditional statistical methods form the backbone of analysis, Artificial Intelligence and Machine Learning are rapidly becoming transformative forces, pushing the boundaries of what is possible. Deep learning models are particularly adept at handling the scale and noise inherent in single-cell data. Autoencoders, a type of neural network, are powerful tools for non-linear dimensionality reduction and denoising. They learn a compressed, latent representation of the data that captures the essential biological signal while filtering out technical noise (Ge et al. 2025). Advanced architectures, including variational autoencoders (VAEs) and generative adversarial networks (GANs), are utilized for tasks such as data imputation (resolving missing values), correction of batch effects, and the in silico generation of realistic single-cell data to augment or expand constrained datasets.

Transfer learning, another Machine Learning type, has proved to be highly valuable. Models pre-trained on massive public single-cell atlases (e.g., from the Human Cell Atlas) can be fine-tuned on new, smaller datasets. This approach can dramatically improve analysis efficiency and accuracy, especially for rare disease samples or clinical biopsies with limited cell numbers, by leveraging knowledge learned from millions of other cells (Lotfollahi et al. 2024). Furthermore, ML classifiers are increasingly used for cell type annotation. Instead of manually labelling clusters based on known marker genes, researchers can now train classifiers on well-annotated reference datasets to automatically and consistently assign cell identities in new experiments, greatly enhancing reproducibility and throughput in R&D pipelines (Inayatullah et al. 2025; Lim et al. 2024).

The integration of AI/ML is also paving the way for predictive biology. Models are being developed to predict how cells will respond to perturbations, such as drug treatments or genetic knockouts, based on their initial state. This has profound implications for drug discovery and personalized medicine, as it could allow clinicians to virtually screen therapies on a patient's own cells computationally before ever administering a treatment in the clinic (Molla and Bitew 2024; Wang and Wang 2023). The computational support for single-cell analysis has thus evolved from a supportive role to a central, driving force that unlocks the full potential of high-resolution biology, turning vast datasets into actionable insights for precision medicine (Ge et al. 2025; Inayatullah et al. 2025).

13.6 Clinical Applications in Precision Medicine

The journey from the technical principles of single-cell analysis to tangible clinical impact is now well underway. The high-resolution cellular data provided by these technologies is moving beyond basic research and beginning to directly influence patient diagnosis, prognosis, and therapeutic strategy. By directly addressing the cellular heterogeneity that underpins treatment response variability, single-cell approaches are refining the very definition of disease and enabling a new era of personalized clinical intervention. The following are the real-world examples that illustrate how this powerful technology is being translated into patient care (Molla and Bitew 2024; Wang and Wang 2023).

13.6.1 Mapping Clonal Evolution in Leukaemia for Relapse Prediction

Acute Myeloid Leukaemia (AML) is an example of intra-tumoral heterogeneity, where relapse is often driven by minor subclones resistant to initial therapy. Bulk sequencing often fails to detect these rare populations at diagnosis. A landmark clinical application involves using scDNA-seq to track clonal evolution in AML patients. By profiling thousands of individual cancer cells at diagnosis and remission, clinicians can identify the complex architecture of founding clones and subclones, including those harbouring mutations associated with therapy resistance (e.g., in *FLT3* or *TP53*) (Naldini et al. 2023; Wegmann et al. 2024). This allows for the early detection of minimal residual disease (MRD) at a cellular resolution far beyond traditional flow cytometry. Patients with high-risk clonal compositions can be identified at diagnosis and stratified towards more aggressive or targeted therapies, such as stem cell transplantation or specific inhibitors, in an effort to proactively eradicate resistant clones and prevent relapse (Kang et al. 2022; Ren et al. 2024). A few of the examples of applications of single-cell analysis in oncology have been provided in Table 13.2.

13.6.2 Identifying Rare Pathogenic Clones in Autoimmunity

Autoimmune diseases like lupus and rheumatoid arthritis are driven by rare, aberrant immune cell clones that escape tolerance mechanisms. Bulk profiling of blood or synovial tissue averages out these critical effectors. scRNA-seq joined with T-cell/B-cell receptor sequencing (scTCR-seq/scBCR-seq) has enabled the identification and tracking of these expanded clonotypes directly from patient samples. Researchers can now pinpoint specific T-cell clones with auto reactive phenotypes or B-cell clones producing pathogenic autoantibodies (Inayatullah et al. 2025; Lim et al. 2024). This not only provides a deeper understanding of disease mechanism but also offers a highly specific biomarker for monitoring disease activity and response to therapy. For instance, the depletion of a specific pathogenic B-cell clone

Table 13.2 Clinical applications of single-cell analysis in precision oncology

Clinical challenge	Bulk analysis limitation	Single-cell approach	Potential clinical utility
Therapy resistance and relapse	Fails to detect rare, pre-existing resistant subclones	scDNA-seq to map clonal architecture and evolutionary trajectories	Identify high-risk patients at diagnosis; guide adjuvant therapy to eradicate resistant clones
Immunotherapy patient selection	Bulk PD-L1 expression is an imperfect biomarker; loses cellular context	scRNA-seq + CITE-seq to profile the working state of immune cells and their interactions inside TME	Develop complex predictive signatures based on specific immune cell types and states to better select responders
Tumour heterogeneity and classification	Provides an averaged molecular profile of the entire tumour	scRNA-seq to identify distinct cellular subtypes and their proportions within a tumour	Refine cancer subtyping beyond organ-of-origin; identify dominant, vulnerable cell states
Minimal residual disease (MRD)	Limited sensitivity and inability to characterize resistant MRD cells	High-sensitivity scDNA/RNA-seq to detect and molecularly characterize residual cells post therapy	Monitor treatment response with extreme sensitivity; detect relapses earlier than old methods

following treatment with a drug like rituximab can be directly correlated with clinical improvement, moving monitoring from a broad measure of inflammation to a targeted assessment of the disease-driving cellular population (Flynn et al. 2023).

13.6.3 Analysis of the Cellular Specificity of Alzheimer's Disease

The brain is a prime example of the cellular complexity, and neurodegenerative diseases have been particularly resistant to targeted therapies, partly because bulk analyses of post-mortem brain tissue homogenize vulnerable and resistant neurons. The scRNA-seq studies on human brain samples have begun to reveal the selective vulnerability of specific neuronal subtypes in Alzheimer's disease (AD). For example, recent work has identified excitatory neurons in specific cortical layers that are among the first to exhibit pathological gene expression signatures related to protein aggregation and stress response, while inhibitory neurons in the same region remain relatively spared (Inayatullah et al. 2025; Woo and Eyun 2025). This cellular-resolution map of pathology is critical for developing targeted therapies. It shifts the focus from broadly targeting amyloid-beta to protecting or supporting these specific, vulnerable neuronal populations, a strategy that could yield more effective and less toxic interventions.

13.6.4 Predicting Response to Checkpoint Inhibition in Cancer

Immune checkpoint inhibitors (ICIs) have modernized cancer care, but they only work in a subset of patients. Bulk RNA sequencing of tumours to find predictive biomarkers like PD-L1 expression has proven imperfect, as it fails to reveal the spatial context and cellular source of the signal. Multi-omics single-cell technologies, particularly CITE-seq and spatial transcriptomics, are now being applied to patient biopsies to deconstruct the tumour microenvironment (TME) with unprecedented detail (Dezem et al. 2024; Quek et al. 2024). These analyses can identify not just the quantity but the functional state of cytotoxic T-cells (e.g., exhausted vs. activated), the immunosuppressive role of specific macrophage subsets, and the complex cellular crosstalk that defines a "hot" versus "cold" tumour. This allows for the development of composite predictive signatures that are far more accurate than any single bulk measurement. A patient's likelihood of responding to anti-PD-1 therapy can now be assessed based on the precise composition and spatial localization of immune cells within their specific tumour ecosystem (Molla and Bitew 2024; Wegmann et al. 2024).

13.6.5 Developmental Disorders and In Vitro Fertilization

Single-cell analysis is making a profound impact even before birth. Preimplantation genetic testing (PGT) during IVF has traditionally relied on bulk sequencing of a few cells biopsied from an embryo, which can be inaccurate and damaging. scRNA-seq is now being explored as a tool to assess embryonic viability non-invasively by analysing spent embryo culture media for RNA transcripts shed by the embryo (Overbey et al. 2022). Furthermore, for couples carrying serious genetic disorders, applying scDNA-seq to embryos allows for the highly accurate detection of mosaic mutations (where only some cells are affected) and aneuploidies, ensuring that only the healthiest embryos are selected for implantation. This application directly uses cellular resolution to prevent the transmission of devastating genetic diseases, representing a powerful form of preventive precision medicine (Inayatullah et al. 2025; Li et al. 2021).

These examples represent just the forefront of a coming wave of clinical applications. As single-cell technologies become more robust, automated, and cost-effective, their integration into routine clinical workflows will redefine pathological diagnosis from a tissue-level art to a precise science of cellular ecosystems, ultimately fulfilling the promise of delivering the accurate treatment to a specific patient at the correct time.

13.7 Challenges and Limitations

While the potential of single-cell analysis (SCA) to revolutionize precision medicine is immense, as illustrated by the clinical applications in the previous section, its widespread adoption faces significant hurdles. Translating these powerful research tools into routine clinical practice requires overcoming a multifaceted array of scientific, technical, legal, regulatory, and social challenges. Acknowledging and addressing these limitations is crucial for responsibly guiding the field forward and ensuring that the benefits of this technology are realized equitably and effectively.

13.7.1 Scientific and Technical Challenges

The foremost obstacles remain scientific and technical. Tissue acquisition and processing pose an immediate blockage. Clinical biopsies are often small, precious, and fraught with artifacts from collection methods. The required dissociation process to create a single-cell suspension is harsh and can alter the transcriptional state of cells, potentially skewing the data and obscuring the truth in in vivo biology. This is particularly problematic for fragile cell types like neurons or certain immune cells (Overbey et al. 2022).

Spatial context loss is another critical limitation. While technologies like scRNA-seq excel at characterizing cellular diversity, they inherently destroy the native tissue architecture. The spatial relationships between cells such as those in which immune cells are directly contacting tumour cells are often crucial for understanding disease mechanisms and predicting therapy response. Although spatial transcriptomics technologies are emerging to address this, they are not yet as high-throughput or widely accessible as droplet-based scRNA-seq, creating a disconnection between cell identity and location.

Furthermore, the sheer complexity and scale of the data generated pose a monumental analytical challenge. Integrating multi-omics datasets (e.g., RNA, ATAC, protein from the same cell) requires sophisticated computational tools and significant expertise. Batch effects, and technical variations between experiments processed at different times or locations can be severe and difficult to correct, potentially confounding biological findings. There is also a persistent trade-off between the number of cells sequenced and the depth of sequencing per cell, meaning that rare but biologically critical transcripts can still be missed in large-scale atlas projects (Flynn et al. 2023; Lähnemann et al. 2020).

13.7.2 Legal, Ethical, and Regulatory Challenges

The clinical use of SCA introduces novel legal and ethical dilemmas. Informed consent becomes vastly more complex. How can patients truly consent to the myriads of potential future discoveries that might arise from the deep analysis of their cellular data? Traditional consent forms are inadequate for explaining the implications of identifying rare somatic mutations or unexpected disease predispositions that could have implications for relatives.

Data ownership and privacy are major concerns. Single-cell data is the ultimate personal identifier because a person's cellular makeup is unique. The risk of re-identification from publicly shared datasets, even when "de-identified," is a serious threat. Robust governance frameworks are needed to determine who owns this data, the patient, the hospital, or the sequencing company and how it can be used for secondary research while protecting individual privacy (Molla and Bitew 2024).

From a regulatory perspective, agencies like the FDA are navigating uncharted territory. How does one approve a diagnostic or therapeutic based on a dynamic cellular signature that may evolve over time or vary within a tumour? Current regulatory pathways are designed for static biomarkers (e.g., a specific genetic mutation), not for complex, multi-dimensional cellular maps. Defining analytical validation standards for these highly multiplexed assays ensuring they are accurate, reproducible, and clinically meaningful presents a significant hurdle for their use in guiding treatment decisions.

13.7.3 Social and Economic Challenges

Finally, significant social and economic barriers threaten to exacerbate existing health disparities. The high cost of single-cell technologies, both in terms of laboratory reagents and the required computational infrastructure and bioinformatics expertise, is currently prohibitive for most routine clinical settings. This creates a real danger of a "genomic divide," where cutting-edge precision medicine is only accessible to wealthy individuals and institutions, further widening global health inequities (Inayatullah et al. 2025). This leads to issues of health equity and justice. If SCA-guided therapies are only available to a privileged few, their development could divert resources from more fundamental, equitable healthcare needs. There is a risk that already marginalized populations will be further left behind, both as beneficiaries of these technologies and as participants in the research that develops them, leading to datasets that are not representative of human diversity and thus biased in their clinical applications.

Successfully navigating the transition of SCA from a research marvel to a clinical mainstay will require a concerted, multidisciplinary effort. Scientists must develop more robust and spatially preserved protocols, computational biologists must create user-friendly and standardized analytical pipelines, ethicists and lawyers must build new consent and data governance models, regulators must adapt pathways for complex signatures, and policymakers must address issues of cost and access. Only by tackling these trials head-on can the promise of single-cell powered precision medicine be ethically and equitably fulfilled.

13.8 Conclusion and Future Directions

This chapter has presented the transformative journey of precision medicine from a broad conceptual framework to a discipline being fundamentally reshaped by single-cell analysis. It began by establishing the core premise of precision medicine tailoring health care to the individual is intrinsically dependent on understanding cellular heterogeneity. The limitations of traditional bulk analysis, which obscures this diversity by providing only an averaged molecular profile, created a critical technological and conceptual gap. This gap has been bridged by the revolutionary principles and methodologies of single-cell analysis, which provided the lens to bring the intricate cellular mosaic of human health and disease into sharp focus.

The advanced techniques now available to us, from high-throughput scRNA-seq to multimodal omics and spatial transcriptomics, allow for the dissection of tissues into their fundamental cellular units. This has moved from simply knowing a tumour containing immune cells to understanding the exact functional states of those cells, their clonal lineages, their epigenetic regulations, and their physical interactions with the malignant ecosystem. However, as explored, this power generates a bulk of data. The field of computational biology has risen to this challenge, developing sophisticated algorithms for dimensionality reduction, clustering, and trajectory inference, and is now being supercharged by artificial intelligence to extract deeper meaning and predictive power from these complex datasets.

The ultimate output of this technological revolution is a growing list of clinical applications that is already impacting patient lives. From tracking clonal evolution in leukaemia to predicting its relapse, to identifying rare autoreactive immune cells in autoimmunity, and mapping the specific neuronal vulnerabilities in Alzheimer's disease, single-cell analysis is refining diagnoses, prognoses, and therapeutic strategies. It is transforming diseases once considered monolithic into spectra of molecularly defined subtypes, each potentially requiring a distinct therapeutic approach.

Yet, the journey from a research-grade discovery to a clinically validated tool is fraught with scientific challenges like the loss of spatial context and analytical complexity, as well as profound ethical, regulatory, and economic hurdles concerning data privacy, equitable access, and clinical validation. These challenges are not dispensable; rather they are central considerations that must be addressed with the same rigour and innovation as the scientific questions themselves.

Looking ahead, the future of precision medicine lies in the seamless integration of these multi-scale, single-cell insights into a coherent clinical picture. The goal is to move beyond a static snapshot to a dynamic, high-resolution view of a patient's disease, a living plan that can be monitored over time and used to predict treatment responses before a therapy is even administered. This will require not only continued technological refinement but also the development of new ethical frameworks, regulatory pathways, and clinical infrastructures.

In conclusion, single-cell analysis has done more than just provide a new set of tools. It has fundamentally changed our understanding of biological complexity. It has been affirmed that heterogeneity is not noise to be averaged out, but the very signal we must decode. By continuing to unravel the cellular and molecular intricacies of each individual, the idea of precision medicine delivering patient-specific treatment at the right time is steadily transitioning from an aspirational idea into an achievable reality, heralding a new era of truly personalized health care.

References

Dezem FS, Arjumand W, DuBose H, Morosini NS, Plummer J (2024) Spatially resolved single-cell omics: methods, challenges, and future perspectives. Annu Rev Biomed Data Sci 7(2024):131–153

Flynn E, Almonte-Loya A, Fragiadakis GK (2023) Single-cell multiomics. Annu Rev Biomed Data Sci 6(1):313–337

Gao G, Deng A, Liang S, Liu S, Fu X, Zhao X, Yu Z (2022) Integration of bulk RNA sequencing and single-cell RNA sequencing to reveal uveal melanoma tumor heterogeneity and cells related to survival. Front Immunol 13:898925

Ge S, Sun S, Xu H, Cheng Q, Ren Z (2025) Deep learning in single-cell and spatial transcriptomics data analysis: advances and challenges from a data science perspective. Brief Bioinform 26(2):bbaf136

He J, Lin L, Chen J (2022) Practical bioinformatics pipelines for single-cell RNA-seq data analysis. Biophys Rep 8(3):158

Hulsen T, Jamuar SS, Moody AR, Karnes JH, Varga O, Hedensted S et al (2019) From big data to precision medicine. Front Med 6:34

Inayatullah M, Dwivedi AK, Tiwari VK (2025) Advances in single-cell omics: transformative applications in basic and clinical research. Curr Opin Cell Biol 95:102548

Jovic D, Liang X, Zeng H, Lin L, Xu F, Luo Y (2022) Single-cell RNA sequencing technologies and applications: a brief overview. Clin Transl Med 12(3):e694

Kang B, Camps J, Fan B, Jiang H, Ibrahim MM, Hu X et al (2022) Parallel single-cell and bulk transcriptome analyses reveal key features of the gastric tumor microenvironment. Genome Biol 23(1):265

Lähnemann D, Köster J, Szczurek E, McCarthy DJ, Hicks SC, Robinson MD et al (2020) Eleven grand challenges in single-cell data science. Genome Biol 21(1):31

Li Y, Ma L, Wu D, Chen G (2021) Advances in bulk and single-cell multi-omics approaches for systems biology and precision medicine. Brief Bioinform 22(5):bbab024

Lim J, Park C, Kim M, Kim H, Kim J, Lee D-S (2024) Advances in single-cell omics and multiomics for high-resolution molecular profiling. Exp Mol Med 56(3):515–526

Lotfollahi M, Hao Y, Theis FJ, Satija R (2024) The future of rapid and automated single-cell data analysis using reference mapping. Cell 187(10):2343–2358

Molla G, Bitew M (2024) Revolutionizing personalized medicine: synergy with multi-omics data generation, main hurdles, and future perspectives. Biomedicine 12(12):2750

Naldini MM, Casirati G, Barcella M, Rancoita PMV, Cosentino A, Caserta C et al (2023) Longitudinal single-cell profiling of chemotherapy response in acute myeloid leukemia. Nat Commun 14(1):1285

Overbey EG, Das S, Cope H, Madrigal P, Andrusivova Z, Frapard S et al (2022) Challenges and considerations for single-cell and spatially resolved transcriptomics sample collection during spaceflight. Cell Rep Methods 2(11):100325

Quek C, Pratapa A, Bai X, Al-Eryani G, da Silva IP, Mayer A et al (2024) Single-cell spatial multiomics reveals tumor microenvironment vulnerabilities in cancer resistance to immunotherapy. Cell Rep 43(7):114392

Ren Y, Wu R, Li C, Liu L, Li L, Weng S et al (2024) Single-cell RNA sequencing integrated with bulk RNA sequencing analysis identifies a tumor immune microenvironment-related lncRNA signature in lung adenocarcinoma. BMC Biol 22(1):69

Wang RC, Wang Z (2023) Precision medicine: disease subtyping and tailored treatment. Cancer 15(15):3837

Wegmann R, Bonilla X, Casanova R, Chevrier S, Coelho R, Esposito C et al (2024) Single-cell landscape of innate and acquired drug resistance in acute myeloid leukemia. Nat Commun 15(1):9402

Weldy CS, Ashley EA (2021) Towards precision medicine in heart failure. Nat Rev Cardiol 18(11):745–762

Woo H, Eyun S-I (2025) Applications and techniques of single-cell RNA sequencing across diverse species. Brief Bioinform 26(4):bbaf354

Single-Cell Analysis in Regenerative Medicine

14

Aneela Mustafa, Shah Fawad, Mustafeez Mujtaba Babar, and Jayakumar Rajadas

Abstract

Single-cell analysis approaches are reshaping modern biological research. Single-cell analysis techniques divulge cellular heterogeneity, differentiation, and repair mechanisms at a single-cell level unlike bulk homogenous mixture analysis that obscure complex details. Single-cell analysis can reveal cellular pathway involved in cell, tissue, and organ repair. Regenerative medicine (RM) is a promising field in medical research. RM holds the potential to repair, replace, and regenerate damaged or worn-out cells, tissues, and organs to restore their normal function. Single-cell analysis techniques (genomics, transcriptomics, epigenomics, proteomics, metabolomics, and multi-omics) reveal individual cell profile at high resolution. This chapter aims to provide different single-cell techniques utilized in

A. Mustafa
Department of Biomedicine (DBM), Atta-ur-Rahman School of Applied Biosciences (ASAB), National University of Sciences and Technology (NUST), Islamabad, Pakistan

S. Fawad
Department of Microbiology & Biotechnology (MBT), Atta-ur-Rahman School of Applied Biosciences (ASAB), National University of Sciences and Technology (NUST), Islamabad, Pakistan

M. M. Babar (✉)
Shifa College of Pharmaceutical Sciences, Shifa Tameer e Millat University, Islamabad, Pakistan

Advanced Drug Delivery and Regenerative Biomaterials Laboratory of Cardiovascular Institute, Stanford University School of Medicine, Stanford University, Palo Alto, CA, USA
e-mail: mustafeez.babar@fulbrightmail.org

J. Rajadas (✉)
Advanced Drug Delivery and Regenerative Biomaterials Laboratory of Cardiovascular Institute, Stanford University School of Medicine, Stanford University, Palo Alto, CA, USA
e-mail: jayraja@stanford.edu

V. Singh (ed.), *Advances in Single Cell Analysis*,
https://doi.org/10.1007/978-981-95-9789-5_14

the regenerative medicine field. Moreover, the chapter also provides comprehensive computational methods and tools to analyze large datasets obtained by single-cell techniques. Additionally, this chapter also covers the applications of single-cell techniques in RM. To conclude, it discusses the current bottlenecks and future perspectives in single-cell analysis techniques in RM.

Keywords

Regenerative medicine · Tissue repair · Omics · Healing · Lineage tracing · Stem cell · Bioinformatics · CITE-seq

14.1 Introduction

Cells are building blocks of all living organisms. The human body is composed of heterogenous cell population containing different types of cells. The study of individual cell genotype, phenotype, and behavior is crucial to understanding the complexities in living organisms. Single-cell techniques can detect differences in DNA sequence, gene expression, and protein and metabolite levels effectively. Single-cell genotypes provide key insights into clonal heterogeneity at the genome level (Mincarelli et al. 2018). Single-cell techniques interpret individual cell profile at high resolution that provide rare cell population driving various signaling pathways. By analyzing the signaling cascades at a single-cell level, scientists can study disease onset and progression. Single-cell techniques help researchers find targeted and effective treatment strategies for different disorders (Lee et al. 2020). Single-cell analysis approaches have progressed the biological research arenas, e.g., regenerative medicine, stem cell biology, oncology, and immunology. Single-cell techniques are crucial in these fields in order to design effective therapeutic strategies (Tellez-Gabriel et al. 2016; Papalexi and Satija 2018).

The bulk analysis methods provide an average signal of cells. The average signals of cells hide the heterogeneity of mixtures of cells under study. The heterogenous mixture masks the rare cell population that may be involved in disease onset, progression, and regeneration process. For example, bulk transcriptomics do not differentiate between different cell forms: for example malignant, immune, and stromal cell types in a tumor microenvironment. This results in loss of crucial biological information required for better understanding of disorder and its suitable treatment (Patel et al. 2014). Bulk analysis also fails to measure cell plasticity and dynamics. When a drug is administrated to cells, it is only taken up by a few cells; when bulk methods are applied to analyze such heterogenous cell populations, these bulk methods only provide an average profile. Thus, the important information such as development transition, lineage hierarchy, and change in states induced by drugs is missed. Therefore, bulk analysis often leads to incorrect conclusion, as bulk analysis does not reveal the cell diversity. Moreover, bulk analysis does not provide information on cell-cell interaction and spatial knowledge that are crucial for tissue function and regeneration (Trapnell 2015).

The cellular heterogeneity is an essential characteristic of biological tissue, and is influenced by pathological and physiological factors. Transcriptional variability is produced even in a cell type that appears to be homogenous because it can be influenced by a variety of factors: cells can be placed in different microenvironments or niches which modulate gene expression, expression patterns can change over different cell-cycle phases, and gene-expression programs themselves exhibit stochasticity. Stem cells possess two properties that define them: the capability to regenerate indefinitely, and the ability to form distinct cells. Stem cells are classified as pluripotent stem cells and tissue-specific stem cells. Pluripotent stem cells give rise to all three germ layers, i.e., ectoderm, mesoderm, and endoderm, whereas tissue-specific stem cells contribute to embryo formation and adult tissue homeostasis. Pluripotent stem cells are present in extremely small quantities in the early development of mammals, and tissue-specific stem cells also represent a very small fraction of the number of cells in any particular tissue or organ. In turn, these unique stem cells are present together with a variety of differentiated and intermediate cell populations, which form highly heterogeneous cellular landscapes.

Single-cell sequencing is a revolutionary approach in defining the molecular phenotype of such heterogeneous populations, such as stem cells. Single-cell sequencing techniques can reveal cellular heterogeneity and reveal common and rare cell types without prior information on population composition. Single-cell analysis is paving its way in regenerative medicine (RM) as it reveals cell heterogeneity at extraordinary resolution. Single-cell DNA sequencing (scDNA-seq), for instance, provides indepth single-cell details at genomic level. Single-cell RNA sequencing (scRNA-seq) reveals single-cell profile at the transcriptome level. Single-cell epigenomics divulge key information of epigenetic alteration of single cell at the epigenetic level. Single-cell proteomics (SCP) describe varying protein levels at a single level; single-cell metabolomics gives metabolite fluctuation at a single level. Whereas multi-omics approach combines two or more techniques to profile an individual cell at the transcriptome or proteome level. Single-cell techniques reveal rare cell population of stem cells or progenitor cell also construct cell lineage and molecular pathway important for cell, tissue and organ repair, healing or regeneration. A previous study utilized scRNA-seq to map stem and progenitor cell lineages involved in skin- and wound-healing processes (Guerrero-Juarez et al. 2019). Single-cell techniques have been applied to study cardiomyocytes and fibroblast cell behavior involved in post-injury heart regeneration (Gladka et al. 2018). Similarly, single-cell analysis approaches have been employed to identify the activation pathways of satellite cells and regenerative differentiation mechanisms in skeletal muscles (Dell'Orso et al. 2019). Single-cell analysis techniques can also provide key details about tissue architecture which is important to study cell-cell communication in regeneration (Mao and Mooney 2015). Altogether, single-cell techniques are crucial in RM to advance stem cell-based medicine, biomaterial, and other regenerative strategies.

14.2 Single-Cell Analysis Techniques

14.2.1 Single-Cell Genomics

Single-cell DNA sequencing (scDNA-seq) allows the scientific community to examine genetic alteration of a single cell at the genome level. It has the ability to determine somatic mutations, single-nucleotide polymorphism variations, copy number alterations, and changes in microsatellites in stem and progenitor cells. Single-cell genomics helps to have a picture of genomic heterogeneity in tissues (Wen and Tang 2016). ScDNA-seq is crucial to monitor genetic stability of the cells utilized in the regenerative therapies to reduce the risk of tumor development following transplantation (Wang et al. 2014).

Genomes of individual cells can also offer an extra source of information that can be used to see how stem cell populations develop and become heterogeneous: the cell lineage. A single stem cell has the capability of producing many specialized cells during development by way of repeated division and differentiation. The probability of replication errors is low in each instance of the cell division, and these errors, although uncommon in normal mammalian cells, can be used as a marker to trace the history of development of individual progeny. The study of these mutations caused by replication allows the researchers to build detailed lineage trees that depict the processes of development (Wen and Tang 2016).

Among the various scDNA-seq techniques, Whole-Genome Amplification (WGA) has resolved the problem of limited starting material. Among WGA methods, multiple displacement amplification (MDA) is the most widely employed technique to produce and characterize DNA fragments with a high degree of fidelity. MDA is widely employed to detect single-nucleotide polymorphisms. Degenerate oligonucleotide-primed PCR (DOP-PCR) is useful in detecting copy number alteration of entire genome; however, its resolution is lower compared to other scDNA-seq techniques.

Multiple annealing and looping-based amplification (MALBAC) is considered more suitable than other scDNA-seq techniques as it allows uniform genome coverage, minimizes amplification bias, and accomplishes a balance between copy number variations and point mutation detection. In contrast, whole-genome amplification (WGAM) is a general umbrella term that covers all these strategies and provides flexibility to choose a method according to study interest, e.g., structural variants, point mutations, and genome-wide copy number variations. Lastly, single-nucleus DNA sequencing is more accurate because it has less cytoplasmic contamination. It also provides a better representation of the nuclear genome by preserving the genomic integrity (Wen and Tang 2016).

14.2.2 Single-Cell Transcriptomics

Single-cell transcriptomics (scRNA-seq) was initially developed by James Eberwine, and independently by Iscove and colleagues. ScRNA-seq have transformed our understanding of cellular diversity within tissues. ScRNA-seq methods can directly measure and address the natural variability in gene expression unlike bulk transcriptomic studies, which assume all cells are part of a homogeneous population and ignore the stochasticity of gene expression (Munsky et al. 2012; Kulkarni et al. 2019). There are various methods to study and analyze single cell isolated in high throughput or low throughput. These involve droplet-based technology, which is a popular non-biased, high-throughput technology. 10X Genomics Chromium, DropSeq, and inDrop are most commonly employed scRNA-seq techniques (Zhang et al. 2019; Zheng et al. 2017; Klein et al. 2015; Macosko et al. 2015). These systems can rely on the microfluidics system to capture single cells with their beads which have a distinct barcode. All mRNA transcripts are labeled with a distinct molecular identifier (UMI) to record the absolute number of transcripts produced in each cell. Moreover, 10X Genomics Chromium detects a high number of transcripts at high accuracy; however, it is an expensive technique when it comes to large-scale research. DropSeq is a cheaper technique; however, it detects fewer transcripts, whereas inDrop allows adjustable parameters, which can be utilized when it comes to detecting lowly expressed genes. Whereas several scRNA-seq techniques cannot detect both poly A + tail RNAs and poly A-RNA in a single cell, however, SUPeR-seq techniques have made it possible to obtain poly(A)-positive and poly(A)-negative RNA species within one cell. A previous study detected thousands of circular RNAs with no poly A tails and hundreds of RNA with poly A-negative tails (Fan et al. 2015).

14.2.3 Single-Cell Epigenomics

Epigenetics refers to the investigations of the regulatory mechanisms which permit the hereditarily modifiable gene expression in genetically equivalent cells. It includes chemical modification of DNA and histone proteins and alterations in DNA and chromatin structure accessibility. Single-cell chromatin condition alters the functional structure of the genome by controlling the availability of gene regulatory elements like enhancers that serve as a binding location of transcriptional regulators. Chromatin accessibility of different cell types is altered by cascade of transcription factors in developmental process. The cell-type-specific transcription factor analysis is beneficial to decipher underlying molecular mechanism deriving cell fate determination.

Single-cell epigenomics (sc-Epi) can profile DNA methylation, histone modification, and chromatin transitions of a single cell. ScATAC-seq provide details of various regulatory elements, transcription biding sites of a single cell in mixture of cells. ScBS-seq measures DNA methylation patterns at the genome-wide level. Similarly, scWGBS also gives DNA methylation patterns of a single cell in a heterogenous cell population. scChIP-seq profiles transcription factors and histone modifications in a single cell providing key insights between chromatin state and regulatory activity. Sc-Epi techniques allowed to profile cell identity, lineage, and epigenetic variation at the individual cell level (Gkountela et al. 2015; Farlik et al. 2015; Gu et al. 2011).

14.2.4 Single-Cell Proteomics

Single-cell proteomics (SCP) has enabled the scientific community to comprehend biological systems at a granularity level. The protein expression levels do not correlate with the transcriptomics level in biological samples (Budnik et al. 2022). A previous study conducted reported that the transcriptomics levels vary as compared to proteomics level. They indicated that proteomics are stable when they analyzed proteome of a single cell using SCP (Chung et al. 2021). Therefore, it is crucial to analyzing translational regulation at the individual cell level to understand and analyze cell heterogeneity and behavior under disease conditions. SCP is a complementary technique applied when scRNA-seq analysis are performed. SCP measures and analyzes protein expression levels in living cell revealing cell-to-cell difference providing real-time information happening within a cell. SCP analysis is important in studying post-translational modifications. SCP is an essential technique owing to its ability to provide reliable information on initial cell signal, cell adaptation in response to environmental and drug therapy. SCP is important in revealing biological information relating to protein binding, modification, degradation, and interaction at a single-cell level. As protein activities are highly regulated by degradation, SOP provides key information on protein-protein interaction (PPI). It has also been employed to determine protein localization within a cell. All these mechanisms are imperative in decoding the complex signaling pathways involved in regenerative processes (Ahmad and Budnik 2023).

SOP has developed at a rapid pace with the development of different techniques. Cytometry by Time-of-Flight (CyTOF) measures over 40 metal-tagged antibodies by mass spectrometry. CyTOF performs the high-dimensional analysis of immune responses and micro-phenotypic heterogeneity of cellular populations in single-cell mass spectrometry (Sahu et al. 2022). Imaging Mass Cytometry combines CyTOF with spatial imaging that allows the researcher not only to establish the degree of expression of the proteins, but also to map their localization to subcellular sites in whole tissues (South et al. 2023). Single-Cell Proteomics by Mass Spectrometry (SCoPE-MS) employs tandem mass tags (TMT) for multiplexed measurement of proteins in individual cells. SCoPE-MS has a sensitive variant SCoPE2 suitable for large scale SOP with improved sensitivity, precise quantification, and high throughput (Budnik et al. 2018). These technologies offer to study and observe cell-to-cell variability in cell protein expression, indicate regulatory processes, and examine the mechanisms of disease processes.

14.2.5 Single-Cell Metabolomics

Metabolome contains the complete set of metabolite content of a cell, tissue, and organisms. Analyzing metabolites at a single-cell level is difficult compared to analyzing metabolite of a whole tissue or a large sample. However, bulk metabolite analysis fails to show intricacies and complexities within a single cell. Single-cell metabolomics (sc metabolomics) is an evolving technique that maps and quantitates the metabolites that can provide meaningful information about cellular function in extremely heterogeneous systems with spatial resolution at a single-cell level. Sc metabolomics techniques are helpful in establishing metabolic phenotype, functional condition of individual cells which are affected by age, genetic status, disease status, and drug exposure (Ali et al. 2022).

The type and dynamic range of metabolites are very diverse, which brings difficulties for their detection, identification, and quantification; however sc metabolomics has become a powerful method to visualize, detect, and identify cellular content on the metabolic scale that provides information that is frequently lost in bulk metabolomics. Mass spectrometry (MS)-based methods, e.g., nano-electrospray ionization (nanoESI), mass guided single-cell matrix-assisted laser desorption/ionization (MALDI), and secondary ion mass spectrometry (SIMS) can detect attomole to zeptomole range metabolite (Xu et al. 2024; Cairns et al. 2025; Taylor et al. 2021). Progress to subsequent single-cell metabolic measurements involving microfluidic devices, single-probe sampling, and capillary sampling is becoming possible. In addition, mass spectrometry imaging (MSI), e.g., MALDI-MSI, microfluidics-MS, and SIMS-MSI also extends these capabilities by preserving anatomic locations and mapping metabolites at subcellular compartments or whole-tissue levels (Zhang et al. 2025; Feng et al. 2022; Tian et al. 2024).

14.2.6 Single-Cell Multi-omics

Single-cell multi-omics (sc multi-omics) is the combination of two or more layers of molecules (e.g., genomics, transcriptomics, epigenomics, and proteomics) within the same cell. Sc multi-omics can be used to give a more global picture of cellular behavior, in contrast with single-modality assays, which only measure one layer of information at a time, by connecting different biomolecular profiles in the same cell. Sc multi-omics allows scientists to reveal genetic variations in transcriptional programs, chromatin accessibility, protein expression, and even metabolic states all at once. Notably, sc multi-omics is capable of solving cellular heterogeneity at the most detailed level, giving a chance to make new findings in regenerative medicine cancer biology, immunology, and developmental biology. Various techniques employed in single-cell omics have been presented in Table 14.1.

Various new methods have been invented to make multi-layer profiling of single cells possible. CITE-seq is a blend of scRNA-seq and antibody-derived tags that simultaneously measures both transcriptomes and surface protein expression. It gives a direct relationship between mRNA level and protein abundance and

Table 14.1 An overview of various singe-cell omics layer techniques and their targets

Omics layer	Techniques	Target	Reference
Single-cell genomics	WGA MALBC DOP-PCR MDA single-cell nucleus DNA seq	Variant identification, clonal structure, lineage heterogeneity	Zong et al. (2012), Carter et al. (1992), Fu et al. (2015), Leung et al. (2015)
Single-cell transcriptomics	Smart-seq2 Dropseq inDROP SUPeR-seq MARS-seq Microfluidic scRNA-seq 10X Genomics Chromium MALBAC-RNA	Transcriptome/cell, mRNA levels	Klein et al. (2015), Hashimshony et al. (2012), Streets et al. (2014), Chapman et al. (2015), Cadwell et al. (2016), Jaitin et al. (2014)
Single-cell proteomics	CyTOF Imaging Mass Cytometry SCoPE-MS SCoPE	Multiplexed markers, PTMs, numerous proteins	Sahu et al. (2022), South et al. (2023), Budnik et al. (2018)
Single-cell epigenomics	scATAC-seq scBS-seq scChIP-seq scWGBS	Chromatin, histone marks, DNA accessibility, and methylation, 5mC	Gkountela et al. (2015), Farlik et al. (2015), Gu et al. (2011), Miura et al. (2012), Guo et al. (2014), Smith et al. (2014), Cusanovich et al. (2015)
Single-cell metabolomics	nanoESI MALDI SIMS MALDI-MSI SIMS-MSI microfluidics-MS	Sensitive metabolite detection, precise singe cell metabolites extraction with spatial information	Xu et al. (2024), Cairns et al. (2025), Taylor et al. (2021), Zhang et al. (2025), Feng et al. (2022), Tian et al. (2024)
Single-cell multi-omics	CITE-seq REAPP-seq Sci-CAR SNARE-seq 10x Multiome	RNA + proteins, RNA + ATAC, and so forth	Lakkis et al. (2022), Peterson et al. (2017), Cao et al. (2018), Chen et al. (2019)

addresses the issue of poor mRNA-protein correlation (Stoeckius et al. 2017; Lakkis et al. 2022). REAP-seq is based on the same principle, combing transcriptomic and protein profiling on a high-throughput scale, and thus it is particularly useful in characterizing the immune system and finding rare cells of interest (Peterson et al. 2017). Sci-CAR is used to investigate the relation between chromatin structure and transcription within a same cell (Cao et al. 2018). Similarly, SNARE-seq can also access chromatin accessibility as well as transcriptomes of single nuclei, which is especially handy in frozen tissue and clinical samples where intact cells are hard to find (Chen et al. 2019). The 10x Multiome platform is a commercial platform that combines scATAC-seq and scRNA-seq to map regulatory elements as well as transcriptional programs at a high throughput.

14.3 In Silico Methods for Single-Cell Analysis

The single-cell analysis techniques provide big data which is difficult to analyze. Therefore, comprehensive and specialized computation methods are developed to analyze the big data to study the biological mechanism and associated disease pathways and regeneration mechanism accurately.

ScDNA-seq data is often noisy due to low coverage and amplification bias. BWA and Bowtie 2 are utilized to process and align data obtained through scDNA-seq. GATK and muTect2 are employed to analyze variant calling data. inferCNV and CopyKAT tools are helpful in processing data of copy number variation. All these tools are utilized for quality control to remove cells with low coverage and excessive noise. Dimensionality reduction and visualization are done like principal component analysis (PCA), t-distributed stochastic neighbor embedding (t-SNE), or uniform manifold approximation and projection (UMAP). The grouping of cells is based on mutation or CNV profiles, which identify the genetic differences between clones. LineageO and Cassiopeia are utilized for clonal evolution construction which helps in tracking stem cell lineage clonal. DNA repair and WNT pathway are mapped to redevelop phenotypes (Baslan and Hicks 2017).

The main aim of scRNA-seq is to identify the transcriptional schemes that define the identity and state of cells. It starts by analyzing quality control in order to eliminate low-quality cells and then proceeds to normalization and scaling to address sequencing depth variations. Visualization of high-dimensional data is then performed by the use of dimensionality reduction approaches like PCA, t-SNE, UMAP. Clustering algorithms are used to determine sets of transcriptionally similar cells, and cells can be annotated using marker genes. Downstream analyses involve differential gene expression testing and the classification of dynamic processes, like cell differentiation, using trajectory analysis. Seurat (R) and Scanpy (Python) are widely used tools which offer major pipelines to combine all those steps and are still the most popular databases to analyze scRNA-seq. Monocle, Slingshot, and RNA velocity are used for tracing dynamic process construction. GSEA is utilized to map differentially expressed genes to reveal cell signaling (Luecken and Theis 2019).

Both ArchR and Signac are used to align ATAC-seq reads, and quality control is performed using transcription start site (TSS) enrichment. Latent Semantic Indexing (LSI) and UMAP are used to regulate the diversity within cells. Motif enrichment analysis is done to access chromatin states. cisTopic is employed for dynamic reconstruction. SCENIC links is used to infer transcription factor activities (Granja et al. 2021; Stuart et al. 2023).

Recent developments in SCP have produced high-dimensional datasets of millions of cells, which give insightful information about basic biology and disease. SCP involves data normalization, batch correction, and noise removal. Spectronaut and MaxQuant are utilized to achieve data normalization, batch correction, and noise removal from data coming from CyTOF or SCoPE-MS. PCA and UMAP are utilized for dimension reduction for protein expression states. Clustering and grouping of protein signatures are based on immune, stromal, and stem cell subpopulations. Pseudotime is utilized to trace cell fate on protein level. STRING and Reactome are helpful in pathway construction at the protein level (Specht et al. 2021; Guldberg et al. 2023).

Sc metabolomics raw data is processed initially with XCMS or MZmine and then normalized to provide sample consistency. PCA and UMAP are important in visualizing and uncovering metabolic diversity in cell states. The grouping and clustering of cells is based on metabolites abundance. Time flux and lineage analysis provide useful information on metabolism rewiring. MetaboAnalyst tool is important in pathway analysis; it links identified metabolites to KEGG and Reactome pathway databases (Zenobi 2013).

Data obtained through multi-omics techniques employ a combined pipeline to combine and analyze different types of data within a single cell. MOFA+ and totalVI tools are used to simplify and interpret the complex multimodal datasets. Weighted nearest neighbor (WNN) tool is employed to cluster and determine different population of cells resulting from multi-model datasets. MATCHER and MultiVI tools are useful in trajectory and lineage construction. Pathway analysis is performed through pathway enrichment analysis (Argelaguet et al. 2021; Hao et al. 2021).

14.4 Application of Single-Cell Analysis in Regenerative Medicine

Single-cell techniques are crucial in studying and analyzing individual-cell behavior providing real-time information on the regeneration process. Single-cell techniques hold the potential to uncover diverse cell populations, lineage tracing and identify crucial and minor cell populations required for the regeneration process. Single-cell techniques also provide details on cell-cell interaction. Scientist has identified common and tissue-specific regenerative mechanisms utilizing single-cell analysis techniques and in silico tools. The single-cell techniques have contributed significantly to the broadening of our knowledge of regenerative medicine as follows.

14.4.1 Germline and Somatic Lineage

ScDNA-seq is employed to investigate meiotic recombination, aneuploidy, and mutation rates in human germ cells (e.g., sperm and oocytes) to map and identify chromosomal abnormalities in the process of gametogenesis (Wang et al. 2012; Hou et al. 2013; Lu et al. 2012). ScDNA-seq of a normal adult somatic tissue makes it possible to trace the lineage of that cell and how it relates to adult tissue architecture. It is also being utilized in cancer research to study clonal evolution by determining the copy number alterations and single-nucleotide variations in a single tumor cell. Irrespective of these developments, it is hard to detect single-nucleotide variants with a low false positive rate (Wang et al. 2014). The advancements in single-cell whole-genome sequencing processes are likely to improve tracking of the lineage of both normal and cancerous stem cells offering an overall map of the genomic changes in single-cell scale.

14.4.2 Limb Regeneration

Gerber and colleagues mapped connective-tissue (CT) cell type lineages at high-throughput sc RNA-seq during axolotl limb regeneration using Cre-lox lineage tracking. They demonstrated that several mature types of CT subtypes (dermal fibroblasts, interstitial fibroblasts, etc.) merged into a reasonably homogeneous blastema progenitor form which revived traits of embryonic limb bud gene expression. This study explained the role of differentiated CT in multipotent skeletal progenitor regrowth. The study gave a cellular atlas and a set of marker genes of blastema populations which could provide a roadmap of functional perturbations (Gerber et al. 2018). ScRNA-seq of adult axolotl limb regeneration in large scales categorized various types of cells and revealed new regeneration-specific clusters. Among them, there was a mitochondria-formed cluster that was related to energy supply, and an ECM-secreting COL2+ cluster that was related to intercellular signaling. Dedifferentiation and re-differentiation in a subpopulation of COL1+/COL2+ was also detected, as well as a COL2-mitochondria subcluster that facilitated musculoskeletal repair. A previous study's results recapitulated the dynamic scRNA-seq profiles of axolotl limb regeneration and found musculoskeletal populations involving mitochondria as important constituents of the regenerative microenvironment (Qin et al. 2021).

14.4.3 Sensory Hair Cell Regeneration

CRNA-seq has been used in zebrafish to study transcriptional programs of sensory hair cell regeneration. Based on the lateral line system, a time-resolved profiling of gene expression in the supporting cells was performed using the combination of scRNA-seq with Pseudotime and RNA velocity analysis. This strategy identified three sequential stages of regeneration. (1) Immature injury and inflammatory response with a down-regulation of progenitor maintenance genes. (2) A temporary activation of regeneration-specific genes absent in homeostasis. (3) A subsequent reinstatement of developmental programs, including cell cycle progression, a hair cell specification gene atoh1a, and metabolic reprogramming. Significantly, it was demonstrated that dynamic regulation of Notch signaling was required to permit progenitor differentiation during their early developmental phases but to inhibit excess differentiation at their later developmental stages in order to maintain a supporting cell pool. Through scRNA-seq, the researchers identified pathways that are important in zebrafish and which can be used as possible therapeutic options to induce hair cell regeneration in the human ear (Baek et al. 2022).

14.4.4 Heart Regeneration

ScRNA-seq was utilized to study ErbB2/Nrg1 signaling with metabolic reprogramming of border-zone cardiomyocytes in heart regeneration. Proliferative cardiomyocytes were found to have glycolytic activation in a single-cell profile; functional experiments revealed that glycolysis is necessary in cardiomyocyte proliferation in the process of regeneration. The experiment links the definition of the single-cell state to a manipulable metabolic process that facilitates cardiac regeneration (Honkoop et al. 2019).

With the help of scRNA-seq and complementary chromatin analysis, a neonatal mouse heart model was studied for regenerative competence. The study identified transcriptional programs and gene regulatory networks that involved cardiomyocytes proliferation and tissue repair. The study addressed cell-type-specific injury responses, determined the border-zone cardiomyocyte patterns of proliferative signatures, and discovered metabolic processes involved in the neonatal regenerative response (Wang et al. 2020).

14.4.5 Whole Body Regeneration

ScRNA-seq analysis identified transcription factor atlas of planarian stem cells. It also provided information on differentiate progeny required to map the lineage process in the regeneration process. The analysis provided planarian stem cell subpopulation along with crucial transcription factor expression regulatory modules required for neural, muscular, and epidermal fate determination. The transcription factor atlas showed genes that are crucial for self-renewal lineage-priming, and terminal-differentiation of stem cells.

14.4.6 Tail Regeneration

Regeneration-initiating cells (RICs) were discovered in *Xenopus laevis* tail through scRNA-seq analysis. RICs grew from basal epidermal cell and had their role in extracellular matrix remodeling and regeneration process. The study discovered that RIC loss was associated with lost regeneration ability in *Xenopus laevis* tail supporting RICs' important role in the regenerative process (Sindelka et al. 2024).

14.4.7 Liver Regeneration

Single-cell transcriptomic profiling of mouse liver following acute injury outlined the spatiotemporal program of zonal regeneration. It was found that hepatocyte proliferation was an uneven process which began mainly in certain periportal and midlobular areas driven by localized signaling pathways. Transcriptional profiles of hepatocytes appeared along the central-portal axis, with regenerative clusters demonstrating the activation of cell cycle, stress response, and metabolic reprogramming pathways. It was found that non-parenchymal populations of cells (immune, endothelial) provided zone-specific paracrine signals coordinating hepatocyte proliferation and tissue healing (Ben-Moshe et al. 2022).

14.4.8 Wound Healing

Combined single-cell and spatial dynamics of murine (and human) skin wound analysis revealed dynamic fibroblast heterogeneity and temporal changes of wound-resident cell states. The researcher found that several types of fibroblasts can transition between transient injury-related programs (such as pro-fibrotic and pro-regenerative states). They discovered that the interactions of macrophages, fibroblasts, and keratinocytes are sequential and organized re-epithelialization and hair-follicle regeneration. In a space-resolved analysis, it was established that the distribution of fibroblast subtypes and the local immune composition are broadly defining factors of scarring versus productive regeneration (Almet et al. 2025).

14.4.9 Lineage Tracing

Integrating expressed barcodes or genetic barcoding with single-cell readouts made it possible to directly rebuild lineage during the regeneration process to make it clear whether new tissue is formed by local dedifferentiation, resident stem cells, or recruited progenitors. Recent lineage-capture techniques compatible with transcriptomic and epigenomic measurements enable simultaneous measurement of the history of clones and cell state. This is an effective method to address issues of progenitor contribution and plasticity in organ regeneration. Jindal et al. reported CellTag-multi, an expressed barcode system, which could be captured in both scRNA-seq and scATAC-seq, providing the ability to track clones and at the same time monitor both transcriptional and epigenomic states. They confirmed the approach to hematopoiesis to demonstrate fate-sensitive gene-regulatory modifications associated with clone history. This study showed that multi-omics single-cell assays can be employed to study lineage tracing to answer questions of clonal origins, priming, and regulatory reprogramming (Jindal et al. 2024).

14.4.10 Brain and Nervous System Regeneration

Nervous system regeneration has been most widely investigated in vertebrates such as fish, amphibians, and mammals. Müller glia (MG) in the retina represent the principal regenerative cell type. Upon damage, they reprogram into Müller glia–derived progenitor cells (MGPCs) (Lenkowski and Raymond 2014). These MGPCs selectively differentiate into the lost retinal neuronal cell types, but not the complete retinal composition. The transcriptomics and multi-omics analysis of single cells has shown that MGPCs have shared transcriptional signatures of late-retinal progenitors, but they also engage context-dependent gene regulatory networks (GRNs) (Lyu et al. 2023). Microglial inflammatory signaling plays a role in the reprogramming of Muller glia to MGPCs in chicks and mice, but microglia-mediated inflammatory responses can suppress the neurogenic and proliferative potential of MG in sustained NF-kB signaling (Palazzo et al. 2022).

The regeneration of the brain has been studied in axolotls, which are capable of completely replacing telencephalon populations in the case of significant injury. Single-cell transcriptomics revealed that regeneration occurs in three consecutive steps: proliferation of adult neural stem cells called ependymoglia, differentiation to neuroblast intermediates, and maturation of neuroblast intermediates into previously lost neuronal populations. A method called Div-Seq, which integrates EdU labeling with scRNA-seq, identified a regeneration-specific ependymoglia population. This population expresses genes involved in cell adhesion and wound healing, including Runx1 and Kazald1, which are also associated with limb blastema regeneration. Spatial transcriptomics (Stereo-seq) analysis demonstrated that regeneration-specific ependymoglia were localized in close-out of injury, and tissue-neighboring neurons have transcriptomic profiles similar to immature neurons with the expression of growth-promoting genes including Atf3. Comparing homeostatic post-embryonic neurogenesis with the regeneration induced by injury showed that the transcriptional patterns were similar, showing conservation of cellular programs to govern neural regeneration in disparate situations (Lust et al. 2022; Wei et al. 2022).

Spinal cord regeneration has been reported in mice, Xenopus tropicalis tadpoles, and zebrafish. ScRNA-seq in zebrafish revealed the existence of a selective population in macrophages that facilitates regenerative neurogenesis through TNF α cues to ependymo-radial glia precursors (Cavone et al. 2021). These precursors enhance the neurogenic factor Hb-EGF which is crucial in regenerating tissue since the lack of this factor hampers tissue repair and functional recovery (Cigliola et al. 2023). The surviving neurons aid in the regeneration process through the activation of injury-responsive transcriptomic programs, and vesicular trafficking facilitates spontaneous plasticity and neurite outgrowth (Saraswathy et al. 2024). The scRNA-seq of *Xenopus tadpoles* showed that Foxm1+ progenitor cells develop which induce neuronal differentiation after dividing; Foxm1 knockout induces glial differentiation, indicating its essentiality in the neuronal replacement (Pelzer et al. 2021). An integrated approach to incorporation of single-cell technologies and advanced computational techniques has been summarized in Fig. 14.1.

14.5 Current Challenges

In spite of the paradigm shift brought by single-cell techniques in RM, the single-cell technologies have serious technical and analytical limitations to their general use in the RM. Isolation of individual cells may cause stress, damage, or selective bias (especially in fragile or rare cell populations), and may also change their native molecular states. The data produced by single-cell assays is sparse and noisy by

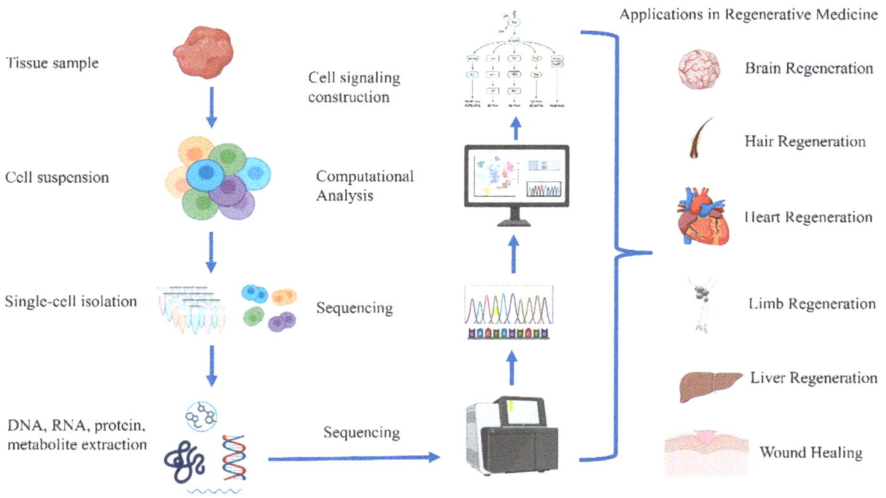

Fig. 14.1 Single-cell techniques, data analysis network construction, and their applications in regenerative medicine

nature, because of small volumes of starting material, leading to dropout events and poor sensitivity to detect lowly expressed genes. In addition, the variability caused by technical reasons and batch effects makes it hard to achieve reproducibility among experiments and between labs, and, therefore, difficult to standardize it. One of the greatest bottlenecks is computational analysis where more sophisticated algorithms are needed to answer such problems as clustering, lineage inference, and multi-omics dataset combination and, in most cases, their results are not consistent. The combination of multi-layered omics data (genomics, transcriptomics, epigenomics, proteomics, and metabolomics) in the same cell is technically difficult, which limits the possibility of a comprehensive view of cell states. Single-cell techniques are costly and resource-intensive. Moreover, these methods require equipment and trained personnel which are not available in resource-limited research settings. Furthermore, results obtained through single-cell analysis techniques need careful interpretation in order to generalize them to the biological and clinical outcomes owing to complexity of the human body (Lähnemann et al. 2020; Kiselev et al. 2019).

14.6 Conclusion and Future Perspectives

The single-cell technologies have transformed the study of RM. RM has been able to acquire a degree of resolution previously unattainable. In addition, advanced computing techniques have allowed improving the capability of unraveling these complex datasets to not only distinguish new regenerative cell subpopulations but also carry out the procedure of determining essential microenvironmental cues.

Single-cell analysis techniques and computational approaches have been applied to identify common and context-dependent regeneration processes.

Although much ground has been covered, there are still technical, analytical, and translational issues like questions of scalability, data integration, and biological interpretation. Nevertheless, they can be lifted because of the existing developments in single-cell multi-omics, spatially resolved technology, and artificial intelligence-based analytics. Collectively, these advances can overcome the gap between scientific findings in basic research and the clinical uses, and can lead to the development of customized RM. To conclude, single-cell strategies are a fundamental solution to RM that does not only present a mechanistic perspective of the problem but provides opportunities to bridge to clinical practice in order to enhance the process of tissue restoration and patient outcomes.

The future of RM is directly related to the development of single-cell analysis methods. Due to always-evolving technologies, the resolution and sensitivity of single-cell genomics, transcriptomics, and proteomics will enable researchers to get a more accurate picture of the cellular states in complex tissues. New multi-omics systems capable of combining genetic, epigenetic, and metabolic data at the single-cell scale are likely to help uncover more about how new developments in stem cell differentiation, tissue repair, and disease progression proceed. Additionally, more comprehensive mapping of cellular interactions in their natural microenvironments will be made possible by the combination of spatially resolved single-cell methods, providing a more holistic view of regeneration. Machine learning and artificial intelligence will also advance further in interpreting data and predicting cell fate decisions and new therapeutic targets. These advances could eventually result in customized regenerative therapies, in which patient-specific cellular profiles are used to design biomaterials, cell therapies, and drug therapies in the long term. With technical issues like scalability, cost, and standardization resolved, single-cell analysis will gradually shift out of research laboratories, into clinical practice, as it will eventually make RM a more specific and customized field.

References

Ahmad R, Budnik B (2023) A review of the current state of single-cell proteomics and future perspective. Anal Bioanal Chem 415(28):6889–6899

Ali A et al (2022) Single cell metabolism: current and future trends. Metabolomics 18(10):77

Almet AA et al (2025) Integrated single-cell analysis reveals spatially and temporally dynamic heterogeneity in fibroblast states during wound healing. J Invest Dermatol 145(3):645–659.e25

Argelaguet R et al (2021) Computational principles and challenges in single-cell data integration. Nat Biotechnol 39(10):1202–1215

Baek S et al (2022) Single-cell transcriptome analysis reveals three sequential phases of gene expression during zebrafish sensory hair cell regeneration. Dev Cell 57(6):799–819. e6

Baslan T, Hicks J (2017) Unravelling biology and shifting paradigms in cancer with single-cell sequencing. Nat Rev Cancer 17(9):557–569

Ben-Moshe S et al (2022) The spatiotemporal program of zonal liver regeneration following acute injury. Cell Stem Cell 29(6):973–989.e10

Budnik B et al (2018) SCoPE-MS: mass spectrometry of single mammalian cells quantifies proteome heterogeneity during cell differentiation. Genome Biol 19(1):161
Budnik B et al (2022) In-depth analysis of proteomic and genomic fluctuations during the time course of human embryonic stem cells directed differentiation into beta cells. Proteomics 22(9):2100265
Cadwell CR et al (2016) Electrophysiological, transcriptomic and morphologic profiling of single neurons using Patch-seq. Nat Biotechnol 34(2):199–203
Cairns JL et al (2025) Mass-guided single-cell MALDI imaging of low-mass metabolites reveals cellular activation markers. Adv Sci 12(5):2410506
Cao J et al (2018) Joint profiling of chromatin accessibility and gene expression in thousands of single cells. Science 361(6409):1380–1385
Carter NP et al (1992) Degenerate oligonucleotide-primed PCR: general amplification of target DNA by a single degenerate primer. Genomics 13(3):718–725
Cavone L et al (2021) A unique macrophage subpopulation signals directly to progenitor cells to promote regenerative neurogenesis in the zebrafish spinal cord. Dev Cell 56(11):1617–1630. e6
Chapman AR et al (2015) Single cell transcriptome amplification with MALBAC. PLoS One 10(3):e0120889
Chen S, Lake BB, Zhang K (2019) High-throughput sequencing of the transcriptome and chromatin accessibility in the same cell. Nat Biotechnol 37(12):1452–1457
Chung H et al (2021) Joint single-cell measurements of nuclear proteins and RNA in vivo. Nat Methods 18(10):1204–1212
Cigliola V et al (2023) Spinal cord repair is modulated by the neurogenic factor Hb-egf under direction of a regeneration-associated enhancer. Nat Commun 14(1):4857
Cusanovich DA et al (2015) Multiplex single-cell profiling of chromatin accessibility by combinatorial cellular indexing. Science 348(6237):910–914
Dell'Orso S et al (2019) Single cell analysis of adult mouse skeletal muscle stem cells in homeostatic and regenerative conditions. Development 146(12):dev174177
Fan X et al (2015) Single-cell RNA-seq transcriptome analysis of linear and circular RNAs in mouse preimplantation embryos. Genome Biol 16(1):148
Farlik M et al (2015) Single-cell DNA methylome sequencing and bioinformatic inference of epigenomic cell-state dynamics. Cell Rep 10(8):1386–1397
Feng D et al (2022) High-throughput single cell metabolomics and cellular heterogeneity exploration by inertial microfluidics coupled with pulsed electric field-induced electrospray ionization-high resolution mass spectrometry. Anal Chim Acta 1221:340116
Fu Y et al (2015) Uniform and accurate single-cell sequencing based on emulsion whole-genome amplification. Proc Natl Acad Sci 112(38):11923–11928
Gerber T et al (2018) Single-cell analysis uncovers convergence of cell identities during axolotl limb regeneration. Science 362(6413):eaaq0681
Gkountela S et al (2015) DNA demethylation dynamics in the human prenatal germline. Cell 161(6):1425–1436
Gladka MM et al (2018) Single-cell sequencing of the healthy and diseased heart reveals cytoskeleton-associated protein 4 as a new modulator of fibroblasts activation. Circulation 138(2):166–180
Granja JM et al (2021) ArchR is a scalable software package for integrative single-cell chromatin accessibility analysis. Nat Genet 53(3):403–411
Gu H et al (2011) Preparation of reduced representation bisulfite sequencing libraries for genome-scale DNA methylation profiling. Nat Protoc 6(4):468–481
Guerrero-Juarez CF et al (2019) Single-cell analysis reveals fibroblast heterogeneity and myeloid-derived adipocyte progenitors in murine skin wounds. Nat Commun 10(1):650
Guldberg SM et al (2023) Computational methods for single-cell proteomics. Annu Rev Biomed Data Sci 6(1):47–71
Guo H et al (2014) The DNA methylation landscape of human early embryos. Nature 511(7511):606–610

Hao Y et al (2021) Integrated analysis of multimodal single-cell data. Cell 184(13):3573–3587. e29
Hashimshony T et al (2012) CEL-Seq: single-cell RNA-Seq by multiplexed linear amplification. Cell Rep 2(3):666–673
Honkoop H et al (2019) Single-cell analysis uncovers that metabolic reprogramming by ErbB2 signaling is essential for cardiomyocyte proliferation in the regenerating heart. elife 8:e50163
Hou Y et al (2013) Genome analyses of single human oocytes. Cell 155(7):1492–1506
Jaitin DA et al (2014) Massively parallel single-cell RNA-seq for marker-free decomposition of tissues into cell types. Science 343(6172):776–779
Jindal K et al (2024) Single-cell lineage capture across genomic modalities with CellTag-multi reveals fate-specific gene regulatory changes. Nat Biotechnol 42(6):946–959
Kiselev VY, Andrews TS, Hemberg M (2019) Challenges in unsupervised clustering of single-cell RNA-seq data. Nat Rev Genet 20(5):273–282
Klein AM et al (2015) Droplet barcoding for single-cell transcriptomics applied to embryonic stem cells. Cell 161(5):1187–1201
Kulkarni A et al (2019) Beyond bulk: a review of single cell transcriptomics methodologies and applications. Curr Opin Biotechnol 58:129–136
Lähnemann D et al (2020) Eleven grand challenges in single-cell data science. Genome Biol 21(1):31
Lakkis J et al (2022) A multi-use deep learning method for CITE-seq and single-cell RNA-seq data integration with cell surface protein prediction and imputation. Nat Mach Intell 4(11):940–952
Lee J, Hyeon DY, Hwang D (2020) Single-cell multiomics: technologies and data analysis methods. Exp Mol Med 52(9):1428–1442
Lenkowski JR, Raymond PA (2014) Müller glia: stem cells for generation and regeneration of retinal neurons in teleost fish. Prog Retin Eye Res 40:94–123
Leung ML et al (2015) SNES: single nucleus exome sequencing. Genome Biol 16(1):55
Lu S et al (2012) Probing meiotic recombination and aneuploidy of single sperm cells by whole-genome sequencing. Science 338(6114):1627–1630
Luecken MD, Theis FJ (2019) Current best practices in single-cell RNA-seq analysis: a tutorial. Mol Syst Biol 15(6):e8746
Lust K et al (2022) Single-cell analyses of axolotl telencephalon organization, neurogenesis, and regeneration. Science 377(6610):eabp9262
Lyu P et al (2023) Common and divergent gene regulatory networks control injury-induced and developmental neurogenesis in zebrafish retina. Nat Commun 14(1):8477
Macosko EZ et al (2015) Highly parallel genome-wide expression profiling of individual cells using nanoliter droplets. Cell 161(5):1202–1214
Mao AS, Mooney DJ (2015) Regenerative medicine: current therapies and future directions. Proc Natl Acad Sci 112(47):14452–14459
Mincarelli L et al (2018) Defining cell identity with single-cell omics. Proteomics 18(18):1700312
Miura F et al (2012) Amplification-free whole-genome bisulfite sequencing by post-bisulfite adaptor tagging. Nucleic Acids Res 40(17):e136–e136
Munsky B, Neuert G, Van Oudenaarden A (2012) Using gene expression noise to understand gene regulation. Science 336(6078):183–187
Palazzo I et al (2022) NFkB-signaling promotes glial reactivity and suppresses Müller glia-mediated neuron regeneration in the mammalian retina. Glia 70(7):1380–1401
Papalexi E, Satija R (2018) Single-cell RNA sequencing to explore immune cell heterogeneity. Nat Rev Immunol 18(1):35–45
Patel AP et al (2014) Single-cell RNA-seq highlights intratumoral heterogeneity in primary glioblastoma. Science 344(6190):1396–1401
Pelzer D et al (2021) Foxm1 regulates neural progenitor fate during spinal cord regeneration. EMBO Rep 22(9):e50932
Peterson VM et al (2017) Multiplexed quantification of proteins and transcripts in single cells. Nat Biotechnol 35(10):936–939

Qin T et al (2021) Single-cell RNA-seq reveals novel mitochondria-related musculoskeletal cell populations during adult axolotl limb regeneration process. Cell Death Differ 28(3):1110–1125
Sahu N, Grandi FC, Bhutani N (2022) A single-cell mass cytometry platform to map the effects of preclinical drugs on cartilage homeostasis. JCI Insight 7(20):e160702
Saraswathy VM, Zhou L, Mokalled MH (2024) Single-cell analysis of innate spinal cord regeneration identifies intersecting modes of neuronal repair. Nat Commun 15(1):6808
Sindelka R et al (2024) Characterization of regeneration initiating cells during Xenopus laevis tail regeneration. Genome Biol 25(1):251
Smith ZD et al (2014) DNA methylation dynamics of the human preimplantation embryo. Nature 511(7511):611–615
South SM et al (2023) Imaging mass cytometry reveals tissue-specific cellular immune phenotypes in the mouse knee following ACL injury. Osteoarthr Cartil Open 5(4):100416
Specht H et al (2021) Single-cell proteomic and transcriptomic analysis of macrophage heterogeneity using SCoPE2. Genome Biol 22(1):50
Stoeckius M et al (2017) Simultaneous epitope and transcriptome measurement in single cells. Nat Methods 14(9):865–868
Streets AM et al (2014) Microfluidic single-cell whole-transcriptome sequencing. Proc Natl Acad Sci 111(19):7048–7053
Stuart T et al (2023) Nanobody-tethered transposition enables multifactorial chromatin profiling at single-cell resolution. Nat Biotechnol 41(6):806–812
Taylor MJ, Lukowski JK, Anderton CR (2021) Spatially resolved mass spectrometry at the single cell: recent innovations in proteomics and metabolomics. J Am Soc Mass Spectrom 32(4):872–894
Tellez-Gabriel M et al (2016) Tumour heterogeneity: the key advantages of single-cell analysis. Int J Mol Sci 17(12):2142
Tian H et al (2024) Multimodal mass spectrometry imaging identifies cell-type-specific metabolic and lipidomic variation in the mammalian liver. Dev Cell 59(7):869–881.e6
Trapnell C (2015) Defining cell types and states with single-cell genomics. Genome Res 25(10):1491–1498
Wang J et al (2012) Genome-wide single-cell analysis of recombination activity and de novo mutation rates in human sperm. Cell 150(2):402–412
Wang Y et al (2014) Clonal evolution in breast cancer revealed by single nucleus genome sequencing. Nature 512(7513):155–160
Wang Z et al (2020) Cell-type-specific gene regulatory networks underlying murine neonatal heart regeneration at single-cell resolution. Cell Rep 33(10):108472
Wei X et al (2022) Single-cell Stereo-seq reveals induced progenitor cells involved in axolotl brain regeneration. Science 377(6610):eabp9444
Wen L, Tang F (2016) Single-cell sequencing in stem cell biology. Genome Biol 17(1):71
Xu T et al (2024) Concentric hybrid nanoelectrospray ionization-atmospheric pressure chemical ionization source for high-coverage mass spectrometry analysis of single-cell metabolomics. Adv Sci 11(16):2306659
Zenobi R (2013) Single-cell metabolomics: analytical and biological perspectives. Science 342(6163):1243259
Zhang X et al (2019) Comparative analysis of droplet-based ultra-high-throughput single-cell RNA-seq systems. Mol Cell 73(1):130–142. e5
Zhang Y et al (2025) Development of a single-cell spatial metabolomics method for the characterization of cell-cell metabolic interactions. Anal Chem 97(14):7986–7994
Zheng GX et al (2017) Massively parallel digital transcriptional profiling of single cells. Nat Commun 8(1):14049
Zong C et al (2012) Genome-wide detection of single-nucleotide and copy-number variations of a single human cell. Science 338(6114):1622–1626

Single-Cell Analysis in Studying Animal Models of Metabolic Disorders

15

Anh-Dao Ngo, Yen-Nhi Khuong, Tien-Dat Pham, and Dinh-Toi Chu

Abstract

Metabolic disorders have a significant impact on human health and quality of life. They not only cause organ dysfunction and inconvenience in daily activities; children whose mothers have metabolic disorders during pregnancy or breast-feeding are also at risk of greater health problems. Historically, many animal experiments have been conducted to understand the cause of diseases and factors related to metabolic disorders in animals in general and humans in particular. To further study physiological processes, single-cell analysis methods were developed to analyze the processes occurring in cells in depth, such as gene expression regulation, transcription, translation, and metabolic pathways. Based on single-cell researches on animal models, scientists have clarified the role of many types of cells and biological processes in the formation and development of metabolic disorders. In this chapter, animal models of metabolic disorders, biomedical applications, and single-cell analysis in animal models of metabolic disorders will be discussed. Despite many challenges, single-cell analysis has emerged as a useful research method, providing reliable evidence in detecting disease markers, developing accurate diagnostic and therapeutic methods.

Keywords

Animal models · Metabolic disorders · Molecular biomedical technique · Single-cell analysis · Single-cell analysis applications

A.-D. Ngo · D.-T. Chu (✉)
The Interdisciplinary Research Group on Biomedicine and Health, International School, Vietnam National University, Hanoi, Vietnam

Faculty of Applied Sciences, International School, Vietnam National University, Hanoi, Vietnam
e-mail: toicd@vnu.edu.vn

Y.-N. Khuong · T.-D. Pham
The Interdisciplinary Research Group on Biomedicine and Health, International School, Vietnam National University, Hanoi, Vietnam

V. Singh (ed.), *Advances in Single Cell Analysis*,
https://doi.org/10.1007/978-981-95-9789-5_15

15.1 Introduction

Metabolic disorders have emerged as a serious health problem. This is a condition in which the body has an abnormal metabolism of nutrients, such as carbohydrates, lipids, and proteins, leading to an excess or deficiency of these substances during life and development. According to a study on the impact of metabolic syndrome in the world, the incidence of diseases such as diabetes and hypertension has significantly increased in the past two decades (Chew et al. 2023). According to the WHO, metabolic disorders are one of the major causes of non-communicable diseases and contribute to tens of millions of deaths worldwide (Noncommunicable Diseases, R.a.D.N. Noncommunicable diseases 2025). The World Health Organization also highlighted the risk of increasing metabolic abnormalities in many countries, which has a significant impact on the sustainable development of the region and the entire human population. Therefore, heads of state and health organizations worldwide need to collaborate to develop plans to control and prevent risk factors that lead to severe health conditions. Despite the decline in mortality, the number of people with metabolic diseases and their long-term effects on the quality of patients' lives are still increasing, especially in developed countries with a high socio-demographic index. In addition, metabolic disorders also contribute directly or indirectly to the health of women of childbearing age and their children (Hoffman et al. 2021). There have been numerous studies demonstrating the relationship between the mother's diet, health status, and habits before and during pregnancy, which affect the nutritional status of the child born. In recent times, the proportion of women of reproductive age suffering from the effects of hypertension, diabetes, or non-alcoholic fatty liver disease tends to increase and become younger (Chen et al. 2025). Reports of the negative impact of these metabolic diseases have raised major and urgent issues for health systems to ensure the healthy development of current populations and future generations.

To detect and diagnose metabolic disorders, the methods commonly used today are to evaluate clinical and paraclinical characteristics based on tests. For example, to determine if a patient has diabetes, the patient needs to do a blood sugar test, or to diagnose obesity, it is necessary to determine the waist/hip ratio combined with the body mass index (BMI) and the distribution of fat in the body (Bergman et al. 2020; Gazarova et al. 2019). With rare, genetic metabolic diseases, diagnosis is even more difficult (Ijaz et al. 2025). These diseases often start very early, or their signs are often unclear or easily confused with other diseases, making treatment difficult. The major difficulty in diagnosing metabolic disorders today is the specificity; the diagnostic methods are often invasive tests that cause pain to the patient, and they cannot monitor the patient's health regularly. It is not only difficult to detect individuals at risk of metabolic disorders early, but accurately and effectively treating metabolic disorders also remains a major challenge. Furthermore, commonly recommended treatments for metabolic disorders, such as dietary modifications, exercise, and medications, often only address the tip of the iceberg (Stefan et al. 2025). They focus on reducing symptoms and maintaining a stable condition, but cannot treat the root cause of the disease. Moreover, these therapies are often long-term, expensive, and can cause frustration and inconvenience for patients with metabolic disorders.

To provide a clearer view of the pathogenesis of metabolic diseases, as well as influencing factors and potential treatments, animal trials have been conducted and they have brought great benefits to science in general and medicine in particular. Animal experiments are conducted on animals that have similar characteristics to humans, and experimental conditions can also be adjusted to explore the relationship between external factors and different diseases. Rodents, typically mouse strains, are the most commonly chosen subjects when performing experiments on metabolic disorders (Kleinert et al. 2018; Zhang et al. 2024). Mouse strains such as the Zucker rat, Wistar rat, and C57BL/6 mouse are commonly used research subjects in the study of obesity, diabetes, and other metabolic diseases (Pandey and Dvorakova 2020; Lanzer et al. 2024). Animal models also contribute to elucidating the influence of negative neuropsychiatric states, such as stress and anxiety, on metabolic disorders through the assessment of metabolic profiles and monitoring of immune cells (Fan et al. 2019). These mouse models may be raised under different feeding conditions, tested with different drug treatments, or genetically and immunologically modified to provide meaningful information about the role of genes, diet in metabolic disease, or to indicate the effectiveness of the drug being tested (Fu et al. 2024). Therefore, animal testing in general or on mice in particular may contribute to narrowing the gap from theoretical research to clinical trials, reducing the cost of drug development or treatments, and enhancing knowledge about the pathogenesis of metabolic diseases.

15.2 Animal Models for Metabolic Disorders

Experimental animal models for studying metabolic disorders have been used since very early times (Fig. 15.1). In 1889, Oskar Minkowski and Josef von Mehring demonstrated the decisive role of the pancreas by removing the entire pancreas in dogs and observed that they developed diabetes (Chew et al. 2023). By the mid-twentieth century, in addition to spontaneous models such as the nonobese diabetic (NOD) mouse and Bio-Breeding (BB) (discovered in 1974) for studying autoimmune diabetes, chemical models with alloxan (since the 1940s) and streptozotocin (STZ, 1960s) were also developed, destroying pancreatic β cells in mice, rabbits, monkeys, and many other species (Noncommunicable Diseases, R.a.D.N. Noncommunicable diseases 2025; Hoffman et al. 2021; Chen et al. 2025). They produce a state of marked hyperglycemia, which is favorable for drug testing, but still has limitations such as extra-pancreatic toxicity and the ability to recover naturally in animals (Bergman et al. 2020). During the 1950s–1960s, genetic models in mice and rats have appeared, typically the ob/ob (leptin-deficient), db/db (leptin receptor-deficient), and Zucker diabetic fatty (ZDF), all of which exhibit obesity, insulin resistance, and type 2 diabetes (Bergman et al. 2020; Gazarova et al. 2019). Subsequently, many other strains such as Goto-Kakizaki (GK) and genetically modified models (knockout, transgenic) from the 1980s–1990s continued to

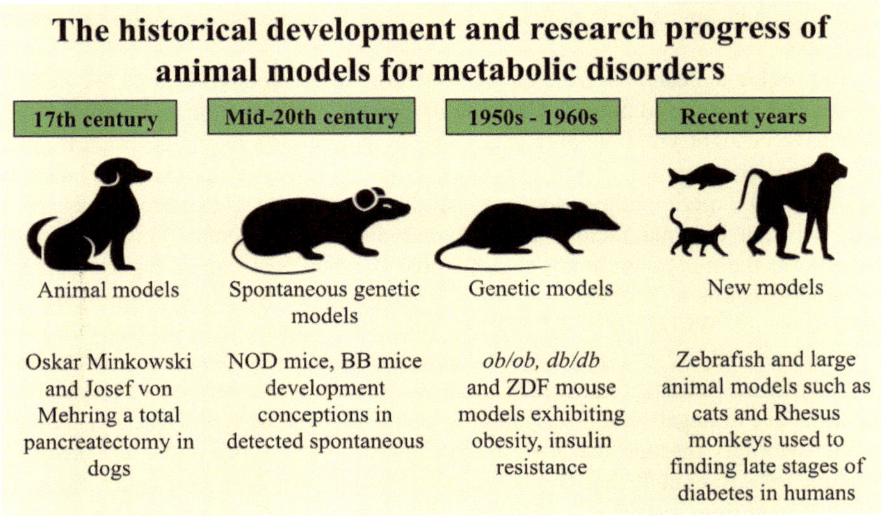

Fig. 15.1 The historical development and research progress of animal models for metabolic disorders

be developed, targeting important genes such as insulin receptor, leptin, leptinR, and Peroxisome Proliferator-Activated Receptor Gamma (PPARγ) to investigate the pathogenesis. In recent years, zebrafish have been increasingly applied for their lipid metabolism, pancreatic structure, and blood glucose regulation mechanism, which have many similarities with humans, helping to effectively simulate obesity and diabetes (Ijaz et al. 2025). In addition, large animal models such as cats and Rhesus monkeys have been used to reproduce the late stages of the disease. In cats, the disease is accompanied by amylin deposition, a 50% reduction in β-cell mass, and neurovascular complications. Meanwhile, ad libitum fed Rhesus monkeys often develop obesity, hyperinsulinemia, progression to diabetes, and pancreatic dysfunction similar to human patients (Bergman et al. 2020). Because of this similarity, they are considered ideal models for complex pathological studies and preclinical trials.

Overall, the diversity of models, from rodents, zebrafish to large animals such as cats and monkeys, has provided an important foundation for exploiting the strengths of each species in metabolic pathology research. Animal models allow for the simulation of many human diseases, such as diabetes, atherosclerosis, and dyslipidemia, and facilitate the evaluation of drug efficacy and safety before clinical trials (Gazarova et al. 2019). For example, diabetic mouse or monkey models have helped to explain the mechanisms of insulin resistance, neurovascular complications and have been used to screen glucose-lowering drugs. According to Pandey et al., the application of animal models not only helps to better understand the disease mechanism but also paves the way for new treatment strategies (Chew et al. 2023). However, it is important to recognize that each model has certain advantages and disadvantages (Table 15.1). What they have in common is that they allow for tight

Table 15.1 Summary of animal models for metabolic disorder research

Type of model	Representative Models	Advantages	Limitations	Applications	Reference
Genetic models in small animals (mice, rats)	ob/ob, db/db, Zucker, Goto-Kakizaki, Koletsky, *Drosophila*	Fast reproduction, low cost. Easy to control, precise gene intervention. Development of obesity, insulin resistance, diabetes when mutated or eating a lot of fat/sugar	Different physiology: high heart rate, fast basal metabolism. Different fat distribution & homeostasis. Many rare mutations (leptin deficiency) in humans	Pathogenesis research. Anti-obesity, insulin resistance, and diabetes drug screening	Wong et al. (2016), Suleiman et al. (2020), Vatashchuk et al. (2022)
Chemical model (β-cell toxicity)	Alloxan, Streptozotocin (STZ)	Simple, low cost. Rapidly increases blood glucose. Good simulation of T1DM in mice & rabbits	Pathogenesis differs from humans (direct toxicity rather than autoimmunity). Some species can recover (e.g., Guinea mice). Induces extra-pancreatic toxicity (liver, kidney, other tissues)	Research on insulin-dependent diabetes mellitus (T1DM). Screening for β-cell protective drugs	Pandey et al. (2023), Dhuria et al. (2015)
Zebrafish model	Zebrafish	Rapid reproduction, low cost. Transparent body, easy observation. Easily genetically modified (CRISPR). The pancreas and liver function similarly to mammals	Physiological differences compared to humans (insulin, leptin, hormones, neurotransmitters). Simplified cardiovascular and endocrine systems. Short lifespan, limiting long-term modeling	Research on obesity, type 2 diabetes mellitus (T2DM), and dyslipidemia. Rapid, low-cost drug screening	Angom and Nakka (2024)
Large animal models	Pig, dog, Rhesus monkey	Anatomical, physiological, and pharmacological features closely resemble humans. Allow application of clinical techniques (biopsy, imaging, laboratory tests)	High maintenance costs. Long growth period. Requirement for specialized housing and equipment. Strict ethical constraints	Research on diabetes mellitus metabolism and complications. Preclinical testing prior to human application	Suleiman et al. (2020), Renner et al. (2020)

control of environmental and genetic conditions, easy tissue sampling, and monitoring of disease progression, but still they cannot fully reflect the diversity and complexity of metabolic diseases in humans (Bergman et al. 2020).

In summary, although no model is absolutely perfect, the continuous development of animal models has played a key role in studying pathogenesis, testing drugs, and evaluating treatment strategies for metabolic disorders before clinical application. In the future, the development of "humanized" or multispecies models is expected to overcome current limitations and enhance the ability to translate results to patients.

15.3 Application of Molecular Biomedical Techniques for Studying Animal Models of Metabolic Disorders

Animal models play an important role in expanding knowledge of the etiology, pathophysiology, and treatment of metabolic disorders (Chew et al. 2023). They offer controlled systems that mimic the state of diseases in humans, enabling researchers to carry out experiments that cannot be applied directly to patients (Noncommunicable Diseases, R.a.D.N. Noncommunicable diseases 2025). In order to achieve these goals, a wide range of molecular biomedical techniques were utilized. To prepare materials for some experiments, researchers apply genetic modifications to induce metabolic syndromes (MetS) in animal models, which can shorten the duration needed to develop the syndromes in comparison with diet-induced MetS (Noncommunicable Diseases, R.a.D.N. Noncommunicable diseases 2025). A study in 2024 used Zinc-finger nuclease technology to delete 16 base pairs on exon 11 of the gene encoding the leptin receptor on chromosome 5 (Hoffman et al. 2021). This frameshift mutation creates SSLepR mutant strain, and rats with a mutation in the leptin receptor exhibit overeating and obesity (Chen et al. 2025). In addition, some molecular biomedical techniques are employed to assess how effective treatments are against metabolic disorders. For instance, polymerase chain reaction (PCR), western blot analysis, and chromatin immunoprecipitation (ChIP) were chosen to evaluate the impact of exendin-4 on improving type 2 diabetes (T2D) (Bergman et al. 2020). The researchers used PCR to analyze the expression levels of genes related to pancreatic β-cell development, and they found that exendin-4 was able to reverse the changes caused by T2D. For example, while the expression levels of Pax4, Ins1, Bcl2, and Pdx1 all reduced in the T2D group, they were shown to increase in the group treated with exendin-4. Besides, a western blot was performed to uncover histone modifications. Under the impact of T2D, the levels of H3K9 di-methylation or H3K9 and H3K23 acetylation were increased, while H3K4 methylation was decreased. However, the opposite results were seen by exendin-4 administration. Lastly, ChIP analyzed histone modifications related to Pdx1 and revealed that the modifications of H3K4 methylation, H3K9 di-methylation, and H3/H3K9 acetylation in Exendin-4-treated mice were opposite to those seen in T2D mice. In short, the results support that Exendin-4 may ameliorate T2D. Other applications of molecular biomedical techniques are given in Table 15.2.

Table 15.2 Application of molecular biomedical techniques in studying animal models of metabolic disorders

Animal	Condition	Technology	Function	Reference(s)
Wild-type and *Crp* knockout (KO) rat	Obesity and insulin resistance	Transcription activator-like effector nucleases (TALENs)	Delete *C-reactive protein (CRP)* gene in *Crp* KO rats since this protein is closely related to the development of diabetes and obesity	Yang et al. (2021)
Lean Dahl salt-sensitive rat and obese SS leptin receptor mutant rat	Obesity	Zinc-finger nuclease	Cause a frame shift mutation in leptin receptor gene	Alhashim et al. (2024)
LEPR knockout rabbits	Obesity	CRISPR/Cas9	Induce *Lepr* mutation by deleting region surrounding *LEPR* exon 10	Silaeva et al. (2024)
C57BL/6J	Type 2 diabetes	Polymerase chain reaction, western blot analysis, and chromatin immunoprecipitation	PCR analyzes the expression of genes associated with pancreatic β-cell development. Western blot analyzes pancreatic histone extracts and discovers histone modifications. ChIP analyzes histone modifications related to Pdx1	Tu et al. (2022)
Wistar Albino rat	Obesity	Enzyme-linked immunosorbent assay (ELISA)	Examine inflammation parameters and NES-1 expression. Measure blood insulin level	Özüdoğru et al. (2023)

(continued)

Table 15.2 (continued)

Animal	Condition	Technology	Function	Reference(s)
C2cd4a and C2cd4b mouse strains	Type 2 diabetes	Fluorescent in situ hybridization (FISH)	Reveal that c2cd4a transcript is present in $gcgb^+$, ins^+ and $sst2^+$ cells	Gharavy et al. (2021)
C57BL/6J mouse and *Hap1* knockout mice	Type 2 diabetes mellitus	Quantitative real-time PCR (RT-qPCR), Western blot, and co-immunoprecipitation (Co-IP) assay	RT-qPCR analysis uncovers that high fat diet-fed group has lower mRNA expression of *Hap1,* compared to normal diet-fed group. Western blot detects HAP1 and GLUT4. Co-IP assay identifies the interactions among proteins: GLUT4, HAP1, and sortilin	Gong et al. (2020)
Wild-type rat and hemizygous human islet amyloid polypeptide transgenic rats	Type 2 diabetes	Fluorescence Lifetime Imaging Microscopy	Monitor glycolysis and oxidative phosphorylation of living islet cells prior to and following glucose stimulation. Provide an approach to study methods to restore the function of α and β cells in diabetes	Wang et al. (2021)
db/*db* mice and control (*db*+) mice	Diabetes	^{1}H magnetic resonance spectroscopy	Provide a noninvasive approach to characterize and monitor metabolic changes of the brain during type 2 diabetes progression	Choi et al. (2024)

Molecular biomedical techniques have greatly enhanced the study of animal models of metabolic disorders, with some undeniable advantages. First, the techniques provide a wide range of tools for researchers to utilize in their investigations, depending on their purposes, including editing genome, exploring gene expression changes, and monitoring cellular and metabolic processes in real time. Second, with the growing prominence of these technologies, efforts are directed toward further development to enhance their performance and quality. Furthermore, collaboration among techniques provides a multi-dimensional picture of disease. However, there are some disadvantages while using molecular biomedical techniques. Firstly, the application of these technologies demands skilled operators and a significant financial investment. Additionally, certain techniques require further exploration since they are relatively new, such as single-cell transcriptomics (Gazarova et al. 2019). Besides that, it is not that all protocols or experimental conditions are universally suitable. Researchers are often required to perform preliminary trials and adjustments to identify the most effective approach for their specific study, ensuring the reliability and relevance of the results obtained. According to these benefits and drawbacks of the molecular biomedical techniques, more and more studies are demanded to strengthen the positives and improve the limitations.

15.4 The Application of Single-Cell Analysis in Studying Animal Models of Metabolic Disorders

Single-cell analysis offers detailed information about individual cells, including cells' molecular contents, function, and heterogeneity (Fu et al. 2024; Wong et al. 2016). Single-cell analysis tools are divided into three groups: genomics, proteomics, and transcriptomics (Suleiman et al. 2020). However, single-cell RNA-sequencing (scRNA-seq) belonging to single-cell transcriptomics is one of the most prevalent technologies to study animal models of metabolic diseases (Table 15.3). In a 2024 study, single-cell transcriptomics was used as the central technology to elucidate the impact of bariatric surgery on the heterogeneity and transdifferentiation of Zucker diabetic rats' pancreatic endocrine cells (Vatashchuk et al. 2022). After analyzing 21,406 cells based on their gene expression variation, the information was illustrated by a t-distributed stochastic neighbor embedding (t-SNE) map including 4 endocrine cell types (α, β, γ, and δ cells). The analysis shows that after surgery, there was an increase in β-cell proportion and a decrease in the ratio of α cells. Moreover, some β-cell-specific transcription factors, such as MAFA, Nkx6.1, and Pdx-1, were found and their function is promoting transdifferentiation of α and δ cells into β cells. Furthermore, scRNA-seq reveals that β cells' heterogeneity is more obvious under the impact of type 2 diabetes mellitus or surgery. Overall, scRNA-seq provides evidence for the influence of bariatric surgery on diabetes treatment. A study in 2021 utilized scRNA-seq to describe vascular remodeling, especially alterations in cell-cell interactions and artery-specific changes during hypertension (Pandey et al. 2023). ScRNA-seq analyzed 21,883 individual cells and grouped them into 4 vascular cell types: smooth muscle cells (SMCs), endothelial

Table 15.3 Application of single-cell analysis in studying animal models of metabolic disorders

Animal	Condition	Technology	Main findings	Reference(s)
Ccr2$^{-/-}$, *Nur77*$^{-/-}$, *R26R-EYFP*, *Cx3cr1*^{CreERT2}, and *Csf1r*^{CreEsr1}, *Plvap*^{tm1Salm}, wild type, C57BL/6J, and C57BL/6N	Obesity	Single-cell mass cytometry	Provide a strategy for white adipose tissue macrophage populations' immune profiling. Reveal the complexity in dividing white adipose tissue subclasses in lean white adipose tissue	Félix et al. (2021)
Zucker diabetic fatty rat	Obesity and diabetes	Single-cell RNA-sequencing	Analyze 21,406 cells based on their gene expression to categorize them into 4 types of cells: α, β, δ, and γ cells with their transcription factors (TFs). Discover novel cell-specific TFs, such as CCND1, HES1, EPAS1, CREM, … Elucidate the increase in the proportion of β cells and decrease in α cells' ratio under the impact of bariatric surgery	Liang et al. (2024)
Spontaneously hypertensive rat (SHR) and Wistar Kyoto rat (WKY)	Hypertension	Single-cell RNA-sequencing	Depict the percentage of major vascular cell types in aortic artery (AA) and mesenteric artery (MA) of SHRs and WKYs to describe vascular remodeling during hypertension at the transcriptomic level. Uncover the heterogeneity of smooth muscle cells and immune cells as the sub-cluster distribution changes were shown in some datasets	Cheng et al. (2021)

(continued)

Table 15.3 (continued)

Animal	Condition	Technology	Main findings	Reference(s)
C57BL/6J mouse	Hypertension	Single-cell RNA-sequencing	Analyze 16,972 cells at the transcriptional level without needing pre-selection. Show the changes in the proportion of smooth muscle cell and endothelial cell subpopulations in order to confirm their contributions to vascular remodeling during hypertension. Uncover that TWEAK protein expression upregulated, which indicates the changes in TWEAK signaling pathways in hypertension	Zhang et al. (2021)
C57BL/6N mouse	Obesity	Single-cell RNA-sequencing	Analyze 375,000 endothelial cells to show the changes in the endothelial cell (EC) population in 6 organs under the affect of obesity. Discover the reduction of EC-pneumocyte population in the lungs, which relates to the innate defense against pathogens	Bondareva et al. (2022)
MCAO mouse and sham mouse	Ischemic stroke and metabolic syndrome	Single-cell RNA-sequencing	Indicate that the level of microglia glycolysis was higher in MCAO mouse than in Sham mouse, while microglia with high glycolysis has higher connection ability to other cells than microglia with lower glycolysis	Li et al. (2024)

(continued)

Table 15.3 (continued)

Animal	Condition	Technology	Main findings	Reference(s)
C57BL/6J mouse	Prediabetes and type 2 diabetes	Single-cell RNA-sequencing	Analyze 364 cells with 1014 ± 942 genes per each cell and reveal the similar cell composition between mouse fed with standard diet (SD) and high-fat diet (HFD). This means that HFD does not affect cell composition. Reveal the impact of HFD on increasing peripheral nervous system macrophages and decreasing mySC	Eid et al. (2023)
Wistar rat	Diabetes	Single-cell RNA-sequencing	Analyze 76,383 cells and reveal the reduction in the number of endocrine cells in islets and increase in hormone secretion under the impact of high fat diet.	Zhao et al. (2022)
NSG mouse	Diabetes	Single-cell RNA-sequencing	Compare transplanted stem cell (SC)-islets to primary cadaveric islets and stage 6 SC-islets to uncover cellular gene expression changes, identity, and maturation. Elucidate that transplantation elevates expression of many β-cell identity and maturation genes that are missing or slow in expression	Augsornworawat et al. (2020)

(continued)

Table 15.3 (continued)

Animal	Condition	Technology	Main findings	Reference(s)
HFD mouse	Obesity and diabetes	Single-cell RNA-sequencing and single-cell proteomics	Analyze 33,999 cells and group them into 8 populations. Uncover that *Ace2*^{low} subpopulation downregulates expression of genes linked to α cell function and upregulates expression of genes involved with β-cell characteristics. Reveal the impact of high-fat diet on impairing function and elevating fragility of *Ace2*^{low}	Fu et al. (2023)

cells (ECs), mesenchymal stromal cells (MSCs), and immune cells. The research shows detailed analysis of vascular changes in SMCs, ECs, and MSCs under hypertension, such as inflammation, contractility, heterogeneity, and extracellular matrix remodeling. Additionally, hypertension seems to enhance some cell-cell interactions, including the TGF signaling pathway, the interaction of Col8a1, Ctgf, and Cyr61 with integrins.

In comparison with bulk RNA-sequencing (bulk RNA-seq), scRNA-seq provides distinct advantages. Firstly, while bulk RNA-seq offers a measurement for the average gene expression across all the samples, scRNA-seq allows analysis of individual cells, even in large datasets. Secondly, this tool can help to discover the genomic and transcriptomic variation among cells. As a result, scRNA-seq can elucidate cellular heterogeneity dynamics and characteristics at the molecular level (Fu et al. 2024). Moreover, scRNA-seq enables the discovery of novel cell types and subpopulations since this technology analyzes all cells without the need for preselection (Dhuria et al. 2015). Furthermore, scRNA-seq analysis data was usually expressed by a heat map, t-SNE map, and other methods, making it easier to interpret and draw the conclusions. However, several obstacles remain in using this technology. It struggles to detect low-abundance transcripts as RNA content can be lost. Besides, the size of the cells can be too big and incompatible with the platform of scRNA-seq (Angom and Nakka 2024). In addition, scRNA-seq gene-expression profiling is naturally noisier than bulk RNA-seq since amplifying small quantities of material and sparse sampling can lead to remarkable distortions. Lastly, expressing transcripts at a high or low level can result in false-negative errors or transcripts appearing artificially abundant. Both cases may cause an artificial increase in the estimated variability between individual cells (Renner et al. 2020).

15.5 Conclusion

The rate of people with metabolic disorders in the world tends to increase and become younger, making this topic one of the major problems in the health system in many countries. Current diagnostic and treatment measures for metabolic diseases have not shown high specificity and often only stop at the level of preventing disease progression rather than treating metabolic disorders at the root. Intending to understand the disease mechanism and develop methods for diagnosing or treating metabolic disorders, many scientists from the past centuries to the present have performed experimental animal models simulating diseases of interest with different influencing conditions. Animals used in biomedical experiments have different sizes and variations to mimic the characteristics of human patients, best serving the discovery of disease mechanisms, factors, or possible treatment methods in humans.

In the present era, biomedical research is increasingly delving into metabolic pathways or changes at a more detailed level, such as the cellular level. Therefore, single-cell analysis has also developed and received much attention from researchers. Single-cell analysis in animals has contributed important, in-depth knowledge about gene expression, transcription processes, and metabolic processes in the formation and progression of metabolic disorders such as obesity and diabetes. However, this technology also faces many difficulties when requiring high-tech equipment in analysis, the implementation techniques also need to be performed accurately, or obstacles in amplifying cell samples for analysis. Despite many difficulties, the topic of single-cell analysis still has a lot of potential in understanding the disease mechanisms hidden deep in each cell, detecting early signs, or developing precise treatment methods for many diseases.

Funding This work was funded by Funded by Vietnam National University, Hanoi (VNU), grant number QG.25.190.

References

Alhashim A et al (2024) Does sex matter in obesity-induced periodontal inflammation in the SS(LepR) mutant rats? Dent J (Basel) 13(1)

Angom RS, Nakka NMR (2024) Zebrafish as a model for cardiovascular and metabolic disease: the future of precision medicine. Biomedicine 12(3)

Augsornworawat P et al (2020) Single-cell transcriptome profiling reveals β cell maturation in stem cell-derived islets after transplantation. Cell Rep 32(8):108067

Bergman M et al (2020) Review of methods for detecting glycemic disorders. Diabetes Res Clin Pract 165:108233

Bondareva O et al (2022) Single-cell profiling of vascular endothelial cells reveals progressive organ-specific vulnerabilities during obesity. Nat Metab 4(11):1591–1610

Chen J et al (2025) Global burden of metabolic disorders among women of child-bearing age, 1990-2021: a population-based study1990-2021. BMC Womens Health 25(1):207

Cheng J et al (2021) Single-cell RNA sequencing reveals cell type- and artery type-specific vascular remodelling in male spontaneously hypertensive rats. Cardiovasc Res 117(4):1202–1216

Chew NWS et al (2023) The global burden of metabolic disease: data from 2000 to 2019. Cell Metab 35(3):414–428 e3

Choi IY et al (2024) Non-invasive in vivo measurements of metabolic alterations in the type 2 diabetic brain by (1)H magnetic resonance spectroscopy. J Neurochem 168(5):765–780

Dhuria RS et al (2015) Current status and patent prospective of animal models in diabetic research. Adv Biomed Res 4(1)
Eid SA et al (2023) Single-cell RNA-seq uncovers novel metabolic functions of Schwann cells beyond myelination. J Neurochem 166(2):367–388
Fan KQ et al (2019) Stress-induced metabolic disorder in peripheral CD4(+) T cells leads to anxiety-like behavior. Cell 179(4):864–879 e19
Félix I et al (2021) Single-cell proteomics reveals the defined heterogeneity of resident macrophages in white adipose tissue. Front Immunol 12:719979
Fu Q et al (2023) Single-cell RNA sequencing combined with single-cell proteomics identifies the metabolic adaptation of islet cell subpopulations to high-fat diet in mice. Diabetologia 66(4):724–740
Fu Y et al (2024) Progress in the study of animal models of metabolic dysfunction-associated Steatotic liver disease. Nutrients 16(18)
Gazarova M, Galsneiderova M, Meciarova L (2019) Obesity diagnosis and mortality risk based on a body shape index (ABSI) and other indices and anthropometric parameters in university students. Rocz Panstw Zakl Hig 70(3):267–275
Gharavy S et al (2021) Sexually dimorphic roles for the type 2 diabetes-associated C2cd4b gene in murine glucose homeostasis. Diabetologia 64
Gong YJ et al (2020) Huntingtin-associated protein 1 plays an essential role in the pathogenesis of type 2 diabetes by regulating the translocation of GLUT4 in mouse adipocytes. BMJ Open Diabetes Res Care 8(1):e001199
Hoffman DJ et al (2021) Developmental origins of metabolic diseases. Physiol Rev 101(3):739–795
Ijaz A et al (2025) Inherited metabolic disorders: presentation, clinical types, laboratory diagnosis and genetic markers. Orphanet J Rare Dis 20(1):422
Kleinert M et al (2018) Animal models of obesity and diabetes mellitus. Nat Rev Endocrinol 14(3):140–162
Lanzer JD et al (2024) Single-cell transcriptomics reveal distinctive patterns of fibroblast activation in heart failure with preserved ejection fraction. Basic Res Cardiol 119(6):1001–1028
Li J et al (2024) Integrated single cell-RNA sequencing and Mendelian randomization for ischemic stroke and metabolic syndrome. iScience 27(7):110240
Liang Y et al (2024) Bariatric surgery induces pancreatic cell transdifferentiation as indicated by single-cell transcriptomics in Zucker diabetic rats. J Diabetes 16(8):e13521
Noncommunicable Diseases, R.a.D.N. Noncommunicable diseases (2025) 25 September 2025 [cited 2025]; Available from: https://www.who.int/news-room/fact-sheets/detail/noncommunicable-diseases#:~:text=Noncommunicable%20diseases%20(NCDs)%2C%20also,deaths%20(32%20million)%20occur
Özüdoğru E et al (2023) Protective effects of swimming exercises and metformin on cardiac and aortic damage caused by a high-fat diet in obese rats with type 2 diabetes, by regulating the Bcl2/Bax signaling pathway. Turk J Med Sci 53(6):1582–1592
Pandey S, Dvorakova MC (2020) Future perspective of diabetic animal models. Endocr Metab Immune Disord Drug Targets 20(1):25–38
Pandey S, Chmelir T, Chottova Dvorakova M (2023) Animal models in diabetic research—history, presence, and future perspectives. Biomedicine 11. https://doi.org/10.3390/biomedicines11102852
Renner S et al (2020) Porcine models for studying complications and organ crosstalk in diabetes mellitus. Cell Tissue Res 380(2):341–378
Silaeva YY et al (2024) Generation of LEPR knockout rabbits with CRISPR/CAS9 system. Dokl Biol Sci 518(1):248–255
Stefan N, Yki-Jarvinen H, Neuschwander-Tetri BA (2025) Metabolic dysfunction-associated steatotic liver disease: heterogeneous pathomechanisms and effectiveness of metabolism-based treatment. Lancet Diabetes Endocrinol 13(2):134–148
Suleiman JB, Mohamed M, Bakar ABA (2020) A systematic review on different models of inducing obesity in animals: advantages and limitations. J Adv Vet Anim Res 7(1):103–114

Tu P et al (2022) Exendin-4 may improve type 2 diabetes by modulating the epigenetic modifications of pancreatic histone H3 in STZ-induced diabetic C57BL/6 J mice. J Physiol Biochem 78(1):51–59

Vatashchuk MV et al (2022) Metabolic syndrome: lessons from rodent and drosophila models. Biomed Res Int 2022:5850507

Wang Z et al (2021) Live-cell imaging of glucose-induced metabolic coupling of β and α cell metabolism in health and type 2 diabetes. Commun Biol 4(1):594

Wong SK et al (2016) Animal models of metabolic syndrome: a review. Nutr Metab (Lond) 13:65

Yang M et al (2021) Genetic ablation of C-reactive protein gene confers resistance to obesity and insulin resistance in rats. Diabetologia 64(5):1169–1183

Zhang K et al (2021) Single-cell analysis of salt-induced hypertensive mouse aortae reveals cellular heterogeneity and state changes. Exp Mol Med 53(12):1866–1876

Zhang Y et al (2024) Single-cell RNA sequencing uncovers pathological processes and crucial targets for vascular endothelial injury in diabetic hearts. Adv Sci (Weinh) 11(47):e2405543

Zhao Y et al (2022) Long-term high-fat high-fructose diet induces type 2 diabetes in rats through oxidative stress. Nutrients 14(11)

Single-Cell Analysis in Pharmacological Research for Blood Lipid Disorders

16

Ha-Le Dang, Duc-Anh Hoang, and Dinh-Toi Chu

Abstract

Dyslipidemia, a widespread and growing chronic blood lipid disorder, poses a significant hurdle in developing nations due to its complex biological variations. The standard approach—measuring blood fats and prescribing traditional drugs such as statins, ezetimibe, and fibrates—often falls short in terms of effectiveness, patient compliance, and side effects. Even newer, highly effective treatments such as *PCSK9* inhibitors are prohibitively expensive. This research aims to find better solutions by applying advanced molecular biology and "Omics" technologies, including genomics, proteomics, and metabolomics. These technologies are crucial for understanding the disorder's intricate genetic underpinnings, interactions between metabolites and protein expression, which in turn facilitates the development of more precise, targeted treatments such as monoclonal antibodies and RNA-based therapies. Specifically, single-cell analysis (SCA), particularly techniques such as single-cell RNA sequencing (scRNA-seq), is becoming an influential tool. It allows researchers to investigate the disease's heterogeneity at the individual cell level, offering unprecedented insights into its development and helping pinpoint new therapeutic targets. While SCA still contends with challenges related to its cost, the complexity of the data, and technical difficulties in isolating cells, ongoing technological progress is rapidly expanding its potential. The ultimate goal is to leverage SCA to create safer, more effective, and personalized drugs for dyslipidemia.

H.-L. Dang · D.-A. Hoang · D.-T. Chu (✉)
Faculty of Applied Sciences, International School, Vietnam National University, Hanoi, Vietnam

The Interdisciplinary Research Group on Biomedicine and Health, International School, Vietnam National University, Hanoi, Vietnam
e-mail: toicd@vnu.edu.vn

V. Singh (ed.), *Advances in Single Cell Analysis*,
https://doi.org/10.1007/978-981-95-9789-5_16

Keywords

Dyslipidemia · Single-cell analysis · Pharmacological research · PCSK9 inhibitors · RNA therapy

16.1 Introduction

Dyslipidemia, also known as blood lipid disorder, is a common and serious chronic disease characterized by abnormal blood lipid levels, including cholesterol (low-density lipoprotein cholesterol—LDL-C, high-density lipoprotein cholesterol—HDL-C) and triglycerides (Berberich and Hegele 2022). Recent studies have shown that the prevalence of dyslipidemia is increasing, especially in developing countries, driven by poor eating habits and sedentary lifestyles, which contribute to rising obesity rates (Vardell 2020). As a result, more cases of atherosclerotic cardiovascular diseases (ASCVD) associated with elevated LDL-C levels are being reported. In addition, decreased HDL-C levels combined with increased triglyceride levels are also associated with other pathologies such as non-alcoholic fatty liver disease (NAFLD), type 2 diabetes (Pirillo et al. 2021).

Despite recent advances in medicine, the diagnosis and treatment of dyslipidemia still face numerous challenges. One of the greatest challenges is the biological heterogeneity of the disease. Dyslipidemia is not a single disease but rather a group of disorders with diverse etiologies, ranging from genetic factors (familial dyslipidemia) to environmental and lifestyle factors (Wazir et al. 2023). This biological heterogeneity leads to significant inter-individual variability in response to the same treatment, a phenomenon known as therapeutic non-response heterogeneity. Current diagnostic methods primarily rely on measuring serum lipid levels. However, measuring only these indices provides just an overview of blood lipid levels, primarily serving to warn of the risk of atherosclerotic cardiovascular disease. It does not offer a deep understanding of molecular mechanisms or cellular-level changes. This lack of insight makes it challenging to select optimal or individualized treatment regimens. In addition, poor patient adherence to the treatment regimen is another significant challenge. Many patients require lifelong medication, yet due to undesirable side effects (e.g., statin-induced myalgia) or lack of understanding, adherence rates are low, which diminishes treatment efficacy and increases the risk of complications (Naderi et al. 2012).

Currently, the most common classes of medications for treating dyslipidemia are as follows:

- Cholesterol-lowering agents: For example, statins, which work by inhibiting the enzyme HMG-CoA reductase, are considered a key enzyme in the process of cholesterol synthesis in the liver (Kosmas et al. 2021). Ezetimibe: The mechanism of action of this medication class is to reduce cholesterol absorption in the small intestine. Ezetimibe is often combined with rosuvastatin to enhance its LDL-C-lowering effect (Chilbert et al. 2022). However, some side effects, such

as elevated liver enzymes, myalgia (muscle pain), and rare cases of rhabdomyolysis, have been reported with the use of this drug class. Furthermore, a significant number of patients fail to achieve their target LDL-C goals despite receiving the maximum tolerated statin dose or due to drug intolerance.
- Triglyceride-lowering agents: Groups of drugs such as fibrates (e.g., fenofibrate, gemfibrozil) are utilized to reduce triglycerides and triglyceride-rich lipoprotein cholesterol (TRL-C) by stimulating the synthesis of lipoprotein lipase (*LPL*) (Canfora and Pierno 2024). However, their efficacy in reducing cardiovascular events remains controversial and is often accompanied by side effects such as flushing (niacin) or gastrointestinal issues.
- Proprotein convertase subtilisin/kexin type 9 (PCSK9) inhibitors, such as evolocumab and alirocumab, represent a major therapeutic advancement. These monoclonal antibodies significantly reduce LDL-C by preventing the degradation of LDL receptors in the liver (Pasta et al. 2020; Zhang et al. 2018). Although this class of drugs is highly effective, their very high cost limits patient access.

The search for novel, more effective, safer, and personalized therapies for dyslipidemia is currently a critical need due to the limitations of existing drugs:

- Suboptimal efficacy: A significant number of patients fail to achieve their desired LDL-C reduction, even with intensive treatment regimens. This increases the risk of cardiovascular events as well as the economic burden.
- Drug resistance and side effects: Adverse drug reactions (ADRs), particularly from statins, have led to discontinuation of treatment in a significant number of patients. Variability in drug response and tolerance is also observed, driven by the biological diversity of the disease, a phenomenon highlighted in recent studies.
- Complex pathogenesis: The pathogenesis of dyslipidemia is more than just a lipid imbalance; it also involves inflammation, oxidative stress, and complex cellular signaling pathways. Current medications primarily target a few specific pathways, thereby neglecting the overall complexity of the disease.

To overcome these limitations, more advanced research methods are being adopted. SCA is emerging as a powerful tool, superseding the traditional study of large cell populations. This approach allows researchers to gain deeper insights into disease mechanisms at the cellular and molecular levels. Specifically, single-cell SCA enables the identification of specific cell populations involved in lipid metabolism disorders, such as hepatocytes (liver cells), vascular endothelial cells, and macrophages. This ultimately helps researchers pinpoint new therapeutic targets, facilitating the development of more precise drugs, minimizing side effects, and boosting treatment efficacy (Huang et al. 2025).

16.2 The Current Medicines and Their Limitations for Blood Lipid Disorders

16.2.1 Drugs Used in the Treatment of Blood Lipid Disorders

In recent decades, lipid-lowering therapies, especially LDL-C, have made significant progress in reducing cardiovascular risk. However, each group of drugs still has certain limitations in effectiveness and safety.

16.2.1.1 Statins

According to guidelines, statins are the first-line choice for dyslipidemia because they effectively control LDL-C and reduce the ASCVD risk (Mach et al. 2019; Pearson et al. 2021; Patel et al. 2025; O'Malley et al. 2020). The cholesterol-atherosclerosis link was identified in the late nineteenth century and reinforced by studies in the 1950s (Endo 2010; Hajar 2011). Prior to that, treatment mainly relied on dietary modification and older agents (Hajar 2011). Akira Endo isolated compactin from *Penicillium citrinum* in 1972, followed by Merck's discovery of lovastatin from *Aspergillus terreus*, which showed strong LDL-C lowering efficacy. This was the first FDA-approved statin. Since then, other statins (e.g., simvastatin, pravastatin) have been developed (Endo 2010; Hajar 2011).

By inhibiting HMG-CoA reductase, statins not only suppress cholesterol synthesis but also enhance hepatic LDL receptor expression, thereby lowering plasma LDL-C by 35–55%. They also reduce myocardial infarction risk by 25–30%, and provide additional benefits in anti-inflammatory effects and improvement of endothelial function (Endo 2010; Hajar 2011). Despite some adverse outcomes, statins remain the cornerstone of dyslipidemia to date.

16.2.1.2 Ezetimibe

The second major drug group for dyslipidemia is cholesterol absorption inhibitors, represented by ezetimibe. Its therapeutic mechanism is to inhibit cholesterol absorption at the intestinal brush border, lowering LDL-C and non-HDL-C (Kim and Wierzbicki 2020).

While statins reduce cholesterol synthesis, ezetimibe inhibits the NPC1L1 protein in the small intestine and liver, reducing cholesterol absorption from bile and dietary sources; this combination provides a synergistic effect (Pearson et al. 2021; Kastelein et al. 2008; Phan et al. 2012). In sitosterolemia caused by *ABCG5/ABCG8* mutations, statins are ineffective, and ezetimibe is an appropriate alternative strategy (Tsubakio-Yamamoto et al. 2010). Ezetimibe lowers LDL-C by 18–20% as monotherapy and provides an additional 15% reduction with statins (Morrone et al. 2012). The SEAS trial demonstrated that ezetimibe reduced cardiovascular events for mild-to-moderate aortic stenosis patients; the SHARP trial showed a 17% reduction in atherosclerotic outcomes in chronic kidney disease (CKD) (Carabello 2010; Sharp Collaborative 2010).

16.2.1.3 Bile Acid Sequestrants

Bile acid sequestrants (BAS) are the oldest and safest lipid-lowering group, with three representatives: cholestyramine (1973), colestipol (1977), and colesevelam (2000) (Ewang-Emukowhate and Wierzbicki 2013; Lewek et al. 2024). These non-absorbed drugs bind bile acids in the intestine for fecal excretion. Due to the depletion of reabsorbed bile acids, the liver increases bile acid synthesis from the cholesterol pool, thereby reducing intracellular cholesterol levels (Ewang-Emukowhate and Wierzbicki 2013; Lewek et al. 2024; Antilipemic Agents 2012). Activation of SREBP2 (Sterol Regulatory Element-Binding Protein 2), a transcription factor that stimulates LDL receptor expression on hepatocytes, enhances hepatic uptake of LDL-C from the blood. This ultimately leads to a reduction in plasma LDL-C (Ewang-Emukowhate and Wierzbicki 2013; Lewek et al. 2024; Antilipemic Agents 2012).

BAS monotherapy lowers LDL-C by 16–30% and provides an additional 10–15% reduction in LDL-C when combined with statins; however, evidence for reduction in cardiovascular events remains limited. BAS are preferred for patients intolerant to statins and are the only FDA-approved lipid-lowering agents during pregnancy (Lewek et al. 2024; Esan et al. 2022).

16.2.1.4 PCSK9 Inhibition

PCSK9, a protease that promotes the degradation of LDL receptors, is a novel therapeutic target in patients intolerant to statins or uncontrolled with statins (Roth and Davidson 2018; Sabouret et al. 2019; Squizzato et al. 2017). Alirocumab and evolocumab block PCSK9 from binding to LDL receptors, thereby increasing receptors and reducing LDL-C levels by up to 50–70% (Roth and Davidson 2018; Sabouret et al. 2019; Barale et al. 2021).

In the GLAGOV trial (2016), adding evolocumab to statin therapy was shown to promote regression of atherosclerotic plaque (Nicholls et al. 2016). The FOURIER trial (2017) showed that evolocumab reduced LDL-C levels by up to 60% and lowered the risk of major adverse cardiovascular events (MACE) by 15% (Sabatine et al. 2017). Results from the ODYSSEY OUTCOMES trial (2018) were also positive: alirocumab reduced LDL-C by 55%, and decreased both MACE and all-cause mortality by 15% (Schwartz et al. 2018).

Figure 16.1 illustrates the mechanisms of the main drug classes currently used in the treatment of lipid disorders (dyslipidemia).

16.2.2 Challenges and Limitations

Overall, statins and other current lipid-lowering drugs are clinically effective and widely prescribed, but all have limitations regarding efficacy and safety.

Statins are linked to a 10–45% higher risk of new-onset diabetes, though the mechanism is still only hypothesized. The incidence is higher in patients with pre-existing diabetes risk factors (Yandrapalli et al. 2019; Wallemacq 2019). Statins also cause statin-associated muscle symptoms (SAMS), which lead to clinical side

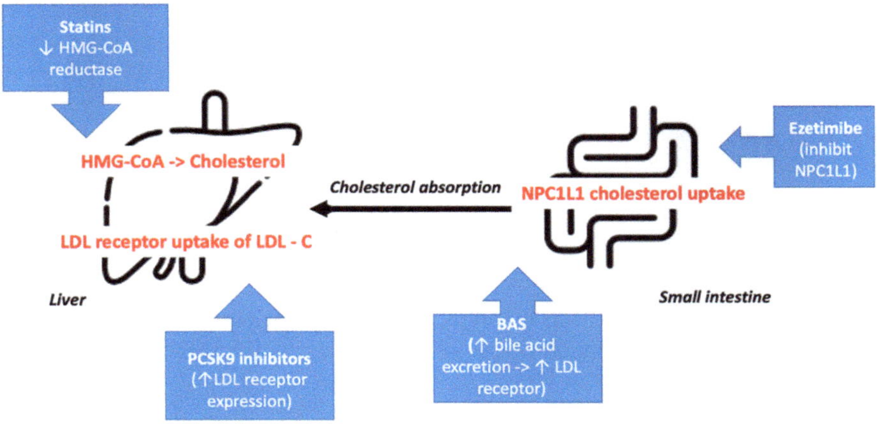

Fig. 16.1 Mechanisms of major lipid-lowering drug classes

effects such as myalgia, myositis, and even rhabdomyolysis, affecting about 1–5% of cases (Tomaszewski et al. 2011). Some high-dose data, such as liver, eye, and reproductive toxicity, have been noted in animal studies but have not been observed at therapeutic doses in humans. There remains some concern about cancer and cataracts, for which there is no consistent evidence from large clinical trials (Hajar 2011; Alves et al. 2018).

The efficacy of ezetimibe monotherapy is limited and remains controversial, especially after the ENHANCE trial (Kim and Wierzbicki 2020; Kastelein et al. 2008; Phan et al. 2012). Therefore, the drug is currently mainly recommended in combination with statins rather than as monotherapy. BAS, although safe, are often associated with gastrointestinal adverse effects, increased triglycerides, and mild elevation of liver enzymes (particularly with cholestyramine) (Ewang-Emukowhate and Wierzbicki 2013; Antilipemic Agents 2012). In addition, these drugs reduce the absorption of fat-soluble vitamins and many cardiovascular drugs, resulting in poor tolerability and restricted clinical utility (Jain 2024). PCSK9 inhibitors are generally well tolerated, with mostly mild adverse effects, but their very high cost and limited accessibility remain significant barriers to widespread application (Roth and Davidson 2018; Sabouret et al. 2019; Squizzato et al. 2017; Barale et al. 2021).

Other options such as lomitapide and mipomersen may be considered in certain cases, though their cardiovascular benefits and safety have not been fully established. Niacin and fibrates—two drugs that can lower triglycerides and reduce LDL-C levels in some patients—have not yet been classified as LDL-lowering drugs by the American College of Cardiology (ACC) and the American Heart Association (AHA) (Mach et al. 2019; Grundy et al. 2019). Research is now focused on advanced approaches, including molecular biomedical techniques and single-cell analysis in pharmacological studies on dyslipidemia, to improve both efficacy and safety in dyslipidemia management.

16.3 Application of Molecular Biomedical Techniques in Pharmacological Research for Blood Lipid Disorders

The current trend in therapy development for dyslipidemia is shifting away from traditional, non-specific drug approaches toward a focus on targeted molecular therapies that precisely address core disease pathways. This shift is primarily driven by the remarkable development of advanced molecular biomedical technologies such as gene sequencing and omics techniques (including genomics, proteomics, and metabolomics). These technologies, which enable the analysis of proteins and metabolites, have provided scientists with a powerful set of tools to study diseases at the molecular level and discover new pharmacological targets.

16.3.1 Genomics

The genes that play a key role in dyslipidemia have now been identified through next-generation sequencing. This technology has not only helped to differentiate single-gene lipid disorders, such as familial hypercholesterolemia (FH), but also to clarify the role of polygenic factors in increasing the risk of disease across the entire population (Berberich and Hegele 2022). Here are some of the techniques currently in use.

16.3.1.1 Next-Generation Sequencing (NGS)

Next-generation sequencing is used to identify genetic variants associated with inherited dyslipidemia, such as mutations in the *LDLR, APOB*, and *PCSK9* genes, which are the main genes causing familial hypercholesterolemia. Early identification of these mutations allows timely diagnosis, treatment, and the development of personalized therapies (Butnariu et al. 2023). This method allows sequencing of the entire genome or specific gene regions at a lower cost and significantly faster than conventional methods. In addition, NGS also facilitates the study of molecular mechanisms by helping to identify novel genes and biological pathways involved in dyslipidemia, such as the role of microRNAs (miRNAs) or long non-coding RNAs (lncRNAs) in the regulation of lipid metabolism (Agnello et al. 2024).

16.3.1.2 CRISPR–Cas9 Gene-Editing Technology

In pharmacology, CRISPR–Cas9 is being investigated as a potential therapy for dyslipidemia. It is a powerful tool that allows precise gene-editing. Some research is investigating the possibility of using CRISPR–Cas9 to "turn off" genes linked to hyperlipidemia, such as the *PCSK9* gene, to reduce cholesterol levels. This is a promising direction for gene therapy (Hoekstra and Van Eck 2024). CRISPR–Cas9 is also used to create animal models of disease, such as mice with *PCSK9* or *APOB* gene mutations. This allows for the study of disease pathogenesis and the testing of novel drug compounds (Zhao et al. 2020).

16.3.2 Proteomics

Proteomics plays an essential role in the comprehensive study of proteins in a diseased biological system. While genomics identifies which genes are involved in disease, proteomics reveals the mechanism of action and interactions of genes with proteins. By analyzing the entire proteome, researchers can discover disease mechanisms; this allows us to understand exactly how genetic mutations affect protein function leading to disease. It also identifies new biomarkers, discovers new protein-based indicators for diagnosing cardiovascular disease, and understands the pathways leading to atherosclerosis. Proteomics also searches for potential drug targets by identifying new proteins to target with drugs and analyzing drug efficacy and any unwanted side effects. This method provides a direct and complete view of ongoing cellular processes, making it an important tool in all stages of drug development (Mayr et al. 2006; Russell et al. 2013).

16.3.3 Metabolomics and Lipidomics

Metabolomics and lipidomics are playing an increasingly important role in the context of diagnosis and drug development. Traditional clinical markers of dyslipidemia, such as TC and LDL-C, are often not sensitive enough for early diagnosis because their concentrations only increase significantly as the disease progresses (Chen et al. 2014). Detecting individuals at risk for this disorder before symptoms manifest would facilitate timely intervention and management to prevent potential complications. Quantitative metabolomics is a method for early detection of diagnostic biomarkers for hyperlipidemia. In the context of drug development, changes in the composition or concentration of lipids and metabolites associated with genetic variants have provided insights into the function and genetic underpinnings of human genetics. This approach also sheds light on potential new targets for therapeutic intervention (Astarita et al. 2023; Alwahsh et al. 2024).

16.3.4 Advanced Molecular Therapies

The successful targeting of the PCSK9 protein has proven that a single molecular target can be treated in multiple ways, leading to two distinct types of advanced therapies:

- Monoclonal antibodies: Drugs like alirocumab and evolocumab are injected to block the PCSK9 protein already circulating in the bloodstream. By stopping PCSK9 from destroying the LDLR receptors on liver cells, these antibodies enable the liver to clear more LDL cholesterol from the blood (Coppinger et al. 2022).
- RNA therapy: A different approach, an RNA interference therapy called Inclisiran, works by preventing the body from producing the PCSK9 protein in

the first place (Chebli et al. 2024). This approach makes it easier for patients to adhere to the treatment regimen, requiring only two injections a year, whereas monoclonal antibody therapy requires frequent injections.

The integration of molecular biomedical techniques has opened the way for the development of innovative therapies that fundamentally change the dyslipidemia treatment landscape, moving from traditional therapies that require long-term adherence to a paradigm focused on precise, sustainable, and potentially curative intervention. Targeting the *PCSK9* gene and related proteins using CRISPR/Cas9 technology has yielded promising results in reducing total cholesterol (Chadwick and Musunuru 2017). While current advances in monoclonal antibody and RNA-based therapies have achieved remarkable results, the emerging field of gene-editing offers future potential as a one-time treatment application to provide lifelong cures for patients.

Advanced biomedical techniques such as NGS and CRISPR–Cas9, while promising for the future, have several limitations. The high cost of these technologies, coupled with the computational complexity of analyzing the large amounts of data they generate, poses a significant barrier to their routine application. Furthermore, although these techniques are effective in elucidating the genetic basis, they do not fully account for the complex interactions between lifestyle and environmental factors that also contribute to dyslipidemia. Finally, the clinical use of genetic information for diagnostic and therapeutic purposes raises complex ethical and legal questions regarding patient privacy and the potential for discrimination.

A compilation of the primary applications in dyslipidemia pharmacology research, along with the advantages and limitations of advanced molecular biomedical techniques, is indicated in Table 16.1.

16.4 The Application of Single-Cell Analysis in Pharmacological Research for Blood Lipid Disorders

16.4.1 Technologies of Single-Cell Analysis

Single-cell analysis (SCA) is a promising strategy to investigate rare and diverse cell populations, resolve cellular complexity, and support diagnosis, pathology, and therapy by optimizing drug efficacy, minimizing adverse events, and promoting personalized treatment—a distinct value that conventional omics approaches have yet to provide (Valihrach et al. 2018). Since the early twenty-first century, this field has expanded widely with over 4000 publications, focusing on scRNA-seq and bioinformatics tools (Valihrach et al. 2018; Tian et al. 2020). Advances in microfluidics, droplet-based systems, and capillary technologies have enabled automation, high throughput, and multimodal analysis for clinical applications, while improvements in devices and software have enhanced multiparametric capabilities and signal quality (Tian et al. 2020; Källberg et al. 2022; Mitra-Kaushik et al. 2021).

Table 16.1 Pharmacological applications of molecular biomedical techniques in dyslipidemia

Technique	Major Applications in Lipid Pharmacology	Outstanding Advantages/ Progress	Technical Limitations	References
RNA therapy (ASO/siRNA)	Inhibition of target protein expression (PCSK9, APOC3)	High specificity, potential for prolonged action and dose reduction (e.g., Inclisiran)	Challenges in RNA delivery and stability, risk of off-target effects, and high manufacturing costs	Coppinger et al. (2022), Merćep et al. (2022)
Gene-editing (CRISPR)	Repair/Disabling of disease-causing genes (*PCSK9*, *ANGPTL3*)	One-time treatment potential, targeting the genetic root, precision, and flexibility in tools	Off-target effects, load-carrying capacity of viral vectors (such as AAV—Adeno-associated virus) or lipid nanoparticles (LNP), immune regeneration, irreversibility	Chadwick and Musunuru (2018), Ding et al. (2014), Jarrett et al. (2017)
Proteomics	New drug target discovery, Biomarker discovery, Research on the mechanism of action of Drugs (MoA), Post-translational modification studies	Discovering new drug targets, Understanding drug mechanism of action (MoA) and effects, Biomarker development and Personalized medicine	Proteome complexity and variants, Difficulty in biomarker detection due to the small amount of some proteins, Data analysis and processing, Sample preparation issues	Gianazza et al. (2023), Nurmohamed et al. (2023)
Metabolomics	Elucidating pathogenesis and Drug action, Biomarker discovery, Precision medicine	Advanced biomarker discovery, Mechanistic mapping of drug action, Enhanced target discovery	Complex data, need strong computational biology, need standardization of analysis process	Astarita et al. (2023), Alwahsh et al. (2024)

In pharmacology, SCA provides tools for drug target discovery, off-target detection, and resistance mechanisms (Lähnemann et al. 2020; Van de Sande et al. 2023). Emerging techniques such as scATAC-seq, multiome, and spatial transcriptomics reveal gene regulatory changes and spatial patterns of drug-resistant cells (Van de Sande et al. 2023; Liu et al. 2024).

In SCA, scRNA-seq is the central tool with various protocols differing in sensitivity, cost, and throughput (Kashima et al. 2020; Hodzic 2016; Haque et al. 2017). Microfluidic and droplet-based platforms can process thousands of cells simultaneously and support automated workflows applicable to clinical settings. Multi-omics approaches, including scATAC-seq, bisulfite sequencing, and single-cell proteomics, expand the analysis to proteins and the epigenome (Kashima et al. 2020; Hodzic 2016). The steps of single-cell isolation, addition of unique molecular identifiers (UMIs), and data amplification determine the quality of the results (Haque et al. 2017).

16.4.2 Pharmacological Applications of SCA in Lipid Disorders Via Therapeutic Targets

Table 16.2 summarizes some key pharmacological applications of SCA in lipid metabolism disorders, highlighting its role in uncovering cell-level disease mechanisms and supporting the development of more precise therapeutic strategies.

16.4.2.1 PCSK9 Inhibition as Revealed by SCA

PCSK9, discovered in 2003, downregulates LDL receptors (LDL-R) in hepatocytes, thereby increasing plasma LDL-C. Gain-of-function (GOF) mutations (e.g., F216L, S127R) cause familial hypercholesterolemia, while loss-of-function (LOF) mutations (e.g., heterozygous Y142X) are protective—forming the basis for the

Table 16.2 Pharmacological applications of SCA in lipid disorders

SCA technique	Main feature	Pharmacological insights in lipid disorders	Target/ Biomarkers	References
ScRNA-seq/ snRNA-seq	Transcriptome mapping Cell-subtype mapping	Basis for cell atlases (adipose) Define foam-macrophage subtypes	PPARγ, foam macrophages	Deutsch et al. (2020), So et al. (2024), De Siqueira et al. (2024), Maniyadath et al. (2023)
ScATAC-seq/ multiome	Chromatin accessibility Regulatory landscape	Epigenetic shifts in drug-resistant cells	*DGAT2*, *PNPLA3*	Van de Sande et al. (2023), Liu et al. (2024)
Spatial transcriptomics	Preserves tissue context	Ligand–receptor mapping Plaque microenvironment	*ITLN1*⁺ foam cells	Lu et al. (2024), Li et al. (2025), Xiong et al. (2023)
Single-cell proteomics/ lipidomics	Protein and lipid diversity	CETP expression, HDL remodeling	CETP	Traeuble et al. (2025), Yu et al. (2025), Sharma et al. (2024)
Single-cell atlases (e.g., Tabula Muris; organ-specific atlases)	Cross-tissue/ cell maps across age/sex	Target prioritization Patient stratification Safety prediction	*PCSK9*, *ANGPTL3/8*	Schaum et al. (2018)

development of monoclonal antibody (mAb), small interfering RNA (siRNA), and CRISPR-based therapies (Seidah 2021).

The Tabula Muris project (2018) conducted single-cell mRNA sequencing on mouse cells to create a comprehensive atlas, and the results showed that PCSK9 is predominantly expressed in hepatocytes, reinforcing the liver's role as a primary target organ (Schaum et al. 2018).

Single-cell studies on aged tissues also revealed that PCSK9 expression varies with age and cell type for personalized therapies and explained different patient responses to alirocumab, evolocumab, or inclisiran (Csiszar et al. 2024; Almanzar et al. 2020).

16.4.2.2 ANGPTL3/8 (Betatrophin Blockade and Triglyceride Modulation)

The Tabula Muris project also demonstrated that ANGPTL3 is predominantly expressed in hepatocytes (Schaum et al. 2018). Recent studies have shown the cooperative inhibition of lipoprotein lipase (LPL) and the modulation of triglyceride metabolism by ANGPTL3 and ANGPTL8. Dual ANGPTL3/8 antibodies significantly reduce plasma triglycerides, opening opportunities for the development of dual inhibition therapies and personalized treatment. A further prospect is the application in systemic lipid metabolic diseases such as metabolic syndrome and non-alcoholic fatty liver disease (NAFLD) and its progressive form, non-alcoholic steatohepatitis (NASH) (Balasubramaniam et al. 2022; Landfors et al. 2023).

Using SCA-based data as a reference, Feng et al. demonstrated that ANGPTL8 deficiency reduces liver injury and improves lipid disorders in lipopolysaccharide (LPS)-induced hepatitis through modulating inflammatory pathways, emphasizing its promise as a therapeutic target (Feng et al. 2024).

16.4.2.3 PPAR/LXR Agonists and Adipose–Cholesterol Regulation

SnRNA-seq has characterized adipose tissue heterogeneity, identifying lipid-laden macrophage populations and hypertrophic adipocytes (Deutsch et al. 2020; So et al. 2024). Trajectory analysis has shown that adipocytes transition from an adaptive to a pathological state along with alterations in PPARγ expression. Meanwhile, scRNA-seq has demonstrated that activation of PPARγ by rosiglitazone can reactivate transcriptional and translational programs for mRNAs regulating adipogenesis and lipid metabolism (So et al. 2024; De Siqueira et al. 2024; Maniyadath et al. 2023). This process is influenced by obesity, leading to dysregulated lipid accumulation (De Siqueira et al. 2024).

These studies conclude that PPARγ is a central node that can guide the pathological and physiological states of adipose tissue. Combining modulation of the PPARγ axis with Liver X receptors (LXR)—another nuclear receptor capable of controlling cholesterol efflux and reducing foam macrophages—is a potential target for developing therapies for lipid disorders, obesity, and other metabolic syndromes (Deutsch et al. 2020; So et al. 2024; De Siqueira et al. 2024; Maniyadath et al. 2023; Hong and Tontonoz 2014; Endo-Umeda and Makishima 2025).

16.4.2.4 Foam Macrophages as Therapeutic Targets in Atherosclerosis

Foam cells, central to atherosclerosis, have become a major focus of SCA because their single-cell heterogeneity remains poorly characterized. An integrative study of multiple SCA datasets classified foam cells into five subtypes based on marker gene expression profiles, while using trajectory analysis and immunostaining to reconstruct the developmental pathway from pro-foam macrophages and to elucidate their heterogeneity. Ligand–receptor interaction analysis and Mendelian randomization revealed strong communication potential among these subpopulations and identified genes associated with regulated cell death pathways, opening new avenues for investigating the atherosclerotic plaque microenvironment and causal roles in pathogenesis (Lu et al. 2024). Heterogeneity and interaction analyses of foam cells have suggested several potential therapeutic targets, including "dysfunctional" foam macrophages and *ITLN1*$^{+}$ foam cells (Li et al. 2025; Xiong et al. 2023).

16.4.2.5 CETP Inhibition and HDL Remodeling Insights

Advances in the analysis of heterogeneity among foam macrophage subpopulations have provided insights into CETP expression in cholesterol pathways and high-density lipoprotein (HDL) remodeling, revealing the reasons for differential therapeutic responses among patients with lipid disorders. CETP has emerged as a key molecule to elucidating disease pathogenesis and enhances therapeutic strategies in these diseases (Traeuble et al. 2025; Yu et al. 2025; Sharma et al. 2024).

16.4.2.6 Other Research Directions

Overall, lipid metabolic disorders still remain a promising area for the development of novel biomarkers and pharmacological targets. Through single-cell analysis, researchers are expanding the understanding of pathogenic mechanisms, thereby developing new pharmacological intervention strategies targeting biomarkers such as *APOC3, APOA5*, and *PNPLA3*, as well as potential drug targets involved in lipid regulation such as *DGAT2* and *GPAM* (Rosenson et al. 2014; Farstad-O'Halloran et al. 2025; van Rijn et al. 2019; Chen et al. 2025).

16.4.3 Challenges of SCA in Pharmacology

In addition, the application of SCA in pharmacological research still faces several difficulties and challenges. Cells may be damaged or altered in certain biological and spatial characteristics due to errors in tissue dissociation, sampling, and isolation (Lähnemann et al. 2020; Hodzic 2016; Ascensión et al. 2022). The amplification of cDNA/mRNA may lead to data loss for lowly expressed genes. For scRNA-seq in particular, the most common difficulty is the dropout phenomenon—loss of RNA signals resulting in missing data (Lähnemann et al. 2020; Hodzic 2016). In single-cell lipidomics, results may be confounded if samples are affected by background noise or lipid contamination from the external environment. Data processing and integration are complex, requiring advanced bioinformatics tools.

The loss of spatial context when cells are detached from tissues reduces the accuracy of measuring cell–cell interactions, affecting the ability to regenerate tissues and interpret drug responses. Another challenge that cannot be ignored is the high cost and limited scalability (Hodzic 2016; Kontiza et al. 2024; Vanderaa and Gatto 2023). Ongoing technological innovations are gradually overcoming these barriers, thereby broadening the applicability of SCA in pharmacological research.

16.5 Conclusion

It is evident that although current treatments for dyslipidemia have significantly reduced the cardiovascular burden, they still face major challenges, especially the heterogeneous treatment responses among individuals. In addition, some existing drugs are still under debate regarding their safety.

With the rapid development of molecular biomedicine, advances in molecular biomedical techniques have created opportunities to develop more accurate diagnostic methods, targeted therapies, and personalized approaches. Gene-editing (CRISPR–Cas9) is expected to provide a treatment strategy for lipid disorders caused by genetic mutations, while RNA-based therapies (such as *inclisiran*), with their infrequent dosing schedules, can reduce the treatment burden for patients (Berberich and Hegele 2022; Hoekstra and Van Eck 2024).

Single-cell analysis, particularly scRNA-seq, shows notable advantages by addressing cellular heterogeneity and identifying rare cell populations, thereby optimizing therapy and minimizing side effects. Novel biomarkers and potential drug targets include *APOC3, PNPLA3*, and *DGAT2*, while focusing on tissue-specific characteristics (liver, vasculature, adipose tissue) helps explain the variability in drug responses among patients (Schaum et al. 2018; Rosenson et al. 2014; Farstad-O'Halloran et al. 2025; van Rijn et al. 2019; Chen et al. 2025).

However, current studies remain relatively new and face several obstacles, such as large and complex datasets, dropout or missing data, high costs, and ethical issues (Lähnemann et al. 2020; Hodzic 2016; Kontiza et al. 2024). If these challenges are addressed, and molecular tools are combined with SCA, dyslipidemia can be eventually treated as its root rather than being managed only as a chronic condition.

Funding This work is funded by Funded by Vietnam National University, Hanoi (VNU), grant number QG.25.190.

References

Agnello F et al (2024) Novel and emerging LDL-C lowering strategies: a new era of dyslipidemia management. J Clin Med 13(5):1251

Almanzar N et al (2020) A single-cell transcriptomic atlas characterizes ageing tissues in the mouse. Nature 583(7817):590–595

Alves C, Mendes D, Batel Marques F (2018) Statins and risk of cataracts: a systematic review and meta-analysis of observational studies. Cardiovasc Ther 36(6):e12480

Alwahsh M et al (2024) The application of metabolomics in hyperlipidemia: insights into biomarker discovery and treatment efficacy assessment. Metabolites 14(8)
Antilipemic Agents. In LiverTox: clinical and research information on drug-induced liver injury. National Institute of Diabetes and Digestive and Kidney Diseases, Bethesda, 2012
Ascensión AM, Araúzo-Bravo MJ, Izeta A (2022) Challenges and opportunities for the translation of single-cell RNA sequencing technologies to dermatology. Life (Basel) 12(1)
Astarita G, Kelly RS, Lasky-Su J (2023) Metabolomics and lipidomics strategies in modern drug discovery and development. Drug Discov Today 28(10):103751
Balasubramaniam D et al (2022) An anti-ANGPTL3/8 antibody decreases circulating triglycerides by binding to a LPL-inhibitory leucine zipper-like motif. J Lipid Res 63(5):100198
Barale C et al (2021) PCSK9 biology and its role in atherothrombosis. Int J Mol Sci 22(11)
Berberich AJ, Hegele RA (2022) A modern approach to dyslipidemia. Endocr Rev 43(4):611–653
Butnariu LI et al (2023) Current data and new insights into the genetic factors of atherogenic dyslipidemia associated with metabolic syndrome. Diagnostics (Basel) 13(14):2348
Canfora I, Pierno S (2024) Hypertriglyceridemia therapy: past, present and future perspectives. Int J Mol Sci 25(17)
Carabello BA (2010) The SEAS trial. Curr Cardiol Rep 12(2):122–124
Chadwick AC, Musunuru K (2017) Treatment of dyslipidemia using CRISPR/Cas9 genome editing. Curr Atheroscler Rep 19(7):32
Chadwick AC, Musunuru K (2018) CRISPR-Cas9 genome editing for treatment of atherogenic dyslipidemia. Arterioscler Thromb Vasc Biol 38(1):12–18
Chebli J, Larouche M, Gaudet D (2024) APOC3 siRNA and ASO therapy for dyslipidemia. Curr Opin Endocrinol Diabetes Obes 31(2):70–77
Chen H et al (2014) Metabolomics in dyslipidemia. Adv Clin Chem 66:101–119
Chen ZJ et al (2025) Single-cell DNA methylome and 3D genome atlas of human subcutaneous adipose tissue. Nat Genet 57(9):2238–2249
Chilbert MR et al (2022) Combination therapy of Ezetimibe and Rosuvastatin for dyslipidemia: current insights. Drug Des Devel Ther 16:2177–2186
Coppinger C et al (2022) A comprehensive review of PCSK9 inhibitors. J Cardiovasc Pharmacol Ther 27:10742484221100107
Csiszar A et al (2024) PCSK9: an emerging player in cardiometabolic aging and its potential as a therapeutic target and biomarker. Geroscience 46(1):257–263
De Siqueira MK et al (2024) PPARγ-dependent remodeling of translational machinery in adipose progenitors is impaired in obesity. Cell Rep 43(12):114945
Deutsch A et al (2020) The impact of single-cell genomics on adipose tissue research. Int J Mol Sci 21(13)
Ding Q et al (2014) Permanent alteration of PCSK9 with in vivo CRISPR-Cas9 genome editing. Circ Res 115(5):488–492
Endo A (2010) A historical perspective on the discovery of statins. Proc Jpn Acad Ser B Phys Biol Sci 86(5):484–493
Endo-Umeda K, Makishima M (2025) Exploring the roles of liver X receptors in lipid metabolism and immunity in atherosclerosis. Biomolecules 15(4):579
Esan O, Viljoen A, Wierzbicki AS (2022) Colesevelam - a bile acid sequestrant for treating hypercholesterolemia and improving hyperglycemia. Expert Opin Pharmacother 23(12):1363–1370
Ewang-Emukowhate M, Wierzbicki AS (2013) Lipid-lowering agents. J Cardiovasc Pharmacol Ther 18(5):401–411
Farstad-O'Halloran K et al (2025) Discovery of novel APOC3 isoforms in hepatic and intestinal cell models using long-read RNA sequencing. Genes (Basel) 16(4)
Feng Y et al (2024) ANGPTL8 deficiency attenuates lipopolysaccharide-induced liver injury by improving lipid metabolic dysregulation. J Lipid Res 65(8):100595
Gianazza E et al (2023) Proteomics and lipidomics to unveil the contribution of PCSK9 beyond cholesterol lowering: a narrative review. Front Cardiovasc Med 10:1191303
Grundy SM et al (2019) 2018 AHA/ACC/AACVPR/AAPA/ABC/ACPM/ADA/AGS/APhA/ASPC/NLA/PCNA guideline on the management of blood cholesterol: a report of the American

College of Cardiology/American Heart Association task force on clinical practice guidelines. Circulation 139(25):e1082–e1143
Hajar R (2011) Statins: past and present. Heart Views 12(3):121–127
Haque A et al (2017) A practical guide to single-cell RNA-sequencing for biomedical research and clinical applications. Genome Med 9(1):75
Hodzic E (2016) Single-cell analysis: advances and future perspectives. Bosn J Basic Med Sci 16(4):313–314
Hoekstra M, Van Eck M (2024) Gene editing for the treatment of hypercholesterolemia. Curr Atheroscler Rep 26(5):139–146
Hong C, Tontonoz P (2014) Liver X receptors in lipid metabolism: opportunities for drug discovery. Nat Rev Drug Discov 13(6):433–444
Huang JXF et al (2025) Recent advances in the management of dyslipidemia: a systematic review. Cureus 17(3):e81034
Jain P (2024) Traditional and novel non-statin lipid-lowering drugs. Indian Heart J 76(Suppl 1):S38–s43
Jarrett KE et al (2017) Somatic genome editing with CRISPR/Cas9 generates and corrects a metabolic disease. Sci Rep 7:44624
Källberg J et al (2022) Frontiers in single cell analysis: multimodal technologies and their clinical perspectives. Lab Chip 22(13):2403–2422
Kashima Y et al (2020) Single-cell sequencing techniques from individual to multiomics analyses. Exp Mol Med 52(9):1419–1427
Kastelein JJP et al (2008) Simvastatin with or without ezetimibe in familial hypercholesterolemia. N Engl J Med 358(14):1431–1443
Kim EJ, Wierzbicki AS (2020) The history of proprotein convertase subtilisin kexin-9 inhibitors and their role in the treatment of cardiovascular disease. Ther Adv Chronic Dis 11:2040622320924569
Kontiza A et al (2024) Single-cell lipidomics: an automated and accessible microfluidic workflow validated by capillary sampling. Anal Chem 96(44):17594–17601
Kosmas CE et al (2021) New and emerging lipid-modifying drugs to lower LDL cholesterol. Drugs Context 10
Lähnemann D et al (2020) Eleven grand challenges in single-cell data science. Genome Biol 21(1):31
Landfors F, Chorell E, Kersten S (2023) Genetic mimicry analysis reveals the specific lipases targeted by the ANGPTL3-ANGPTL8 complex and ANGPTL4. J Lipid Res 64(1):100313
Lewek J et al (2024) Dyslipidaemia management in pregnant patients: a 2024 update. Eur Heart J Open 4(3):oeae032
Li Y et al (2025) Single-cell and spatial analysis reveals the interaction between ITLN1(+) foam cells and SPP1(+) macrophages in atherosclerosis. Front Cardiovasc Med 12:1510082
Liu B, Hu S, Wang X (2024) Applications of single-cell technologies in drug discovery for tumor treatment. iScience 27(8):110486
Lu Y et al (2024) Integrated single-cell analysis revealed novel subpopulations of foamy macrophages in human atherosclerotic plaques. Biomolecules 14(12)
Mach F et al (2019) 2019 ESC/EAS guidelines for the management of dyslipidaemias: lipid modification to reduce cardiovascular risk: the task force for the management of dyslipidaemias of the European Society of Cardiology (ESC) and European Atherosclerosis Society (EAS). Eur Heart J 41(1):111–188
Maniyadath B et al (2023) Adipose tissue at single-cell resolution. Cell Metab 35(3):386–413
Mayr M et al (2006) Proteomics-based development of biomarkers in cardiovascular disease: mechanistic, clinical, and therapeutic insights. Mol Cell Proteomics 5(10):1853–1864
Merćep I et al (2022) Advantages and disadvantages of inclisiran: a small interfering ribonucleic acid molecule targeting PCSK9-A narrative review. Cardiovasc Ther 2022:8129513
Mitra-Kaushik S et al (2021) The evolution of single-cell analysis and utility in drug development. AAPS J 23(5):98

Morrone D et al (2012) Lipid-altering efficacy of ezetimibe plus statin and statin monotherapy and identification of factors associated with treatment response: a pooled analysis of over 21,000 subjects from 27 clinical trials. Atherosclerosis 223(2):251–261
Naderi SH, Bestwick JP, Wald DS (2012) Adherence to drugs that prevent cardiovascular disease: meta-analysis on 376,162 patients. Am J Med 125(9):882–7.e1
Nicholls SJ et al (2016) Effect of Evolocumab on progression of coronary disease in statin-treated patients: the GLAGOV randomized clinical trial. JAMA 316(22):2373–2384
Nurmohamed NS et al (2023) Proteomics and lipidomics in atherosclerotic cardiovascular disease risk prediction. Eur Heart J 44(18):1594–1607
O'Malley PG et al (2020) Management of dyslipidemia for cardiovascular disease risk reduction: synopsis of the 2020 updated U.S. Department of Veterans Affairs and U.S. Department of Defense Clinical Practice Guideline. Ann Intern Med 173(10):822–829
Pasta A et al (2020) PCSK9 inhibitors for treating hypercholesterolemia. Expert Opin Pharmacother 21(3):353–363
Patel SB et al (2025) American Association of Clinical Endocrinology Clinical Practice Guideline on pharmacologic management of adults with dyslipidemia. Endocr Pract 31(2):236–262
Pearson GJ et al (2021) 2021 Canadian Cardiovascular Society Guidelines for the Management of Dyslipidemia for the prevention of cardiovascular disease in adults. Can J Cardiol 37(8):1129–1150
Phan BA, Dayspring TD, Toth PP (2012) Ezetimibe therapy: mechanism of action and clinical update. Vasc Health Risk Manag 8:415–427
Pirillo A et al (2021) Global epidemiology of dyslipidaemias. Nat Rev Cardiol 18(10):689–700
van Rijn JM et al (2019) DGAT2 partially compensates for lipid-induced ER stress in human DGAT1-deficient intestinal stem cells. J Lipid Res 60(10):1787–1800
Rosenson RS et al (2014) Genetics and causality of triglyceride-rich lipoproteins in atherosclerotic cardiovascular disease. JACC 64(23):2525–2540
Roth EM, Davidson MH (2018) PCSK9 inhibitors: mechanism of action, efficacy, and safety. RCM 19(S1):31–46
Russell C, Rahman A, Mohammed AR (2013) Application of genomics, proteomics and metabolomics in drug discovery, development and clinic. Ther Deliv 4(3):395–413
Sabatine MS et al (2017) Evolocumab and clinical outcomes in patients with cardiovascular disease. N Engl J Med 376(18):1713–1722
Sabouret P, Farnier M, Puymirat E (2019) Inhibiteurs de PCSK9: quelle place dans la prise en charge actuelle des dyslipidémies? Presse Med 48(3, Part 1):227–237
Schaum N et al (2018) Single-cell transcriptomics of 20 mouse organs creates a Tabula Muris. Nature 562(7727):367–372
Schwartz GG et al (2018) Alirocumab and cardiovascular outcomes after acute coronary syndrome. N Engl J Med 379(22):2097–2107
Seidah NG (2021) The PCSK9 discovery, an inactive protease with varied functions in hypercholesterolemia, viral infections, and cancer. J Lipid Res 62:100130
Sharma D et al (2024) Comprehensive integration of multiple single-cell transcriptomic data sets defines distinct cell populations and their phenotypic changes in murine atherosclerosis. Arterioscler Thromb Vasc Biol 44(2):391–408
Sharp Collaborative G (2010) Study of Heart and Renal Protection (SHARP): randomized trial to assess the effects of lowering low-density lipoprotein cholesterol among 9,438 patients with chronic kidney disease. Am Heart J 160(5):785–794.e10
So J et al (2024) Robust single nucleus RNA sequencing reveals depot-specific cell population dynamics in adipose tissue remodeling during obesity. bioRxiv
Squizzato A et al (2017) PCSK9 inhibitors for treating dyslipidemia in patients at different cardiovascular risk: a systematic review and a meta-analysis. Intern Emerg Med 12(7):1043–1053
Tian H et al (2020) The trends of single-cell analysis: a global study. Biomed Res Int 2020:7425397
Tomaszewski M et al (2011) Statin-induced myopathies. Pharmacol Rep 63(4):859–866
Traeuble K et al (2025) Integrated single-cell atlas of human atherosclerotic plaques. Nat Commun 16(1):8255

Tsubakio-Yamamoto K et al (2010) Current therapy for patients with sitosterolemia--effect of ezetimibe on plant sterol metabolism. J Atheroscler Thromb 17(9):891–900
Valihrach L, Androvic P, Kubista M (2018) Platforms for single-cell collection and analysis. Int J Mol Sci 19(3)
Van de Sande B et al (2023) Applications of single-cell RNA sequencing in drug discovery and development. Nat Rev Drug Discov 22(6):496–520
Vanderaa C, Gatto L (2023) Revisiting the thorny issue of missing values in single-cell proteomics. J Proteome Res 22(9):2775–2784
Vardell E (2020) Global health observatory data repository. Med Ref Serv Q 39(1):67–74
Wallemacq C (2019) Statins and new-onset diabetes: benefit-risk balance. Rev Med Suisse 15(659):1454–1457
Wazir M et al (2023) Lipid disorders and cardiovascular risk: a comprehensive analysis of current perspectives. Cureus 15(12):e51395
Xiong J et al (2023) Bulk and single-cell characterisation of the immune heterogeneity of atherosclerosis identifies novel targets for immunotherapy. BMC Biol 21(1):46
Yandrapalli S et al (2019) Statins and the potential for higher diabetes mellitus risk. Expert Rev Clin Pharmacol 12(9):825–830
Yu B, Sopic M, Sluimer JC (2025) Single-cell RNA sequencing (scRNA-seq) and its insights into cellular heterogeneity in atherosclerosis. Vasc Pharmacol 159:107499
Zhang J et al (2018) Usefulness of alirocumab and evolocumab for the treatment of patients with diabetic dyslipidemia. Proc (Bayl Univ Med Cent) 31(2):180–184
Zhao H et al (2020) In vivo AAV-CRISPR/Cas9–mediated gene editing ameliorates atherosclerosis in familial hypercholesterolemia. 141(1):67–79

Nanotechnology in the Generation and Analysis of Single Cells

17

Aparna Jayachandran, Suresh Ramakrishna, Prashanth Prithviraj, Viswanathaiah Matam, George Kannourakis, and Arpita Poddar

Abstract

Cellular heterogeneity characterises normal physiology as well as driving disease progression, therapeutic resistance, and variable clinical outcomes. Single-cell technologies address such complexities by enabling molecular and functional analysis at the level of individual cells, but are fundamentally limited by low analyte abundance, perturbation during handling, and loss of spatial context. Nanotechnology provides powerful solutions to these challenges by enabling precise single-cell generation, manipulation, and ultra-sensitive analysis at biologically relevant length scales. This chapter reviews advances in nanotechnology-enabled single-cell platforms, spanning nano-engineered cell isolation and handling, single-cell genomics, transcriptomics, proteomics, and spatial and

A. Jayachandran
Fiona Elsey Cancer Research Institute, Ballarat Central, VIC, Australia

S. Ramakrishna
Graduate School of Biomedical Science and Engineering, Hanyang University, Seoul, South Korea

College of Medicine, Hanyang University, Seoul, South Korea

P. Prithviraj · G. Kannourakis
Fiona Elsey Cancer Research Institute, Ballarat Central, VIC, Australia

Federation University, Ballarat, VIC, Australia

V. Matam
Department of Biotechnology, Alliance School of Applied Engineering, Alliance University, Bengaluru, Karnataka, India

A. Poddar (✉)
Fiona Elsey Cancer Research Institute, Ballarat Central, VIC, Australia

Federation University, Ballarat, VIC, Australia

RMIT University, Melbourne, VIC, Australia

The University of Melbourne, Parkville, VIC, Australia
e-mail: arpita@fecri.org.au

V. Singh (ed.), *Advances in Single Cell Analysis*,
https://doi.org/10.1007/978-981-95-9789-5_17

functional multi-omics. We highlight translational applications across oncology, immunology, and therapeutic development, including nano-enabled drug response profiling, imaging, and targeted delivery. Together, these approaches position nanotechnology as a key driver of next-generation single-cell biology, enabling mechanistic insight and clinical translation from bench to bedside.

Keywords

Single cell · Nanotechnology · Nanomaterials · Nanobiotechnology · Cancer · Medical biology

17.1 Introduction

Biological systems are inherently heterogeneous. Even cells that share the same genetic background can differ significantly in gene expression, protein levels, metabolism, mechanical properties, and responses to external signals. Such variabilities arise from dynamic transcriptional activity, epigenetic regulation, complex signalling networks, and differences in the local microenvironment. Rather than representing random noise, cellular heterogeneity is essential for normal biological processes such as development, immune function, and tissue repair (Mendenhall et al. 2021). In disease, particularly cancer, this heterogeneity becomes a major challenge, driving tumour progression, metastasis, immune evasion, and resistance to therapy (Ottaiano et al. 2023). Traditional bulk measurements average signals across several cells and therefore obscure rare but clinically important cell states and markers, motivating the widespread adoption of single-cell technologies in modern biology and precision medicine (Baysoy et al. 2023).

In aggressive cancers such as triple-negative breast cancer (TNBC) and advanced prostate cancer, heterogeneity is evident across multiple regulatory layers (Poddar et al. 2022b, 2024b). TNBC tumours often contain subpopulations with distinct activation of pathways such as androgen receptor (AR), STAT3, and EGFR, as well as diverse epithelial–mesenchymal transition (EMT) states, and metabolic and inflammatory programs that contribute to immune suppression and poor treatment response. In prostate cancer, variation in AR signalling, lineage plasticity, and neuroendocrine differentiation underlies resistance to androgen-deprivation therapies (Poddar et al. 2020; Jayachandran et al. 2016; Lange-Sperandio et al. 2007). Single-cell transcriptomic and proteomic studies have shown that many of these phenotypic changes are driven by dynamic, non-genetic mechanisms that are not detectable in bulk analyses, underscoring the need for tools that can resolve functional states at the level of individual cells (Wen et al. 2022).

Recent advances in single-cell genomics, transcriptomics, proteomics, and metabolomics have enabled detailed mapping of cellular states within complex tissues. However, the accuracy of these approaches depends strongly on how single cells are generated, isolated, and analysed. Common techniques such as limiting dilution, fluorescence-activated cell sorting (FACS), and micromanipulation have enabled important discoveries but are limited by trade-offs in throughput, mechanical and oxidative stress, and disruption of native cell–cell and cell–matrix interactions (Jovic et al. 2022; Wen et al. 2022). In oncology, these stresses can alter signalling programs or lead to loss of fragile yet biologically relevant cell populations, complicating data interpretation (Mendenhall et al. 2021).

Nanotechnology offers a complementary solution by using materials and devices engineered at scales comparable to cellular membranes and subcellular structures (Ramos et al. 2017). At the nano scale, materials can be designed to precisely control cell adhesion, membrane interactions, receptor organisation, and molecular transport. Nanostructured surfaces and nano-fluidic devices enable reliable capture and confinement of single cells while maintaining viability and physiological signalling. By reducing reaction volumes and increasing local analyte concentrations, nano-enabled platforms can also improve the sensitivity of detecting low-abundance transcripts and proteins in individual cancer cells, which is critical for identifying signalling states associated with therapy resistance (Fu et al. 2024).

In single-cell cancer research, nanotechnology plays two closely linked roles. First, it enables controlled manipulation of individual tumour and immune cells, including spatial organisation, mechanical stimulation, and exposure to well-defined biochemical gradients. Nano-patterned substrates and nanochannels can mimic aspects of the tumour microenvironment, allowing researchers to study how physical cues influence processes such as EMT or immune escape at single-cell resolution. Second, nanotechnology enhances analytical sensitivity by enabling detection of nucleic acids, proteins, metabolites, and biophysical properties from extremely small cellular volumes. Nano-enabled sensors, probes, and electrodes allow direct measurement of signalling dynamics, metabolic activity, cytokine release, and drug response in single cells, revealing functional heterogeneity that is not accessible with bulk or conventional microscale methods (Wang et al. 2025; Zhu et al. 2025; Kim et al. 2024).

This chapter reviews nanotechnology-enabled approaches for the generation and analysis of single cells, with an emphasis on their application in biomedical and translational research. Section 17.2 introduces the biological basis of cellular heterogeneity and reviews conventional single-cell isolation and handling methods, highlighting technical limitations that motivate nanoscale solutions. Section 17.3 focuses on nanomaterials and nano-engineered platforms for single-cell generation, manipulation, and controlled interrogation. Section 17.4 subsequently examines nano-enabled analytical technologies spanning single-cell genomics, transcriptomics, proteomics, and spatial and multi-omics profiling, followed by functional applications including drug response assessment, tumour–microenvironment interactions, immune profiling, and biophysical characterisation at the single-cell level. The chapter concludes by discussing emerging nano-enabled platforms, data integration challenges, and opportunities for clinical translation, positioning nanotechnology as a central enabler of next-generation single-cell science.

17.1.1 Historical Context

The completion of the Human Genome Project in 2003 provided a comprehensive reference for interpreting the human genome, while the introduction of massively parallel next-generation sequencing platforms between 2005 and 2007 made high-depth sequencing faster, cheaper, and broadly accessible. Together, these advances laid the foundation for single-cell omics, beginning with the first demonstration of single-cell RNA sequencing (scRNA-seq) in 2009. Method development progressed rapidly over the following years, with improvements in RNA capture and cDNA amplification, and a major increase in scalability as microwell- and droplet-based microfluidic platforms developed around 2015. These innovations enabled routine, high-throughput profiling of thousands to tens of thousands of individual cells per experiment. In parallel, combinatorial indexing and in situ strategies further increased throughput while reducing per-cell cost, thereafter establishing single-cell sequencing as a practical approach for systematic mapping of cellular heterogeneity (Wang et al. 2023; Aldridge and Teichmann 2020).

As the field matured, single-cell analysis expanded beyond transcriptomics to include genomic and epigenomic layers, reflecting the multi-level regulation of cell state. Single-cell chromatin conformation and DNA methylation profiling were introduced around 2013, followed shortly by the adaptation of chromatin accessibility assays (scATAC-seq) from bulk to single-cell formats in 2015. These methods were subsequently integrated into droplet- and indexing-based workflows to improve throughput and robustness. In parallel, single-cell genome sequencing advanced through refinements in whole-genome amplification strategies, improving coverage uniformity and reducing technical artefacts such as allele dropout and false variant calls. Importantly, the field moved towards integrated measurement: single-cell multi-omics assays capable of jointly profiling combinations of genome, DNA methylation, chromatin accessibility, and transcriptome emerged from 2015 onwards. At the same time, spatially resolved transcriptomic approaches, such as multiplexed FISH, in situ sequencing, and in situ capture, began to restore tissue context and local cell–cell interactions that are otherwise lost during dissociation (Mazan-Mamczarz et al. 2022; Minnoye et al. 2021; Wen and Tang 2022).

Nanotechnology has played a critical enabling role throughout this progression by improving both single-cell handling and analytical sensitivity under the low-input constraints of single-cell biology. Nano-engineered surfaces and nanostructured interfaces allow more controlled capture, organisation, and maintenance of individual cells within defined microenvironments, while nano-enabled probes and sensors enhance detection of low-abundance nucleic acids, proteins, and metabolites by reducing dilution volumes and improving signal transduction at the cell–material interface (Poddar et al. 2023). These capabilities are particularly important in cancer, where clinically relevant behaviours often arise from rare or transient cell states that bulk assays fail to resolve. In aggressive diseases, single-cell approaches have uncovered subpopulations with distinct signalling dependencies, such as variable androgen receptor activity, inflammatory and metabolic pathways, and differential therapy responses, as well as complex interactions with immune and stromal cells. By enabling higher fidelity isolation, functional studies, and integration with multi-omics and spatial analyses, nanotechnology substantially strengthens single-cell platforms for dissecting tumour ecosystems and informing functionally guided strategies in cancer research (Zhu et al. 2025; Wang et al. 2025; Ramos et al. 2017).

17.1.2 Fundamentals of Single-Cell Generation and Isolation

In multicellular organisms, genetically identical cells can adopt distinct phenotypes due to stochastic gene expression, epigenetic regulation, and microenvironmental cues. For instance, rare subpopulations of drug-tolerant cancer cells can survive therapy and seed relapse, while small subsets of immune cells can dictate the magnitude and quality of immune responses (Hirsch et al. 2024; Van Heyningen 2024).

Capturing this heterogeneity requires robust methods to generate and isolate individual cells while preserving their viability, molecular integrity, and functional state. Importantly, the method of isolation itself can influence cellular behaviour. Mechanical stress, temperature fluctuations, and exposure to non-physiological surfaces can induce transcriptional and metabolic changes within minutes. The methods utilised for single-cell generation are not merely a preparatory step, but a critical determinant of data quality and biological relevance (Yuan et al. 2017b; Huang et al. 2024).

(a) *Conventional Single-Cell Isolation Techniques*

A variety of techniques have been developed to isolate single cells from heterogeneous populations (Hu et al. 2016). Limiting dilution, one of the earliest methods, involves diluting a cell suspension such that wells contain, on average, one cell. While conceptually simple, this approach is inefficient for rare cell populations or high-throughput applications (Ye et al. 2021).

Micromanipulation techniques, including micropipette aspiration and laser capture microdissection, enable precise isolation of individual cells from tissues or cultures. These methods offer high spatial resolution but are labour-intensive, low-throughput, and prone to operator-dependent variability. Moreover, physical contact and laser exposure can perturb cellular physiology (Hohnadel et al. 2018).

Fluorescence-activated cell sorting (FACS) remains one of the most widely used single-cell isolation techniques. FACS enables high-throughput sorting based on cell surface markers or fluorescent reporters and can achieve single-cell resolution with high purity. However, cells are subjected to significant shear stress and pressure, which can affect viability and gene expression. In addition, FACS requires fluorescent labelling and large input cell numbers, limiting its applicability for certain samples such as primary tissues or rare cell populations (Miwa et al. 2022).

Microfluidic technologies have addressed some of these limitations by enabling controlled manipulation of cells in microscale channels and droplets. Droplet-based microfluidics has revolutionised single-cell genomics by encapsulating individual cells with barcoded beads for downstream sequencing. Despite their success, microfluidic systems often rely on passive hydrodynamic forces and channel geometries that limit fine control at the single-cell and subcellular level. Furthermore, scaling down from micro- to nano-dimensions is essential to achieving more biomimetic interfaces and enhanced analytical sensitivity (Ruan et al. 2026; Wang et al. 2024).

(b) *Technical Challenges in Single-Cell Generation*

The generation and isolation of single cells present several technical challenges that motivate the integration of nanotechnology. First, scalability and throughput must be balanced against precision. Techniques that offer high throughput often sacrifice control over individual cells, while highly precise methods tend to be slow and low throughput (Yuan et al. 2017a; Ryoo et al. 2024).

Second, cell viability and perturbation are critical concerns. Physical forces, chemical exposure, and surface interactions can rapidly alter cellular states. Methods that minimise stress and maintain physiological conditions are essential for functional single-cell studies (Bao et al. 2010).

Third, compatibility with downstream analysis is a major consideration. Single-cell isolation platforms must integrate seamlessly with genomic, proteomic, imaging, and functional assays. Inefficient transfer between platforms can result in cell loss and signal degradation (Skylaki et al. 2016).

Finally, heterogeneity in cell size, shape, and mechanical properties complicates isolation strategies. A universal method suitable for all cell types remains elusive, highlighting the need for adaptable and tuneable technologies (Xie et al. 2018).

Nanotechnology addresses these challenges by enabling precise control over surface chemistry, topography, and molecular interactions at scales relevant to cellular processes. By bridging the gap between molecular biology and microscale engineering, nanotechnology provides a foundation for next-generation single-cell platforms (Abdullah et al. 2025; Liang et al. 2014).

17.2 Nanomaterials for Single-Cell Generation and Manipulation

Nanomaterials characterise dimensions that are comparable to those of cellular membranes, cytoskeletal elements, and macromolecular complexes (Murthy 2007). Such size compatibility allows nanostructures to interact with individual cells in a controlled and minimally invasive manner, overcoming several limitations of conventional microscale techniques. Unlike bulk or purely microfluidic approaches, nano-enabled platforms can form precise physical interfaces with single cells, allowing for deterministic capture, positioning, membrane access, and molecular exchange while preserving cell viability and functional states (Kaladharan et al. 2021). These properties are particularly important in single-cell workflows, where

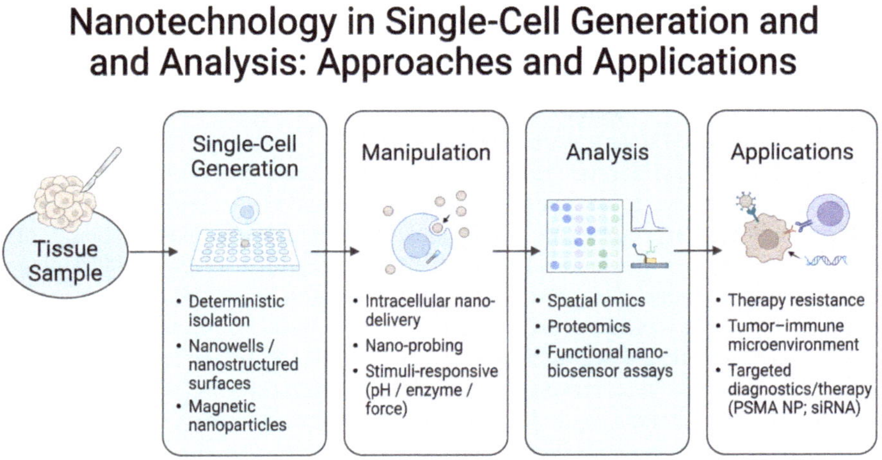

Fig. 17.1 Nanotechnology-enabled single-cell generation, manipulation, and analysis, and their translational applications. The schematic outlines the progression from tissue samples to deterministic single-cell isolation using nano-engineered platforms, followed by intracellular nano-delivery and probing, ultra-sensitive single-cell analytical profiling, and downstream biomedical applications. Created in BioRender. Poddar, A. (2026) https://BioRender.com/sn8bozs

mechanical stress or perturbation during isolation can rapidly alter gene expression, signalling activity, and cellular behaviour.

By matching the physical and functional interfaces of biological systems, nanotechnology provides a mechanistically distinct framework for single-cell generation and manipulation. Nano-engineered materials can modulate cell-surface adhesion, membrane curvature, receptor organisation, and transport pathways in ways that are inaccessible to larger scale tools (Casanellas et al. 2022). As a result, nanomaterial-based approaches reduce mechanical and biochemical stress during isolation and handling, improve compatibility with multi-omics and functional assays, and enable reproducible single-cell workflows across heterogeneous populations (Woodland et al. 2025) (Fig. 17.1).

17.2.1 Nano-engineered Platforms

A diverse range of nano-enabled platforms has been developed to support controlled access to single cells, including nanostraws, nanoneedles, nanopipettes, and nanoelectrode-based systems, each offering complementary modes of interaction. Nanostraw arrays, comprising vertically aligned hollow nanostructures fabricated from materials such as alumina or silicon nitride, can spontaneously and non-lethally penetrate adherent cell membranes to form stable conduits to the cytosol. This architecture enables direct intracellular delivery of biomolecules, including plasmid DNA, RNA, and proteins, without reliance on endocytosis (Liu et al. 2022; Poddar et al. 2019). Voltage-assisted nanostraw systems further enhance delivery efficiency and temporal control, allowing synchronised manipulation of thousands of cells in parallel (Polash et al. 2024). Importantly, nanostraws also enable longitudinal intracellular sampling, supporting repeated extraction of mRNA or proteins from the same living cells to track dynamic cellular responses such as differentiation or drug-induced state transitions (Quek and Tay 2024).

Nanoneedle-based approaches provide higher spatial precision and are often derived from atomic force microscopy technologies. With tip diameters below 100 nm and high aspect ratios, nanoneedles can penetrate the plasma membrane, and in some cases the nuclear envelope, with minimal mechanical damage. Functionalised nanoneedles have been used to deliver nucleic acids directly into the nucleus, bypassing inefficiencies associated with conventional transfection, as well as to extract specific intracellular proteins using antibody or probe-based capture. These features make nanoneedles particularly well suited for interrogating heterogeneous cell populations, including rare or therapy-resistant cancer cells that are difficult to access using conventional methods (Yum et al. 2011).

Nanopipette and nanoelectrode systems further expand the toolkit for single-cell manipulation by integrating fluidic, electrical, and sensing modalities (Xu et al. 2023). Glass nanopipettes with nanoscale apertures enable controlled injection or aspiration of picolitre-scale volumes via electroosmotic or electrophoretic forces. When combined with high-resolution positioning methods such as scanning ion conductance microscopy, nanopipettes can target defined subcellular regions within living cells. In parallel, vertically aligned nanoelectrode arrays and nanowire-based probes enable intracellular electrical recording or stimulation, transforming single-cell manipulation from a disruptive intervention into a programmable and scalable operation (Xu et al. 2020; Bulbul et al. 2018).

17.2.2 Nanostructured Environments

Beyond direct membrane access, nanomaterials enable the creation of engineered microenvironments that support deterministic single-cell capture, organisation, and functional interrogation. Nanoparticles can be precisely engineered in terms of size, shape, surface charge, and chemical functionalisation, enabling selective interactions with specific cell types or intracellular compartments. Magnetic nanoparticles, for example, allow gentle single-cell separation and spatial positioning using external magnetic fields, making them particularly attractive for handling fragile or rare cell populations (Stueber et al. 2021; Zhang et al. 2024).

Metal–organic frameworks (MOFs), including zeolitic imidazolate frameworks (ZIFs), represent an emerging class of nanomaterials for single-cell manipulation. Their high porosity, tuneable chemistry, and pH-responsive degradability enable efficient loading and controlled release of biomolecules or drugs at the single-cell level. Notably, ZIF nanoparticles have been shown to exhibit cell-state-dependent uptake and intracellular processing, providing opportunities for selective manipulation and functional profiling of individual cancer cells based on membrane properties and endocytic pathways (Pyreddy et al. 2023; Polash et al. 2024; Poddar et al. 2022a).

Nanostructured surfaces, such as nanopillars, nanowires, nanopores, and nanowells, provide physical interfaces that interact with cells at subcellular resolution. Compared with flat substrates, these surfaces better mimic features of the extracellular matrix, supporting improved cell adhesion, viability, and preservation of physiological behaviour. Arrays of nanopits or nanowells can be engineered to accommodate individual cells, enabling deterministic single-cell capture and spatially organised single-cell arrays suitable for high-throughput analysis. Such platforms overcome the randomness of limiting dilution and enable parallel manipulation of thousands of individual cells under uniform conditions (Torres et al. 2014; Ctortecka et al. 2023).

Nano-fluidic platforms further refine single-cell control by operating at length scales comparable to biomolecules. Nanochannels and nanopores exploit size exclusion, electrokinetic forces, and surface charge interactions to isolate and confine single cells without the need for labels. Carefully tuned channel dimensions allow nutrients and signalling molecules to diffuse freely while maintaining stable confinement, creating well-controlled microenvironments suitable for long-term single-cell studies. Confinement within nanolitre or picolitre volumes also enhances analytical sensitivity by reducing dilution effects, allowing secreted proteins or metabolites from a single cell to accumulate to detectable concentrations for real-time functional monitoring (Haywood et al. 2015; Iarossi et al. 2025).

Finally, nanomaterials enable active manipulation and stimulation of single cells with high spatial and temporal precision. Nanoparticles can deliver mechanical, chemical, optical, or thermal stimuli to individual cells or specific subcellular regions; for example, gold nanoparticles convert light into heat, enabling localised photothermal stimulation. Nanoelectrodes and nanowire-based devices similarly allow electrical interrogation and modulation of membrane potential with minimal invasiveness. Together, these nano-enabled capabilities transform single-cell generation and manipulation into controlled, reproducible, and functionally informative processes, providing the physical foundation for advanced single-cell analysis and downstream translational applications (Haywood et al. 2015; Iarossi et al. 2025).

17.3 Nano-enabled Analytical Platforms for Single-Cell Profiling

A central challenge in single-cell biology is extracting meaningful molecular and functional information from extremely small sample volumes while preserving cellular context and heterogeneity. Bulk and microscale analytical techniques often lack the sensitivity required to detect low-abundance molecules and inherently obscure cell-to-cell variability, particularly in heterogeneous diseases such as cancer. Nano-enabled single-cell analysis addresses these limitations by operating at length scales comparable to biomolecules, thereby enhancing detection sensitivity, reducing dilution effects, and enabling quantitative interrogation of individual cells

(Poddar et al. 2024a). As a result, nanotechnology has become essential for resolving heterogeneous molecular states, tracking intracellular dynamics, and linking nano–bio interactions to functional outcomes at single-cell resolution (Heath 2015).

One of the most widely adopted nano-enabled analytical strategies is flow-based single-cell analysis of nanoparticle–cell interactions. Conventional flow cytometry has been adapted to quantify nanoparticle uptake by exploiting nanoparticle-induced changes in light scattering properties, enabling label-free, high-throughput assessment of nano–bio interactions at the single-cell level (Shin et al. 2025). These approaches have consistently revealed substantial cell-to-cell variability in nanoparticle internalisation that is masked in bulk measurements. For example, magnetite nanoparticles interact more strongly with PC-3 prostate cancer cells than with non-malignant BPH1 epithelial cells under identical conditions, highlighting disease-specific nano–bio interaction signatures that are directly relevant to nanomedicine design (Youhannayee et al. 2019). Similar studies using gold, silver, silica, and polymeric nanoparticles demonstrate that nanoparticle size, surface charge, and surface chemistry drive highly heterogeneous uptake patterns within cancer cell populations, reinforcing the need for single-cell analytical resolution (Singh et al. 2025).

17.3.1 Advanced Nano-enabled Cytometry and Intracellular Nanosampling

Recent technological advances have significantly expanded flow-based single-cell analysis through nano-flow cytometry (nFCM) and mass cytometry (CyTOF) (Liu et al. 2026). Nano-flow cytometry improves sensitivity by reducing interrogation volumes and enhancing optical resolution, enabling detection of nanoparticles and extracellular vesicles as small as 7–10 nm (Bokun et al. 2025). This capability has enabled direct quantification of nanoparticle binding and uptake at the single-particle and single-cell levels, as well as detailed profiling of extracellular vesicle heterogeneity in cancer systems. Notably, nFCM has revealed highly uneven distributions of extracellular vesicle surface proteins across individual tumour cells, providing insights into intercellular communication and phenotypic diversity that are inaccessible with ensemble measurements. Mass cytometry further extends these capabilities by combining metal-tagged antibodies and nanoparticles with time-of-flight mass spectrometry, enabling simultaneous measurement of dozens of protein markers alongside nanoparticle content in individual cells and directly linking nano–bio interactions to cellular phenotype (Delgado-Gonzalez and Sanchez-Martin 2021).

Beyond flow-based approaches, nanoscale intracellular sampling and sensing technologies enable direct molecular analysis from living single cells with minimal disruption. Nanopipettes, hollow atomic force microscopy tips, carbon nanotubes, and nanostraw arrays can extract picolitre-scale volumes of cytosolic material, allowing repeated sampling from the same cell over time (Haywood et al. 2015). This longitudinal capability is critical for capturing temporal heterogeneity in signalling responses, differentiation trajectories, and therapy adaptation. The nanostraw extraction (NEXT) platform, for example, has been used to repeatedly sample mRNA and proteins from individual living cells, revealing dynamic and asynchronous signalling behaviours in cancer cells exposed to therapeutic stress. In parallel, nanoelectrode-based electrochemical sensors enable intracellular detection of metabolites such as reactive oxygen and nitrogen species, NADH, and glucose, directly linking metabolic state to disease progression and treatment response at single-cell resolution (Quek and Tay 2024; Liu et al. 2022).

17.3.2 Nano-enabled Nucleic Acid Analysis, Imaging, and Functional Biosensing

Nanotechnology has also transformed single-cell nucleic acid analysis through the development of nanopore sequencing platforms. Nanopore-based systems detect individual DNA or RNA molecules in real time by measuring changes in ionic current as nucleic acids traverse nanoscale pores. This enables direct, amplification-free detection of nucleotide modifications, alternative splicing events, and chromosomal rearrangements. In oncology, nanopore sequencing has revealed oncogenic translocations and kinase domain mutations, such as BCR-ABL1 rearrangements in leukaemia, bridging genomic and epigenomic information within individual cells and providing critical insights into clonal evolution and tumour heterogeneity (Singh et al. 2025; Ramos et al. 2017).

Complementing molecular readouts, nano-enabled imaging approaches provide spatial and structural context to single-cell analysis (Minnoye et al. 2021). Expansion microscopy (ExM) uses swellable polymer hydrogels to physically expand biological specimens, achieving effective nanoscale resolution using conventional optical microscopes. ExM and related techniques enable visualisation of proteins, RNA, and chromatin organisation at single-cell and subcellular resolution, revealing nanoscale features of transcriptional regulation and tumour architecture. In cancer, expansion-based imaging has advanced understanding of intratumour heterogeneity by resolving spatial relationships between malignant cells, stromal components, and immune infiltrates that are otherwise obscured using standard imaging modalities (Wang et al. 2025).

Finally, nano-biosensors form a cornerstone of functional single-cell analysis. Nanomaterials such as carbon nanotubes, graphene, quantum dots, and metallic nanostructures transduce biological interactions into electrical or optical signals with exceptional sensitivity due to their high surface-to-volume ratios (Ramesh et al. 2022). Electrochemical nano-biosensors based on nanowires or nanoparticles enable real-time monitoring of cytokines, metabolites, and enzymatic activity from individual cells without amplification. Optical nano-biosensors, including plasmonic and fluorescence-based systems, offer multiplexed detection and long-term stability. Nanoparticle probes engineered to respond to intracellular cues, such as pH, redox state, calcium flux, and enzymatic activity, provide real-time functional readouts of signalling dynamics, exposing heterogeneous cellular responses that are invisible in bulk assays (Ramesh et al. 2022).

Nano-enabled analytical platforms thus transform single-cell analysis from a largely descriptive endeavour into a quantitative, multi-parametric, and mechanistically informative discipline. By enhancing sensitivity, preserving heterogeneity, and integrating molecular, functional, and spatial information, nanotechnology provides a critical analytical foundation for next-generation single-cell biology and its translation into cancer research and precision medicine.

17.4 Nanotechnology in Single-Cell Genomics, Transcriptomics, and Proteomics

17.4.1 Nano-enabled Single-Cell Genomics and Transcriptomics

Single-cell genomic and transcriptomic analyses operate at the limits of molecular sensitivity, where extremely low input material makes measurements vulnerable to amplification bias and technical noise. Nanotechnology has been central to overcoming these constraints by enabling precise cell handling, reaction miniaturisation, and highly efficient molecular capture. By reducing reaction volumes and diffusion losses, nano- and microfabricated platforms increase effective analyte concentrations and improve the sensitivity and reproducibility of DNA and RNA detection at the single-cell level (Wang et al. 2025).

In transcriptomic workflows, nano-engineered beads and surfaces functionalised with oligonucleotides enable efficient mRNA capture while preserving cellular integrity. These principles underpin widely used droplet-based scRNA-seq systems, where nanoscale control of emulsions allows thousands of individual cells to be processed in parallel with high reproducibility (Tieng et al. 2025). Nanostructured interfaces also reduce mechanical and oxidative stress during cell isolation, limiting dissociation-induced transcriptional artefacts that are particularly problematic in fragile clinical samples. Nanowell-based platforms such as BD Rhapsody and ICELL8 exemplify this balance between throughput and control. BD Rhapsody integrates imaging, barcoding, and UMIs to achieve high single-cell purity (~80%) and is well-suited for low-input, targeted studies such as immune profiling, despite bias towards highly expressed genes. ICELL8 enables image-guided selection of up to 5184 cells across a wide size range and supports both 3′ and full-length chemistries; although its standard workflows show lower singlet efficiency and UMI inflation, improved protocols such as 3′ DE-UMI have strengthened its quantitative accuracy for targeted applications (Wang et al. 2025).

Nanotechnology has also driven major advances in spatially resolved transcriptomics, which integrates transcriptome-wide RNA sequencing with histological context to preserve cellular and subcellular organisation within intact tissues. Since being named Method of the Year by Nature Methods in 2020, spatial transcriptomics has expanded to include laser-based (LCM-seq), imaging-based (smFISH, EEL FISH), and barcoding-based in situ sequencing approaches. For example, EEL FISH combines electrophoretic RNA transfer onto indium tin oxide nanosurfaces with combinatorial DNA barcoding, while BARseq and BaristaSeq use in situ barcode sequencing to resolve neuronal projections at cellular or near-subcellular resolution (Wang et al. 2025). More recent nano-enabled platforms such as Stereoseq, STARmap, Slide-seqV2, and Seq-Scope exploit nanostructured capture surfaces, DNA nanoballs, and dense spatial barcodes to achieve near single-cell or submicron resolution. These technologies have transformed single-cell transcriptomics by linking molecular states to spatial context, enabling detailed mapping of tumour microenvironments in cancers such as breast and prostate cancer, where spatial organisation of tumour, immune, and stromal cells critically shapes disease progression and therapeutic response (Wang et al. 2025).

17.4.2 Nano-enabled Single-Cell Proteomics

Single-cell proteomic analysis poses challenges that differ fundamentally from nucleic acid-based assays. Proteins cannot be amplified, span a wide dynamic range, exhibit diverse chemistries, and undergo extensive post-translational modification, all of which demand exceptionally sensitive and low-loss workflows (Carbonara et al. 2021). As a result, conventional approaches, including antibody arrays, electrophoretic or chromatographic separations, and mass spectrometry, have historically required large cell populations, relied heavily on labels, and preferentially detected only abundant proteins (Carbonara et al. 2021).

Nanotechnology has addressed these constraints through extreme miniaturisation and low-loss sample handling at the scale of individual cells. Nano-enabled preparation strategies perform lysis, protein capture, digestion, and cleanup in picolitre- to nanolitre-scale volumes, often within a single reaction vessel (Wang et al. 2025). Methods such as nanoparticle-assisted protein capture, paramagnetic bead-based processing, and nano-droplet workflows reduce surface adsorption and improve protein recovery. These advances underpin modern mass-spectrometry-based single-cell proteomics platforms, including SCoPE-type workflows, which can now quantify hundreds to thousands of proteins per cell across large populations (Ctortecka et al. 2023).

Nanotechnology has also transformed spatial proteomics by enabling high-resolution mapping of protein expression while preserving tissue architecture. Imaging- and sampling-based approaches such as laser capture microdissection (LCM), Multiplexed Ion Beam Imaging (MIBI), and CO-Detection by IndEXing (CODEX) allow multiplexed, single-cell-resolved protein profiling directly in tissues. LCM enables precise isolation of defined cells or subcellular regions for downstream mass spectrometry, supporting analysis of tumour–immune interactions and subclonal heterogeneity. MIBI combines metal-tagged antibodies with time-of-flight secondary ion mass spectrometry to image dozens of proteins at subcellular resolution and has been widely applied to map immune checkpoints and spatial immune organisation in tumours. CODEX uses DNA-barcoded antibodies and cyclic fluorescence imaging to generate highly multiplexed 2D and 3D protein maps across cancer, autoimmune, and infectious disease tissues (Wang et al. 2025).

Complementary nano-enabled mass spectrometry workflows further expand the depth and scalability of spatial proteomics. Platforms such as nanoPOTS minimise reaction volumes to the nanolitre scale, enabling recovery of thousands of proteins from tens of cells, while ProteomEx and nano-DESI MSI provide quantitative, label-free spatial proteome maps at micrometre to subcellular resolution. Integrated pipelines including single-cell Deep Visual Proteomics (scDVP) and DISCO-MS combine AI-guided imaging, nanoscale tissue extraction, and ultrahigh-sensitivity mass spectrometry to directly link spatial phenotype with proteomic state across complex tissues (Wang et al. 2025).

In cancer research, these nano-enabled single-cell proteomic approaches have revealed functional heterogeneity that is invisible to transcriptomic analysis alone. Protein-level profiling of tumour immune microenvironments has uncovered diverse activation states of T cells and macrophages, including variable immune checkpoint expression that shapes response to immunotherapy. In prostate cancer, single-cell proteomics has demonstrated substantial variability in androgen receptor signalling output and downstream pathway activation among cells with similar transcriptomes, highlighting the importance of proteotype-level regulation in therapeutic resistance (Woo et al. 2024). Beyond mass spectrometry, nano-engineered biosensing platforms such as the Nano-in-Micro Array (NiMA)-based 1CellOmes concept integrate nanofabricated wells with ultra-sensitive SIMS and scanning probe microscopy to achieve label-free, high-throughput quantification of proteomes and glycomes from thousands of individual cells in parallel (Misevic et al. 2012). Collectively, these nano-enabled proteomic strategies overcome the "average cell" limitation of classical omics and provide a robust pathway for translating single-cell heterogeneity into clinically actionable diagnostics and prognostics in cancer and other complex diseases.

17.4.3 Integrating Genomics, Transcriptomics, and Proteomics Through Nano-interfaces

A major strength of nanotechnology lies in its ability to bridge multiple molecular layers within the same single-cell workflow. Nanofabricated and nano-fluidic platforms can confine individual cells or tissue regions while enabling controlled reagent exchange, supporting sequential or integrated measurement of DNA, RNA, and proteins without repeated transfers. This capability is particularly important in complex diseases such as cancer, where genomic alterations, transcriptional programs, and protein activity are often uncoupled and cannot be reliably inferred from a single molecular layer alone (Wang et al. 2025).

Spatial multi-omics exemplifies this integrative potential by resolving RNA and protein information within the same tissue context. Early platforms, such as the automated spatial multi-omics system developed by Vickovic and Lötstedt, demonstrated that sequencing-based spatial transcriptomics could be overlaid with protein detection using DNA-barcoded antibodies or immunofluorescence, enabling parallel profiling of dozens of proteins alongside regional transcriptomes. The GeoMx

Digital Spatial Profiler (DSP) extended these concepts to both fresh frozen and FFPE samples, allowing whole-transcriptome analysis (~18,000 RNA targets) together with >96 proteins from defined regions of interest. GeoMx DSP has been widely adopted in oncology to study tumour–stromal interactions and paracrine signalling, although its resolution remains limited to multicellular regions (Wang et al. 2025).

More recent nano-enabled spatial multi-omics technologies move closer to true cellular-scale integration. DBiT-seq uses intersecting microfluidic channels to barcode tissues in situ, enabling simultaneous spatial transcriptomic and proteomic profiling without tissue dissociation and with compatibility for FFPE samples. Spatial-CITE-seq further increases protein multiplexing by integrating hundreds of antibody-derived DNA tags with whole-transcriptome sequencing, albeit with reduced sensitivity for low-abundance transcripts and no subcellular resolution (Wang et al. 2025). In parallel, imaging-based approaches such as MOSAICA achieve nanoscale spatial resolution by combining combinatorial fluorescent barcoding with spectral and lifetime imaging, enabling co-detection of RNA and protein targets in 3D tissue contexts, including colorectal cancer and melanoma samples. Together, these platforms, built on nanofabricated microfluidics, DNA barcodes, and nanoscale imaging probes, enable integrated molecular maps that link transcriptional programs to protein-level function within intact tissues (Wang et al. 2025).

Nanotechnology-driven advances across single-cell genomics, transcriptomics, and proteomics have shifted the field from descriptive cataloguing towards functional and mechanistic insight. By enhancing molecular capture efficiency, reducing technical noise, and enabling integration across omics layers, nano-enabled platforms provide a more faithful representation of cellular state and heterogeneity. As these technologies mature and become more compatible with clinical specimens, their application in oncology will be central to identifying rare but clinically decisive cell populations, informing precision therapies, and translating single-cell discoveries into actionable biomedical outcomes.

17.5 Applications of Single-Cell Technologies and Nanotechnology in Biomedical and Translational Research

The integration of single-cell analysis with nanotechnology is transforming biomedical research by enabling functional, mechanistic, and clinically actionable insights that are inaccessible to bulk approaches. A key application lies in decoding therapy resistance, where nano–cell interactions themselves can serve as diagnostic and functional readouts. In functional assays, nano-engineered drug delivery and sensing platforms now allow single-cell drug response profiling, revealing heterogeneous sensitivity patterns even within ostensibly homogeneous cancer cell populations. By coupling controlled nanoscale delivery with single-cell readouts, these approaches uncover resistance pathways, identify predictive biomarkers, and support rational testing of combination therapies and dosing strategies that account for cellular diversity rather than population averages (Huang et al. 2024; Heath 2015; Baysoy et al. 2023).

A striking example is the use of nanomaterials themselves as functional probes of cellular state. A fluorescent nanoparticle library spanning inorganic (SiO_2), organic (PS-co-PAA), and hybrid metal–organic framework (MOF; ZIF-8) materials was developed to probe drug-resistant cancer cells at single-cell resolution. Distinct nanoparticle accumulation patterns, quantified by flow cytometry, revealed that reduced uptake of ZIF-8 nanoparticles, driven by P-glycoprotein (P-gp)-associated membrane rigidity and altered endocytosis, constituted a single-cell "fingerprint" of multidrug resistance. This nano-enabled single-cell strategy enabled sensitive detection of drug-resistant subpopulations, including high-MDR1 cells, directly in clinical lung cancer and leukaemia samples, illustrating how nanomaterials can function as both probes and reporters of clinically relevant cellular states (Liu et al. 2023).

In oncology, nano-enabled single-cell and spatial multi-omics are increasingly central to translating molecular targets into personalised therapies. Uveal melanoma (UM) exemplifies this need, as its core drivers (GNAQ/GNA11 mutations, YAP/TAZ signalling, BAP1 loss, and metastatic programs involving c-Met, CXCR4, and SF3B1) operate in highly heterogeneous tumour ecosystems. Single-cell transcriptomic and proteomic analyses have shown that MAPK/ERK and YAP/TAZ pathway activation varies markedly across tumour subpopulations, explaining variable responses to MEK inhibitors and agents such as verteporfin that are obscured in bulk profiling. Importantly, nano-enabled spatial transcriptomics revealed that BAP1-null tumour regions exhibit fibrotic, TGFβ-dominated immunosuppression, while adjacent BAP1-intact regions retain immune infiltration, critical information for immunotherapy stratification. These findings underscore how nano-fluidic platforms, nanoscale barcoding, and spatially resolved single-cell technologies provide the resolution needed to align genomic alterations with chromatin state, signalling output, and therapeutic vulnerability in UM (Fu and Li 2025).

Single-cell analysis has similarly transformed immunology, where functional heterogeneity rather than cell identity alone determines clinical outcomes. Nanotechnology enhances immune profiling by enabling sensitive, multiplexed functional measurements at the single-cell level. Nano-biosensors integrated into single-cell platforms can measure cytokine secretion in real time, revealing that only a subset of phenotypically similar T cells actively produces effector cytokines upon stimulation, an observation with direct implications for immunotherapy and vaccine design (An and Varadarajan 2018; Szabo et al. 2019). Nanoparticle-based probes further enable tracking of immune cell activation, metabolism, and cytotoxic function, providing insight into functional exhaustion and dysregulation in tumours, chronic infections, and autoimmune disease. In translational settings, such nano-enabled single-cell immune profiling supports patient stratification, prediction of response to checkpoint inhibitors, and optimisation of personalised immunotherapeutic strategies (An and Varadarajan 2018; Szabo et al. 2019).

Finally, nano-enabled single-cell approaches have also directly informed therapeutic design and delivery. In osteosarcoma, single-cell transcriptomics identified VEGFR2–JMJD3–high quiescent stem-like cells as key mediators of chemotherapy resistance. While co-inhibition of VEGFR2 and JMJD3 suppressed tumour growth, complete eradication of resistant cells required glutathione-scavenging nanoparticles that amplified endoplasmic reticulum stress and apoptosis specifically in stem/progenitor compartments (Wang et al. 2020). Similarly, in prostate cancer, PSMA-targeted semiconducting polymer nanoparticles bridged single-cell specificity with whole-organ imaging: transient absorption microscopy quantified probe uptake and kinetics in individual PSMA-positive cells, while NIR-II photoacoustic tomography enabled deep, non-invasive tumour detection in vivo (Wu et al. 2020). In brain tumours, nano-engineered lipid–polymer nanoparticles combined with focused ultrasound overcame delivery barriers for RNA therapeutics, achieving >ten-fold enhancement in siRNA uptake and target knockdown as confirmed by single-cell image analysis (Guo et al. 2021). Collectively, these examples show how the integration of nanotechnology with single-cell analysis enables precise diagnostics, functional dissection of resistance mechanisms, rational combination therapies, and effective drug delivery. By resolving heterogeneity across tumour, immune, and microenvironmental compartments, nano-enabled single-cell platforms are accelerating the translation of molecular discoveries into clinically actionable strategies across a broad range of biomedical applications.

17.6 Conclusion and Future Directions

Despite rapid progress, nano-enabled single-cell technologies remain an evolving field with significant potential to further transform biomedical and clinical research. Continued advances in materials science and device engineering are expected to yield smart, responsive nanomaterials that dynamically interact with cellular states relevant to disease. Stimuli-responsive nanoparticles and nano-engineered interfaces capable of sensing pH, enzymatic activity, redox balance, or mechanical forces will enable adaptive, closed-loop systems in which single-cell responses directly modulate sensing, stimulation, or drug delivery. Such platforms are particularly relevant for oncology and immunology, where real-time adaptation to cellular heterogeneity could support more precise interrogation of resistance mechanisms and inform the design of adaptive, patient-specific therapeutic strategies (Yuan et al. 2017b; Xu et al. 2020).

As nano-enabled single-cell platforms increasingly generate high-dimensional datasets that integrate molecular, functional, spatial, and temporal information, advanced computational approaches will be essential for clinical interpretation. Integration with artificial intelligence and machine learning will enable identification of clinically relevant cellular states, prediction of treatment response, and stratification of patients based on single-cell phenotypes rather than bulk averages. AI-driven analyses can bridge nanoscale measurements with clinical outcomes, accelerating biomarker discovery and supporting decision-making in precision medicine. To ensure clinical utility, standardisation of data formats, analytical pipelines, and metadata across platforms will be critical for reproducibility, regulatory validation, and cross-cohort comparison (Wen et al. 2022; Murthy 2007; Misevic et al. 2012).

A key translational frontier lies in extending nano-enabled single-cell technologies from in vitro and ex vivo studies to in vivo and clinical settings. Advances in biocompatible nanomaterials, minimally invasive sensors, and implantable or injectable nano-devices may enable single-cell analysis within living tissues, capturing disease-relevant cellular behaviour in native microenvironments. For clinical adoption, challenges related to scalability, manufacturing consistency, and regulatory approval must be addressed, including robust quality control of nanomaterials and devices (Min et al. 2015). Achieving this transition will require close collaboration between engineers, clinicians, and regulatory bodies, positioning nano-enabled single-cell technologies as powerful enablers of next-generation diagnostics, therapy monitoring, and personalised treatment strategies in clinical practice.

Acknowledgements The authors acknowledge support provided by the team at Fiona Elsey Cancer Research Institute towards this research. A.P. acknowledges the Irene Tellefson Research Fellowship. A. P. and A.J. acknowledge the Victorian Government's Women's Health Research Catalyst Grant.

References

Abdullah M, Obayedullah M, Shariful Islam Shuvo M, Abul Khair M, Hossain D, Nahidul Islam M (2025) A review on multifunctional applications of nanoparticles: analyzing their multiphysical properties. Results Surf Interfaces 21:100635

Aldridge S, Teichmann SA (2020) Single cell transcriptomics comes of age. Nat Commun 11:4307

An X, Varadarajan N (2018) Single-cell technologies for profiling T cells to enable monitoring of immunotherapies. Curr Opin Chem Eng 19:142–152

Bao G, Kamm RD, Thomas W, Hwang W, Fletcher DA, Grodzinsky AJ, Zhu C, Mofrad MR (2010) Molecular biomechanics: the molecular basis of how forces regulate cellular function. Mol Cell Biomech 3:91–105

Baysoy A, Bai Z, Satija R, Fan R (2023) The technological landscape and applications of single-cell multi-omics. Nat Rev Mol Cell Biol 24:695–713

Bokun V, Strang BL, Pantazi P, Liu Y, Holder B (2025) Nano-flow cytometry-guided discrimination and separation of human cytomegalovirus virions and extracellular vesicles. J Extracell Vesicles 14:e70060

Bulbul G, Chaves G, Olivier J, Ozel RE, Pourmand N (2018) Nanopipettes as monitoring probes for the single living cell: state of the art and future directions in molecular biology. Cells 7:55

Carbonara K, Andonovski M, Coorssen JR (2021) Proteomes are of proteoforms: embracing the complexity. Proteomes 9:38

Casanellas I, Samitier J, Lagunas A (2022) Recent advances in engineering nanotopographic substrates for cell studies. Front Bioeng Biotechnol 10:1002967

Ctortecka C, Hartlmayr D, Seth A, Mendjan S, Tourniaire G, Udeshi ND, Carr SA, Mechtler K (2023) An automated nanowell-array workflow for quantitative multiplexed single-cell proteomics sample preparation at high sensitivity. Mol Cell Proteomics 22:100665

Delgado-Gonzalez A, Sanchez-Martin RM (2021) Mass cytometry tags: where chemistry meets single-cell analysis. Anal Chem 93:657–664

Fu S, Li C (2025) Integrating single-cell omics and materials science for uveal melanoma: from mechanistic insights to precision therapeutics. Front Oncol 15:1661889

Fu Y, Liu T, Wang H, Wang Z, Hou L, Jiang J, Xu T (2024) Applications of nanomaterial technology in biosensing. J Sci: Adv Mater Devices 9:100694

Guo Y, Lee H, Fang Z, Velalopoulou A, Kim J, Thomas MB, Liu J, Abramowitz RG, Kim Y, Coskun AF, Krummel DP, Sengupta S, Macdonald TJ, Arvanitis C (2021) Single-cell analysis reveals effective siRNA delivery in brain tumors with microbubble-enhanced ultrasound and cationic nanoparticles. Sci Adv 7

Haywood DG, Saha-Shah A, Baker LA, Jacobson SC (2015) Fundamental studies of nanofluidics: nanopores, nanochannels, and nanopipets. Anal Chem 87:172–187

Heath JR (2015) Nanotechnologies for biomedical science and translational medicine. Proc Natl Acad Sci USA 112:14436–14443

Hirsch MG, Pal S, Mehrabadi FR, Malikic S, Gruen C, Sassano A, Pérez-Guijarro E, Merlino G, Sahinalp C, Molloy EK, Day CP, Przytycka TM (2024) Stochastic modelling of single-cell gene expression adaptation reveals non-genomic contribution to evolution of tumor subclones. bioRxiv

Hohnadel M, Maumy M, Chollet R (2018) Development of a micromanipulation method for single cell isolation of prokaryotes and its application in food safety. PLoS One 13:e0198208

Hu P, Zhang W, Xin H, Deng G (2016) Single cell isolation and analysis. Front Cell Dev Biol 4:116

Huang K, Xu Y, Feng T, Lan H, Ling F, Xiang H, Liu Q (2024) The advancement and application of the single-cell transcriptome in biological and medical research. Biology (Basel) 13

Iarossi M, Verma NC, Bhattacharya I, Meller A (2025) The emergence of nanofluidics for single-biomolecule manipulation and sensing. Anal Chem 97:8641–8653

Jayachandran A, Dhungel B, Steel JC (2016) Epithelial-to-mesenchymal plasticity of cancer stem cells: therapeutic targets in hepatocellular carcinoma. J Hematol Oncol 9:74

Jovic D, Liang X, Zeng H, Lin L, Xu F, Luo Y (2022) Single-cell RNA sequencing technologies and applications: a brief overview. Clin Transl Med 12:e694

Kaladharan K, Kumar A, Gupta P, Illath K, Santra TS, Tseng FG (2021) Microfluidic based physical approaches towards single-cell intracellular delivery and analysis. Micromachines (Basel) 12:631

Kim C, Kang MS, Raja IS, Joung YK, Han DW (2024) Advancements in nanobiosensor technologies for in-vitro diagnostics to point of care testing. Heliyon 10:e40306

Lange-Sperandio B, Trautmann A, Eickelberg O, Jayachandran A, Oberle S, Schmidutz F, Rodenbeck B, Homme M, Horuk R, Schaefer F (2007) Epithelial and mesenchymal cell biology-leukocytes induce epithelial to mesenchymal transition after unilateral ureteral obstruction in neonatal mice. Am J Pathol 171:861–871

Liang J, Cai W, Sun Z (2014) Single-cell sequencing technologies: current and future. J Genet Genomics 41:513–528

Liu B, Yi X, Zheng Y, Yuan Z, Yang J, Yang J, Yu X, Jiang L, Wang C (2022) A review of nano/micro/milli needles fabrications for biomedical engineering. Chin J Mech Eng 35:106

Liu L, Zhang Q, Wang C, Guo H, Mukwaya V, Chen R, Xu Y, Wei X, Chen X, Zhang S, Zhou M, Dou H (2023) Single-cell diagnosis of cancer drug resistance through the differential endocytosis of nanoparticles between drug-resistant and drug-sensitive cancer cells. ACS Nano 17:19372–19386

Liu X, Wang C, Guan X (2026) Technologic advances in flow cytometry. Clin Chim Acta 578:120567

Mazan-Mamczarz K, Ha J, De S, Sen P (2022) Single-cell analysis of the transcriptome and epigenome. Methods Mol Biol 2399:21–60

Mendenhall AR, Martin GM, Kaeberlein M, Anderson RM (2021) Cell-to-cell variation in gene expression and the aging process. Geroscience 43:181–196

Min Y, Caster JM, Eblan MJ, Wang AZ (2015) Clinical translation of nanomedicine. Chem Rev 115:11147–11190

Minnoye L, Marinov GK, Krausgruber T, Pan L, Marand AP, Secchia S, Greenleaf WJ, Furlong EEM, Zhao K, Schmitz RJ, Bock C, Aerts S (2021) Chromatin accessibility profiling methods. Nat Rev Methods Primers 1

Misevic GN, Benassayag G, Rasser B, Sales P, Simic-Krstic J, Misevic N, Popescu O (2012) New nanotechnology applications in single cell analysis: why and how? In: Karamanos Y (ed) Expression profiling in neuroscience. Humana Press, Totowa

Miwa H, Dimatteo R, De Rutte J, Ghosh R, Di Carlo D (2022) Single-cell sorting based on secreted products for functionally defined cell therapies. Microsyst Nanoeng 8:84

Murthy SK (2007) Nanoparticles in modern medicine: state of the art and future challenges. Int J Nanomedicine 2:129–141

Ottaiano A, Ianniello M, Santorsola M, Ruggiero R, Sirica R, Sabbatino F, Perri F, Cascella M, Di Marzo M, Berretta M, Caraglia M, Nasti G, Savarese G (2023) From chaos to opportunity: decoding cancer heterogeneity for enhanced treatment strategies. Biology (Basel) 12

Poddar A, Ahmady F, Prithviraj P, Luwor RB, Shukla R, Polash SA, Li H, Ramakrishna S, Kannourakis G, Jayachandran A (2024a) Chapter seven - advances in CRISPR/Cas systems-based cell and gene therapy. In: Singh V (ed) Progress in molecular biology and translational science. Academic Press

Poddar A, Ahmady F, Rao SR, Sharma R, Kannourakis G, Prithviraj P, Jayachandran A (2024b) The role of pregnancy associated plasma protein-A in triple negative breast cancer: a promising target for achieving clinical benefits. J Biomed Sci 31:23

Poddar A, Banerjee R, Shukla R (2023) Editorial: non-viral vectors for gene therapy/nucleic acid delivery. Front Bioeng Biotechnol 11:1304769

Poddar A, Joglekar MV, Hardikar AA, Shukla R (2019) A novel gene delivery approach using metal organic frameworks in human islet-derived progenitor cells. In: Joglekar MV, Hardikar AA (eds) Progenitor cells: methods and protocols. Springer New York, New York

Poddar A, Pyreddy S, Carraro F, Dhakal S, Rassell A, Field MR, Reddy TS, Falcaro P, Doherty CM, Shukla R (2020) ZIF-C for targeted RNA interference and CRISPR/Cas9 based gene editing in prostate cancer. Chem Commun (Camb) 56:15406–15409

Poddar A, Pyreddy S, Polash SA, Doherty CM, Shukla R (2022a) A quest for cytocompatible metal organic frameworks in non-viral gene therapy: relevance of zeolitic imidazolate framework-8. Biomater Biosyst 8:100065

Poddar A, Rao SR, Prithviraj P, Kannourakis G, Jayachandran A (2022b) Crosstalk between immune checkpoint modulators, metabolic reprogramming and cellular plasticity in triple-negative breast cancer. Curr Oncol 29

Polash SA, Poddar A, Ahmady F, Kannourakis G, Jayachandran A, Shukla R (2024) Impact of ligand concentration on the properties of nucleic-acid-encapsulated MOFs and inflammation modulation in prostate cancer cells. ACS Appl Bio Mater 7:7635–7645

Pyreddy S, Poddar A, Carraro F, Polash SA, Dekiwadia C, Murdoch B, Nasa Z, Reddy TS, Falcaro P, Shukla R (2023) Targeting telomerase utilizing zeolitic imidazole frameworks as non-viral gene delivery agents across different cancer cell types. Biomater Adv 149:213420

Quek YJ, Tay A (2024) Nanoscale methods for longitudinal extraction of intracellular contents. Adv Mater 36:2314184

Ramesh M, Janani R, Deepa C, Rajeshkumar L (2022) Nanotechnology-enabled biosensors: a review of fundamentals, design principles, materials, and applications. Biosensors (Basel) 13

Ramos AP, Cruz MAE, Tovani CB, Ciancaglini P (2017) Biomedical applications of nanotechnology. Biophys Rev 9:79–89

Ruan Q, Guo W, Yang R, Sun T, Yang Q, Ren Y (2026) Empowering single-cell analysis with emerging active microfluidic devices. Biomaterials 325:123617

Ryoo H, Kimmel H, Rondo E, Underhill GH (2024) Advances in high throughput cell culture technologies for therapeutic screening and biological discovery applications. Bioeng Transl Med 9:e10627

Shin HJ, Kwak M, Kwon IH, Kim SH, Lee JY (2025) Quantification of cellular uptake of gold nanoparticles via scattering intensity changes in flow cytometry. Nanoscale Adv 7:3558–3567

Singh P, Pandit S, Balusamy SR, Madhusudanan M, Singh H, Amsath Haseef HM, Mijakovic I (2025) Advanced nanomaterials for cancer therapy: gold, silver, and iron oxide nanoparticles in oncological applications. Adv Healthc Mater 14:e2403059

Skylaki S, Hilsenbeck O, Schroeder T (2016) Challenges in long-term imaging and quantification of single-cell dynamics. Nat Biotechnol 34:1137–1144

Stueber DD, Villanova J, Aponte I, Xiao Z, Colvin VL (2021) Magnetic nanoparticles in biology and medicine: past, present, and future trends. Pharmaceutics 13

Szabo PA, Levitin HM, Miron M, Snyder ME, Senda T, Yuan J, Cheng YL, Bush EC, Dogra P, Thapa P, Farber DL, Sims PA (2019) Single-cell transcriptomics of human T cells reveals tissue and activation signatures in health and disease. Nat Commun 10:4706

Tieng FYF, Lee LH, Ab Mutalib NS (2025) Single-cell RNA-sequencing of circulating tumour cells: a practical guide to workflow and translational applications. Cancer Metastasis Rev 44:75

Torres AJ, Hill AS, Love JC (2014) Nanowell-based immunoassays for measuring single-cell secretion: characterization of transport and surface binding. Anal Chem 86:11562–11569

Van Heyningen V (2024) Stochasticity in genetics and gene regulation. Philos Trans R Soc Lond Ser B Biol Sci 379:20230476

Wang C, Qiu J, Liu M, Wang Y, Yu Y, Liu H, Zhang Y, Han L (2024) Microfluidic biochips for single-cell isolation and single-cell analysis of multiomics and exosomes. Adv Sci (Weinh) 11:e2401263

Wang L, Huang X, You X, Yi T, Lu B, Liu J, Lu G, Ma M, Zou C, Wu J, Zhao W (2020) Nanoparticle enhanced combination therapy for stem-like progenitors defined by single-cell transcriptomics in chemotherapy-resistant osteosarcoma. Signal Transduct Target Ther 5:196

Wang R, Hastings WJ, Saliba JG, Bao D, Huang Y, Maity S, Kamal Ahmad OM, Hu L, Wang S, Fan J, Ning B (2025) Applications of nanotechnology for spatial omics: biological structures and functions at nanoscale resolution. ACS Nano 19:73–100

Wang S, Sun ST, Zhang XY, Ding HR, Yuan Y, He JJ, Wang MS, Yang B, Li YB (2023) The evolution of single-cell RNA sequencing technology and application: progress and perspectives. Int J Mol Sci 24

Wen L, Li G, Huang T, Geng W, Pei H, Yang J, Zhu M, Zhang P, Hou R, Tian G, Su W, Chen J, Zhang D, Zhu P, Zhang W, Zhang X, Zhang N, Zhao Y, Cao X, Peng G, Ren X, Jiang N, Tian C, Chen ZJ (2022) Single-cell technologies: from research to application. Innovation (Camb) 3:100342

Wen L, Tang F (2022) Recent advances in single-cell sequencing technologies. Precis Clin Med 5:pbac002

Woo J, Loycano M, Amanullah M, Qian J, Amend S, Pienta K, Zhang H (2024) Single-cell proteomic and transcriptomic characterization of drug-resistant prostate cancer cells reveals molecular signatures associated with morphological changes. bioRxiv

Woodland B, Farrell LA, Brockbals L, Rezcallah M, Brennan A, Sunnucks EJ, Gould ST, Stanczak AM, O'rourke MB, Padula MP (2025) Sample preparation for multi-omics analysis: considerations and guidance for identifying the ideal workflow. Proteomics 25:76–101

Wu J, Lee HJ, You L, Luo X, Hasegawa T, Huang KC, Lin P, Ratliff T, Ashizawa M, Mei J, Cheng JX (2020) Functionalized Nir-Ii semiconducting polymer nanoparticles for single-cell to whole-organ imaging of PSMA-positive prostate cancer. Small 16:e2001215

Xie K, Yang Y, Jiang H (2018) Controlling cellular volume via mechanical and physical properties of substrate. Biophys J 114:675–687

Xu X, Jia J, Guo M (2020) The most recent advances in the application of nano-structures/nanomaterials for single-cell sampling. Front Chem 8:718

Xu X, Valavanis D, Ciocci P, Confederat S, Marcuccio F, Lemineur JF, Actis P, Kanoufi F, Unwin PR (2023) The new era of high-throughput nanoelectrochemistry. Anal Chem 95:319–356

Ye M, Wilhelm M, Gentschev I, Szalay A (2021) A modified limiting dilution method for monoclonal stable cell line selection using a real-time fluorescence imaging system: a practical workflow and advanced applications. Methods Protocols 4:16

Youhannayee M, Nakhaei-Rad S, Haghighi F, Klauke K, Janiak C, Ahmadian MR, Rabenalt R, Albers P, Getzlaff M (2019) Physical characterization and uptake of iron oxide nanoparticles of different prostate cancer cells. J Magn Magn Mater 473:205–214

Yuan G-C, Cai L, Elowitz M, Enver T, Fan G, Guo G, Irizarry R, Kharchenko P, Kim J, Orkin S, Quackenbush J, Saadatpour A, Schroeder T, Shivdasani R, Tirosh I (2017a) Challenges and emerging directions in single-cell analysis. Genome Biol 18:84

Yuan GC, Cai L, Elowitz M, Enver T, Fan G, Guo G, Irizarry R, Kharchenko P, Kim J, Orkin S, Quackenbush J, Saadatpour A, Schroeder T, Shivdasani R, Tirosh I (2017b) Challenges and emerging directions in single-cell analysis. Genome Biol 18:84

Yum K, Yu MF, Wang N, Xiang YK (2011) Biofunctionalized nanoneedles for the direct and site-selective delivery of probes into living cells. Biochim Biophys Acta 1810:330–338

Zhang X, Dai X, Wan H, Jiao Z, Chen W, Wei W, Li J, Liu K, Yang Z, Yan L-T (2024) Interfaces between nanoparticle and biomacromolecular network: dynamic behaviors, confinement and entropy. Adv Funct Mater 34:2314134

Zhu J, Lee H, Huang R, Zhou J, Zhang J, Yang X, Zhou W, Jiang W, Chen S (2025) Harnessing nanotechnology for cancer treatment. Front Bioeng Biotechnol 12:1514890

Social and Economic Aspects of Single-Cell Analysis

18

Thuc-Quyen Nguyen, Nhat Le Bui, Minh Hanh Nguyen, and Dinh-Toi Chu

Abstract

Single-cell analysis (SCA) has become a popular tool in biomedical research, helping to expand the understanding of cellular mechanisms and treatment responses at unprecedented high resolution. In contrast to traditional bulk approach, single-cell approaches preserve cellular heterogeneity, thereby determining the status and role of individual cells in the population. SCA technology has had many specific applications in biomedicine, such as constructing transcriptome maps of tissues and organs, deciphering the mechanisms of relapse and drug resistance in cancers, and optimizing targeted therapies. In terms of economics, due to many technological obstacles and investment costs for facilities, the results of SCA are still limited in basic and preclinical research. In addition, the obvious differences in resources invested in science and technology lead to noticeable gaps in technological conditions and infrastructure, thus preventing

Thuc-Quyen Nguyen and Nhat Le Bui contributed equally with all other contributors.

T.-Q. Nguyen · D.-T. Chu (✉)
The Interdisciplinary Research Group on Biomedicine and Health, International School, Vietnam National University, Hanoi, Vietnam

The Interdisciplinary Research Group on Science, Technology, Innovation and Society, International School, Vietnam National University, Hanoi, Vietnam

Faculty of Applied Sciences, International School, Vietnam National University, Hanoi, Vietnam
e-mail: toicd@vnu.edu.vn

N. Le Bui
The Interdisciplinary Research Group on Biomedicine and Health, International School, Vietnam National University, Hanoi, Vietnam

The Interdisciplinary Research Group on Science, Technology, Innovation and Society, International School, Vietnam National University, Hanoi, Vietnam

M. H. Nguyen
The Interdisciplinary Research Group on Biomedicine and Health, International School, Vietnam National University, Hanoi, Vietnam

V. Singh (ed.), *Advances in Single Cell Analysis*,
https://doi.org/10.1007/978-981-95-9789-5_18

some developing countries from updating new research techniques. The ethical aspects and standards in developing this technology are still controversial. In this book chapter, we will delve into the economic and social advantages and challenges of SCA, which are rarely discussed when developing a new technology.

Keywords

Single-cell analysis · Social · Economic · Cost · Molecular biology · Multi-omics

Abbreviations

CARG	Compound annual growth rate
HCA	Human Cell Atlas
PCR	Polymerase chain reaction
SCA	Single-cell analysis
scDNA-seq	Single-cell DNA sequencing
scRNA-seq	Single-cell RNA sequencing

18.1 Introduction

Single-cell analysis (SCA) is a group of molecular biological techniques to study the characteristics of individual cells in a heterogeneous population, instead of accessing average signals like bulk approaches. Data obtained by SCA can reveal a wide range of cell types and hidden complexities in the cellular landscape, as well as uncover rare cell populations that were masked within bulk measurements (Stuart and Satija 2019; Heumos et al. 2023). The idea of studying the unique characteristics of a specific cell has been around since the advent of cell separation techniques. For a long time afterward, SCA relied primarily on flow cytometry, which classified cells based on their fluorescence characteristics, or laser capture microdissection, which allowed precise cutting and collecting of individual cells in fixed tissue (Czechowska et al. 2008; Müller and Nebe-von-Caron 2010; Suarez-Quian et al. 1999). However, SCA only really achieved a breakthrough when microfluidic technology was successfully developed and became popular (Yin and Marshall 2012). Microfluidic devices adjust solutions from picoliter to nanoliter volumes, which minimizes sample loss and improves assay sensitivity and parallelization ability (Gao et al. 2019). Another advantage is that the integration of different techniques like fluorescence, spectroscopy, and even polymerase chain reaction (PCR) allows simultaneous analysis of multiple biological molecules in each cell automatically for both online and real-time processing (Gao et al. 2019). Currently, the development of SCA is also not outside the general trend of multi-omics integration. The combination of multi-omics dimensions within a single cell, including genome, transcriptome, proteome, chromosome, and epigenetic regulation, facilitates the clarification of underlying molecular mechanisms, complex cellular interactions, and regulatory networks (Lim et al. 2024).

Over the recent decade, single-cell research has gradually become popular and brought various potential results in biomedicine, especially in the fields of oncology and drug development research. For example, a study using SCA conducted by Devon et al. identified a population of early metastatic cells in breast cancer that exhibited a stem-like signature of gene expression, thereby clarifying the hierarchical mechanism in the metastatic process and suggesting the potential for therapeutic intervention by targeting metastasis-initiating cells (Lawson et al. 2015). Another study by Haibin et al. utilized single-cell RNA sequencing to construct the transcriptome map of glioblastoma, which explained the roles of intracellular heterogeneity and tumor microenvironment in drug resistance and cancer recurrence (Wu et al. 2023). With the help of the cell barcoding technique, Bernd et al. successfully performed a high-throughput measurement over more than 18.000 phosphorylation levels from 14 types of human peripheral blood mononuclear cells using mass cytometry, thus recategorized inhibitors and suggested off-target effects (Bodenmiller et al. 2012). In general, SCA offers a great potential for precision medicine, from target discovery, compound screening, drug-resistant mechanism revealing to personalized immunotherapy design, thanks to its ability to decode cellular differentiation at the highest resolution (Heath et al. 2016).

Although this approach provides a holistic view of cellular systems, SCA still faces many challenges, especially in the areas of technology, economics, and social aspects. Regarding technology, the lack of spatial context is a major limitation of traditional SCA, causing the unfaithful reflection of the original biological state (Vandereyken et al. 2023). In addition, SCA requires specialized equipment and complex experimental procedures. Even commercial systems like 10× Genomics or Fluidigm C1 still require high costs and specialized operations (Freytag et al. 2018; Ashton et al. 2021). In addition, each single-cell sequencing experiment can generate hundreds of gigabytes of data, requiring powerful computing infrastructure, large storage systems, and automated standardization pipelines (Heumos et al. 2023; Adil et al. 2021). Integrating spatial and multi-omics data into SCA is necessary; however, on the other hand, it leads to more difficulties in interpreting biological meaning, especially for research units lacking experienced interdisciplinary experts. These challenges hinder the accessibility of individual laboratories and even scientists in under-resourced countries, thereby increasing the technology gap between developed and developing countries. Although various issues surrounding SCA still remain, in this book chapter, we focused mainly on analyzing the impacts that SCA can bring to society and economy, as well as the challenges it may encounter in low- and middle- income countries, which are important but are less often considered when referring to the development of a novel technology.

18.2 Social Aspects of Single-Cell Analysis

SCA has a positive impact on society by enabling early detection of cancer as well as developing the treatment plan for patients by identifying the earliest indications of cancer by looking at cellular and molecular alterations that may develop into cancer. More precisely, SCA studies such as single-cell RNA sequencing (scRNA-seq) and single-cell DNA sequencing (scDNA-seq) have allowed scientists to investigate gene expression variations and mutations in individual cells (Huang et al. 2024). The use of scRNA-seq helped physicians predict patients' response to medication, develop individualized treatment plans for each patient, and improve treatment outcomes (Huang et al. 2024). Ivanecski et al. introduced a method called scTherapy. ScTherapy is a precision medicine approach that treats cancer by targeting only malignant cells while preserving healthy cells. Based on the single-cell RNA sequencing data, physicians can recommend treatment options to cancer patients (Ianevski et al. 2024). In another study, comboSC, which utilized single-cell RNA sequencing to analyze cancer patients' immune microenvironments and predict effective treatments and immunotherapy strategies based on the patients' physiology, was introduced (Tang et al. 2023).

The growing use of animals in scientific research has raised both practical and ethical questions (Park et al. 2024). Nevertheless, ethical concerns regarding animal testing remain. Studies have shown that testing procedures are invasive and can result in death (Park et al. 2024; Luce and Duclos-Vallee 2025). Furthermore, animals also experience pain and have emotional capacities similar to humans. In response, initiatives have sought to limit animal testing. For example, by 2013, Europe had banned animal testing in the cosmetics industry (Daneshian et al. 2015). Researchers are progressively creating human-relevant models in response to these moral and practical issues that could reduce reliance on animal testing (Park et al. 2024). No existing literature states that SCA alone could replace animal testing with the use of human cells. However, they are part of a broader trend in biomedical research that is moving toward human-relevant models, with the potential to decrease the reliance on animal experimentation in the future (Luce and Duclos-Vallee 2025). Advanced biomedical technologies such as stem cell organoids have the potential to replace animal testing. Organoids can serve as an alternative because they can model human physiology and replicate how patients respond to medications and treatments, particularly for rare diseases where full clinical trials are difficult or impossible to conduct (Park et al. 2024; Luce and Duclos-Vallee 2025).

Despite the benefits that SCA technologies can bring, they also faces challenges especially the infrastructure required to implement these technologies: the high cost to obtain the technologies and the issues of ethnic diversity (Fig. 18.1). First is the underdevelopment and slow adoption of scRNA-seq technologies. This underdevelopment is linked to the lack of investment efforts from private companies and governments (Saloojee and Pettifor 2024; Amaeshi et al. 2025) to invest in scRNA-seq technologies. In addition, low- and middle-income countries often pay disproportionately higher prices due to the exclusivity of a few distributors (Boakye Serebour et al. 2024) compared to countries with advancements in biomedical science. The high cost of scRNA-seq technologies becomes a burden to labs with limited funding as it hinders their ability to carry out scRNA-seq research. Setting up a next-generation sequencing facility would cost around US$100,000–700,000 (Amaeshi et al. 2025). Furthermore, low- and middle-income countries also face high import taxes, which further hinder access to lab equipment.

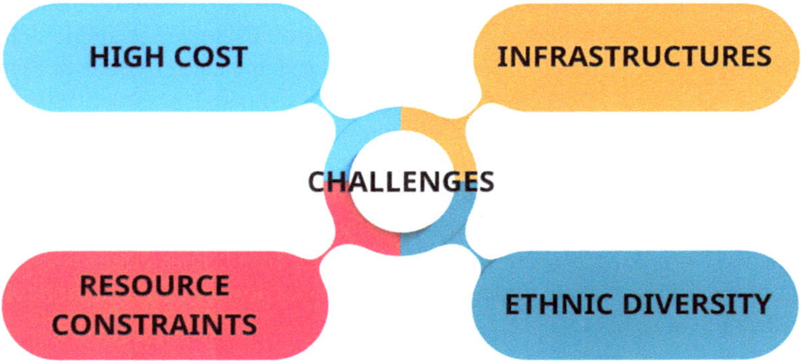

Fig. 18.1 Key challenges in implementing single-cell analysis

Institutionally, laboratories face high costs for technologies and lab materials. At the individual level, the high cost of genetic testing (Vu et al. 2025) remains out of reach for many people from low- and middle-income countries. For example, in sub-Saharan Africa, the cost of genetic testing for cancer can cost between 100$ and 1000$ (Amaeshi et al. 2025). Given the fact that most Africans pay out of pocket for their healthcare expenses while living below the poverty, access to such testing is limited (Boakye Serebour et al. 2024). As a results, SCA is accessible only to the wealthier population, and patients with low income are at risk of being excluded.

In addition to the high cost of obtaining technologies and lab materials, low- and middle-income countries also face resource limitations due to resource-constrained settings. First is the power supply needed to run laboratories. Due to inconsistent power supply, frequent power outages can compromise the preservation of tissues and reagents as well as computational workflows (Amaeshi et al. 2025; Boakye Serebour et al. 2024). While most laboratories have backup power supplies, running a center with backup generators during power outages can be prohibitively expensive (Boakye Serebour et al. 2024). Second is the limited access to supplies to conduct research. On average, it can take up to 3–4 months to obtain reagents through local distributors (Dara et al. 2022). This delay hinders research progression and competitiveness in the international research community. Regarding the limitations of clinical application, another issue that needs to be addressed is that SCA is currently mainly performed at the basic and preclinical research level. SCA requires a complicated implementation process, a long return time, and a high cost for testing, which is not suitable for routine diagnosis. The vast majority of SCA analysis platforms still lack ISO standards or other clinical certifications. Data from SCA contains high-resolution genetic information; thus, the protection of personal data also needs to be considered by the government, and related legal frameworks should be developed.

The SCA faces the issue of under representation of ethnic groups (Kosaji et al. 2023). More population-specific samples can help scientists to detect how the disease varies across different populations, regions, and genetic backgrounds (Peterson et al. 2019). Despite this need, most research samples of single-cell transcriptomic studies come from individuals of Caucasian ancestry (Boakye Serebour et al. 2024). As a result, therapies created using this data are less effective for underrepresented groups. The underrepresentation of ethnic groups in research samples leads to an incomplete outlook of cell types as well as a holistic understanding of the disease (Bentley et al. 2017). The potential for single-cell research to exacerbate pre-existing health inequities, restrict precision medicine's worldwide applicability, and impede efforts toward more equitable healthcare outcomes could be mitigated with more participation from underrepresented populations (Boakye Serebour et al. 2024; Kosaji et al. 2023).

18.3 Economic Aspects of Single-Cell Analysis

The economic aspects of SCA are a significant research direction, as SCA is rapidly developing and has a direct impact on medicine, molecular biology, and the pharmaceutical industry. In a study of Shalek et al., the authors pointed out the potential for economic development in biotechnology based on SCA techniques, stating that the average cost to develop a new drug can reach approximately 2.6 billion USD. The application of SCA helps reduce the failure rate and limit the use of ineffective medications, thereby potentially saving a significant portion of this enormous cost (Shalek and Benson 2017). One of the projects that cannot be ignored in SCA is the international project Human Cell Atlas (HCA) with the goal of mapping all types of cells in the human body at single-cell and multi-omics resolution. The purpose of the project is to build a reference platform for cell states, serving cancer research, immunology, and drug development research. To carry out this project, the Chan Zuckerberg Initiative has confirmed a grant of 68 million USD to support collaborative groups and expand the network (Chan Zuckerberg Initiative awards $68 million to support the growth of the Human Cell Atlas 2019). The official website of HCA also announced receiving a grant of about 13 million USD from the Leona M. and Harry B. Helmsley Charitable Trust for mapping intestinal cells called the Gut cell atlas, which is a part of HCA (Helmsley Trust Announces $13m funding towards creating a Gut Cell Atlas 2019). Thus, it can be observed that the use of advanced techniques such as SCA not only incurs high investment costs but also reduces R&D expenses, thereby opening up significant long-term economic benefits in the future.

According to the report, the global SCA market size was estimated at USD 4.34 billion in 2023 and is projected to reach USD 13.69 billion by 2030, growing at a compound annual growth rate (CAGR) of 18.7% from 2024 to 2030 (Research, G.V 2025). In addition, the SCA market is not only experiencing robust growth on a global scale, but there are also significant differences in revenue, market share, and compound annual growth rates among countries and regions (Research, G.V., UK single cell analysis market size & outlook; Research, G.V., Canada single cell analysis market size & outlook; Research, G.V., China single cell analysis market size

& outlook; Research, G.V., Germany single cell analysis market size & outlook; Research, G.V., India single cell analysis market size & outlook; Research, G.V., Latin America single cell analysis market size & outlook; Research, G.V., Middle East & Africa single cell analysis market size & outlook).

According to reports from Grand View Research, the revenue achieved in 2024 and the projected revenue in 2030 for several countries/regions when applying SCA techniques varied between countries/regions (Fig. 18.2). In Canada, the market revenue reached approximately 125 million USD in 2024 and is expected to increase to 320 million USD by 2030. In Europe, particularly in the United Kingdom and Germany, significant revenue has also been recorded with the implementation of SCA, amounting to 347 million USD and 276 million USD, respectively. Both the United Kingdom and Germany are confident that by 2030, their revenues will exceed 1 billion USD for the UK and more than 810 million USD for Germany. In the Asia region, China stands out with a revenue of approximately 178 million USD and may reach up to 540 million USD by 2030. Meanwhile, India is projected to grow from 131 million USD in 2024 to 389 million USD by 2030. Other countries in Latin America, the Middle East, and Africa regions also demonstrate strong growth trends, with projected revenues of 405 million USD and 201 million USD in 2024, respectively, and expected revenues of 1182 million USD and 580 million USD by 2030. The figures mentioned above are indeed substantial in terms of revenue generated from the application of advanced biomedical techniques, including single-cell analysis.

However, despite the encouraging economic benefits mentioned above, there are still economic barriers for some countries in researching, developing, and applying single-cell analysis techniques. SCA offers many optimal applications in advanced

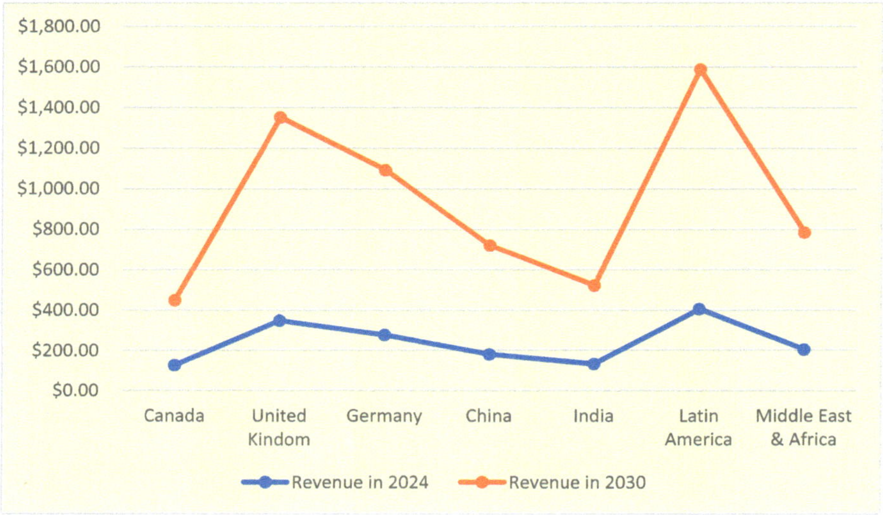

Fig. 18.2 Single-cell analysis market revenue by country/region from 2024 to 2030

biomedical engineering; however, it is also a barrier for some places to access these technologies because the cost of SCA is too expensive and requires a significant investment. In an experiment on Single Cell Core at Harvard Medical School, the cost of services with platforms like 10× Genomics, Parse Biosciences, multiome, etc. is listed, showing that a single run with 8 samples, including the preparation of 8 gene-expression libraries, typically costs around 15,000–20,000 USD or more, depending on sample characteristics and experimental design (School, H.M., Single cell core). This high expenditure hinders the accessibility of some countries to this technology due to limited resources, while developed countries such as the United States, the United Kingdom, and China, which have strong research infrastructure and investment capital, find the application of advanced biomedical techniques, including single-cell analysis techniques, is very easily accessible.

18.4 Conclusion

Advances in core technical platforms such as microfluidics and next-generation sequencing technologies are favorable steps to increase the accuracy and high-throughput scale of SCA research. In addition to technological challenges, SCA also faces many challenges in economic and social aspects. The high cost of each experiment and strict requirements for equipment systems are the main factors hindering the development of SCA in countries with limited resources for investment in science and technology. This leads to an increase in the technology gap between developed and developing countries. In addition, the issue of data security requires more detailed and strict regulations, especially for SCA research that contain genetic data with a variety of personal information. However, the development of research directions using single-cell data is an inevitable trend in the era of personalized medicine. Especially in the future, when integrating omics and spatial data, and even artificial intelligence in data processing, the development of SCA promises to bring a detailed picture down to the cellular level of disease pathogenesis and treatment therapies.

References

Adil A et al (2021) Single-cell transcriptomics: current methods and challenges in data acquisition and analysis. Front Neurosci 15:591122

Amaeshi LC, Okunade KS, Anorlu RI (2025) Current landscape of cancer genomics research in sub-Saharan Africa – a review of literature. Front Oncol 15:1512005

Ashton JM et al (2021) Comparative analysis of single-cell RNA sequencing platforms and methods. 32(4):3fc1f5fe. 3eccea01

Bentley AR, Callier S, Rotimi CN (2017) Diversity and inclusion in genomic research: why the uneven progress? J Community Genet 8(4):255–266

Boakye Serebour T et al (2024) Overcoming barriers to single-cell RNA sequencing adoption in low- and middle-income countries. Eur J Hum Genet 32(10):1206–1213

Bodenmiller B et al (2012) Multiplexed mass cytometry profiling of cellular states perturbed by small-molecule regulators. Nat Biotechnol 30(9):858–867

Chan Zuckerberg Initiative awards $68 million to support the growth of the Human Cell Atlas. 2019.; Available from: https://www.humancellatlas.org/news/chan-zuckerberg-initiative-awards-68-million-to-support-the-growth-of-the-human-cell-atlas-693/
Czechowska K, Johnson DR, van der Meer JR (2008) Use of flow cytometric methods for single-cell analysis in environmental microbiology. Curr Opin Microbiol 11(3):205–212
Daneshian M et al (2015) Animal use for science in Europe. ALTEX 32(4):261–274
Dara A et al (2022) Tackling malaria transmission at a single cell level in an endemic setting in sub-Saharan Africa. Nat Commun 13(1):2679
Freytag S et al (2018) Comparison of clustering tools in R for medium-sized 10x Genomics single-cell RNA-sequencing data [version 2; peer review: 3 approved]. 7(1297)
Gao D et al (2019) Recent advances in single cell manipulation and biochemical analysis on microfluidics. Analyst 144(3):766–781
Heath JR, Ribas A, Mischel PS (2016) Single-cell analysis tools for drug discovery and development. Nat Rev Drug Discov 15(3):204–216
Helmsley Trust Announces $13m funding towards creating a Gut Cell Atlas. 2019
Heumos L et al (2023) Best practices for single-cell analysis across modalities. Nat Rev Genet 24(8):550–572
Huang K et al (2024) The advancement and application of the single-cell transcriptome in biological and medical research. Biology (Basel) 13(6)
Ianevski A et al (2024) Single-cell transcriptomes identify patient-tailored therapies for selective co-inhibition of cancer clones. Nat Commun 15(1):8579
Kosaji N et al (2023) Lack of ethnic diversity in single-cell transcriptomics hinders cell type detection and precision medicine inclusivity. Med 4(4):217–219
Lawson DA et al (2015) Single-cell analysis reveals a stem-cell program in human metastatic breast cancer cells. Nature 526(7571):131–135
Lim J et al (2024) Advances in single-cell omics and multiomics for high-resolution molecular profiling. Exp Mol Med 56(3):515–526
Luce E, Duclos-Vallee JC (2025) Stem cells and organoids: a paradigm shift in preclinical models toward personalized medicine. Pharmaceuticals (Basel) 18(7)
Müller S, Nebe-von-Caron G (2010) Functional single-cell analyses: flow cytometry and cell sorting of microbial populations and communities. FEMS Microbiol Rev 34(4):554–587
Park G et al (2024) Replacing animal testing with stem cell-organoids : advantages and limitations. Stem Cell Rev Rep 20(6):1375–1386
Peterson RE et al (2019) Genome-wide association studies in ancestrally diverse populations: opportunities, methods, pitfalls, and recommendations. Cell 179(3):589–603
Research, G.V (2025) Single-cell analysis market (2024–2030)
Research, G.V., UK single cell analysis market size & outlook
Research, G.V., Canada single cell analysis market size & outlook
Research, G.V., China single cell analysis market size & outlook
Research, G.V., Germany single cell analysis market size & outlook
Research, G.V., India single cell analysis market size & outlook
Research, G.V., Latin America single cell analysis market size & outlook
Research, G.V., Middle East & Africa single cell analysis market size & outlook
Saloojee H, Pettifor JM (2024) Maximizing access and minimizing barriers to Research in low- and middle-income countries: open access and health equity. Calcif Tissue Int 114(2):83–85
School, H.M., Single cell core
Shalek AK, Benson M (2017) Single-cell analyses to tailor treatments. Sci Transl Med 9(408)
Stuart T, Satija R (2019) Integrative single-cell analysis. Nat Rev Genet 20(5):257–272
Suarez-Quian CA et al (1999) Laser capture microdissection of single cells from complex tissues. BioTechniques 26(2):328–335
Tang C et al (2023) Personalized tumor combination therapy optimization using the single-cell transcriptome. Genome Med 15(1):105
Vandereyken K et al (2023) Methods and applications for single-cell and spatial multi-omics. Nat Rev Genet 24(8):494–515

Vu T-D et al (2025) Effects of high fat diet on blood lipids and liver enzymes in murine model: the systemic and experimental study. p 100614
Wu H et al (2023) Single-cell RNA sequencing reveals tumor heterogeneity, microenvironment, and drug-resistance mechanisms of recurrent glioblastoma. Cancer Sci 114(6):2609–2621
Yin H, Marshall D (2012) Microfluidics for single cell analysis. Curr Opin Biotechnol 23(1):110–119

MIX
Papier aus verantwortungsvollen Quellen
Paper from responsible sources
FSC® C105338

If you have any concerns about our products,
you can contact us on
ProductSafety@springernature.com

In case Publisher is established outside the EU,
the EU authorized representative is:
Springer Nature Customer Service Center GmbH
Europaplatz 3, 69115 Heidelberg, Germany

Printed by Libri Plureos GmbH
in Hamburg, Germany